Ergonomics in the Automotive Design Process

Automotive design continues to evolve at a rapid pace. As electric cars become ever more commonplace on the roads to the advent of the driverless vehicle, understanding the ergonomics behind automotive engineering becomes ever more paramount. Vehicle attributes must be considered early during the new vehicle development program by coordinated work of multi-disciplinary teams to begin creating vehicle specifications and development of vehicle attribute requirements.

In *Ergonomics in the Automotive Design Process: Concepts, Issues and Methods*, Vivek D. Bhise covers the need-to-know fundamentals as to what makes an ergonomically sound vehicle. This book covers the entire range of ergonomics issues involved in designing a car or truck and offers evaluation techniques to avoid costly mistakes and assure high customer satisfaction. Across 13 chapters, vehicle design and the attributes of vehicle handling, appearance (interior and exterior styling), safety and security, infotainment, noise and vibrations, emissions, costs and process compatibility are considered in the context of ergonomics. New material to this edition includes coverage of ergonomics in the systems engineering process, decision-making and risks in automotive product programs and ergonomic considerations in electric vehicle development.

This book will allow the reader to develop a more comprehensive knowledge of issues facing the developers of automotive products and delivers methods to manage communication, coordination and integration processes. It provides more tools in implementing systems engineering to minimize the risks of delays and cost overruns, and most importantly, creates the right product for its customers. The reader will develop a knowledge of future in-vehicle devices that are easy to program and use, safe, cheap to manufacture and assemble and are eco-friendly From an author with over forty years of experience in automotive design, this title is an ideal read for students and practitioners of ergonomics, human factors, automotive design, civil engineering, product design, work design and mechanical engineering.

Vivek D. Bhise is currently a LEO Lecturer/Visiting Professor and a Professor in post-retirement of Industrial and Manufacturing Systems Engineering at the University of Michigan-Dearborn. He received his B.Tech. in Mechanical Engineering (1965) from the Indian Institute of Technology, Bombay, India, M.S. in Industrial Engineering (1966) from the University of California, Berkeley, and PhD in Industrial and Systems Engineering (1971) from the Ohio State University, Columbus, Ohio. During 1973 to 2001, he held several management and research positions at the Ford Motor Company in Dearborn, Michigan.

Ergonomics in the Automotive Design Process

Concepts, Issues and Methods

Volume 1

Second Edition

Vivek D. Bhise

CRC Press is an imprint of the
Taylor & Francis Group, an **informa** business

Designed Cover Image: Vivek D. Bhise

First published 2024
by CRC Press
2385 NW Executive Center Drive, Suite 320, Boca Raton FL 33431

and by CRC Press
4 Park Square, Milton Park, Abingdon, Oxon, OX14 4RN

CRC Press is an imprint of Taylor & Francis Group, LLC

© 2024 Vivek D. Bhise

The right of Vivek D. Bhise be identified as author of this work has been asserted in accordance with sections 77 and 78 of the Copyright, Designs and Patents Act 1988.

All rights reserved. No part of this book may be reprinted or reproduced or utilised in any form or by any electronic, mechanical, or other means, now known or hereafter invented, including photocopying and recording, or in any information storage or retrieval system, without permission in writing from the publishers.

Trademark notice: Product or corporate names may be trademarks or registered trademarks, and are used only for identification and explanation without intent to infringe.

ISBN: 9781032739120 (hbk)
ISBN: 9781032779621 (pbk)
ISBN: 9781003485582 (ebk)

DOI: 10.1201/9781003485582

Typeset in Times
by Newgen Publishing UK

Access the Instructor and Student Resources/Support Material: www.routledge.com/9781032739120

Contents

Preface to the Second Edition

This new two volume book was created to provide additional material needed by ergonomics engineers to expand their effectiveness in implementation of ergonomics in future automotive products. It covers advanced topics such as prediction of visibility of targets and glare to evaluate vehicle lighting systems, driver performance and workload measurements, new technology implementations in creating advanced driver assistance systems, electric vehicles and autonomous vehicles. It also includes material to help implement ergonomics during detailed engineering design phase, management of interfaces between different vehicle systems, designing for special driver and user populations, methods for vehicle evaluation in verification and validation, models for conducting cost-benefit analyses, and involvement of ergonomics engineers during litigations associated with product liability cases.

It is hoped that this new two volume book will enable practitioners and students to understand the expanded scope of ergonomic considerations and inputs in the development of future automotive products and makes the two volume set of books more comprehensive and useful.

I would like to thank Suriya Rajasekar and Stuart A.P. Murray from Newgen Knowledge Works, and Ann Chapman and James Hobb from Taylor and Francis for providing many valuable suggestions and guidance in creating this two-volume set of books.

Vivek D. Bhise
March 13, 2024

Preface to the First Edition

The purpose of this book is to provide a thorough understanding of ergonomic issues and to provide background information, principles, design guidelines, tools and methods used in designing and evaluating automotive products. This book has been written to satisfy the needs for both students and professionals who are genuinely interested in improving the usability of automotive products. Undergraduate and graduate students in engineering and industrial design will gain an understanding of the ergonomics engineer's work and the complex coordination and teamwork of many professionals in the automotive product development process. Students will learn the importance of timely information and recommendations provided by the ergonomics engineers and the methods and tools that are available to improve user acceptance. The professionals in the industry will realize that the days of considering ergonomics as a "commonsense" science and simply "winging-in" quick fixes to achieve user-friendliness are over. The auto industry is facing tough competition and severe economic constraints. Their products need to be designed "right the first time" with the right combinations of features that not only satisfy the customers but continually please and delight them by providing increased functionality, comfort, convenience, safety and craftsmanship.

The book is based on my over forty years of experience as a human factors researcher, engineer, manager and teacher, who has performed numerous studies and analyses designed to provide answers to designers, engineers and managers involved in designing car and truck products, primarily for the markets in the United States and Europe. The book is not like many ergonomics textbooks, which compile a lot of information from a large number of references reported in the human factors and ergonomics literature. I have included only the topics and materials that I found to be useful in designing car and truck products and concentrated on the ergonomic issues generally discussed in the automotive design studios and product development teams. The book is really about what an ergonomics engineer should know and do after he or she becomes a member of an automotive product development team and is asked to create an ergonomically superior vehicle.

The book begins with the definitions and goals of ergonomics, historical background and ergonomics approaches. It covers important human characteristics, capabilities and limitations considered in vehicle design in key areas such as anthropometry, biomechanics and human information processing. Next, the reader is led in understanding how the driver and the occupants are positioned in the vehicle space and package drawings and/or CAD models are created from key vehicle dimensions used in the automobile industry. Various design tools used in the industry for occupant packaging, driver vision and applications of other psychophysical methods are described. The book covers important driver information processing concepts and models and driver error categories to understand key considerations and principles used in designing controls, displays, and their usages including current issues related to driver workload and driver distractions.

A vehicle's interior dimensions are related to its exterior dimensions in terms of the required fields of view from the driver's eye points through various window openings and other in-direct vision devices (e.g., mirrors, cameras). Various field of view measurement and analysis techniques and visibility requirements and design areas such as windshield wiper zones, obscurations caused by the pillars, and the required in-direct fields of views are described along with many trade-off considerations. Human factors considerations and night visibility issues are presented to understand basics of headlamp beam pattern design and signal lighting performance, and their photometric requirements.

Other customer/user concerns and comfort issues related to entering and exiting from the vehicle, seating, loading and unloading cargo and other service related issues (in engine and trunk compartment, refueling the vehicle, etc.) are covered. They provide insights into user considerations in designing vehicle body and mechanical packaging in terms of important vehicle dimensions related to body/door openings, roof, rocker panels, and clearances for the user's hands, legs, feet, torso and head, and so forth.

A chapter on craftsmanship covers a relatively new technical and increasingly important area for ergonomics engineers. The whole idea behind the craftsmanship is that the vehicle should be designed and built such that the customers will perceive the vehicle to be built with a lot of attention-to-details by craftsmen who apply their skills to enhance the pleasing perceptual characteristics of the product related to its appearance, touch feel, sounds and ease during operations. Several examples of research studies on measurement of craftsmanship and relating product perception measures to physical characteristics of interior materials are presented.

In addition for the researchers, the second part of the book includes chapters on driver behavioral and performance measurement, vehicle evaluation methods, modeling of driver vision – which illustrates how the target detection distances and legibility of displays can be predicted to evaluate vehicle lighting and display systems, and driver workload to evaluate in-vehicle devices. Discussions on ergonomic issues for the development of new technological features in areas such as telematics, night vision and other driver safety and comfort related devices are included. The second part of the book also presents data and discusses many unique issues associated with designing for different populations segments, such as older drivers, women drivers, drivers in different geographic parts of the world. Finally, the last chapter is focused on various key issues related to future research needs in various specialized areas of ergonomics as well as vehicle systems, and on implementation of available ergonomic design guidelines and tools at different stages of the automotive product design process.

The book can be used to form the basis of two courses in Vehicle Ergonomics. The first course would cover the basic ergonomic considerations needed in designing and evaluating vehicles that are included in Part I – the first eleven chapters of this book. The remaining chapters covered in Part II can be used for an advanced and more research-oriented course.

Vivek D. Bhise

December 26, 2010

MATERIAL FOR WEBSITE

A. Computer Programs and Models

1. Computations of Percentile Value of Normal Distribution
2. Driver Package Parameters Computations
3. Reaction Time Measurement Program
4. Vehicle Program Costs and Revenues Computation Model
5. Visibility Prediction Model
6. Discomfort Glare and Dimming Request Prediction Model

About the Author

Vivek D. Bhise is currently a LEO lecturer and professor in post-retirement of Industrial and Manufacturing Systems Engineering at the University of Michigan-Dearborn. He received his B.Tech. in Mechanical Engineering (1965) from the Indian Institute of Technology, Bombay, India, M.S. in Industrial Engineering (1966) from the University of California, Berkeley, California and Ph.D. in Industrial and Systems Engineering (1971) from the Ohio State University, Columbus, Ohio.

From 1973 to 2001, he held a number of management and research positions at the Ford Motor Company in Dearborn, Michigan. He was the manager of Consumer Ergonomics Strategy and Technology within the Corporate Quality Office and the manager of the Human Factors Engineering and Ergonomics in the Corporate Design of the Ford Motor Company where he was responsible for the ergonomics attribute in the design of car and truck products.

Dr. Bhise has taught graduate courses in Vehicle Ergonomics, Vehicle Package Engineering, Automotive Systems Engineering, Human Factors Engineering, Total Quality Management and Six Sigma, Product Design and Evaluations, and Safety Engineering over the past 43 years (1980–2001 as an Adjunct Professor, and 2001–2009 as a professor, and 2009-present as a lecturer) at the University of Michigan–Dearborn. He also worked on a number of research projects on human factors with late Prof. Thomas Rockwell at the Driving Research Laboratory at the Ohio State University (1968–1973). His publications include over 100 technical papers in the design and evaluation of automotive interiors, vehicle lighting systems, field of view from vehicles, and modeling of human performance in different driver/user tasks.

He received the Human Factors Society's A. R. Lauer Award for Outstanding Contributions to the Understanding of Driver Behavior in 1987. He has served on many committees of the Society of Automotive Engineers, Inc., Vehicle Manufacturers Association, Human Factors Society, and Transportation Research Board of the National Academies. He is a member of the Human Factors and Ergonomics Society, the Society of Automotive Engineers, and the Alpha Pi Mu. He has published the following books:

Bhise, Vivek D. 2011. *Ergonomics in the Automotive Design Process*. ISBN: 9781439842102. Boca Raton, FL: CRC Press. (Also translated in Chinese language and published by China Machine Press in China, 2016).

Bhise, Vivek D. 2013. *Designing Complex Products with Systems Engineering Processes and Techniques*. ISBN: 13: 978-1466507036. Boca Raton, FL: CRC Press.

Bhise, Vivek D. 2017. *Automotive Product Development: A Systems Engineering Implementation*. ISBN: 978-1-4987-0681-0. Boca Raton, FL: CRC Press.

Bhise, Vivek D. 2022. *Decision-Making in Energy Systems*. ISBN: 978-0-367-62015-8. Boca Raton, FL: CRC Press.

Bhise, Vivek D. 2023. *Designing Complex Products with Systems Engineering Processes and Techniques*. Second Edition, ISBN: 978-1032203690. Boca Raton, FL: CRC Press.

Acknowledgments

This book is a culmination of my education, experience and interactions with many individuals from the automotive industry, academia and government agencies. While it is impossible for me to thank all the individuals who influenced my career and thinking, I must acknowledge the contributions of the following individuals.

My greatest thanks go to Prof. Thomas H. Rockwell of the Ohio State University. Tom got me interested in human factors engineering and driving research. He was my advisor and mentor during my doctoral program. I learnt many skills on how to conduct research studies, analyze data; and more importantly, he got me introduced to the technical committees of the Transportation Research Board and the Society of Automotive Engineers, Inc.

I would like to thank Lyman Forbes, Dave Turner and Bob Himes from the Ford Motor Company. Lyman Forbes, manager of the Human Factors Engineering and Ergonomics (HFEE) Department at the Ford Motor Company in Dearborn, Michigan, spent hours with me discussing various approaches and methods to conduct research studies on various crash-avoidance research issues related to the development of motor vehicle safety standards. Dave Turner from Advanced Design Studios helped anchor ergonomics in the automotive design process and also created an environment to establish a human factors group in Europe. Bob Himes of the Advanced Vehicle Engineering staff helped in incorporating "Ergonomics and Vehicle Packaging" as a "Vehicle Attribute" in the vehicle development process.

The University of Michigan-Dearborn campus provided me with the unique opportunities to develop and teach various courses. Our Automotive Systems Engineering and Engineering Management Programs allowed me to interact with hundreds of students who in-turn implemented many of the techniques taught in our graduate programs in solving problems within many other automotive OEMs and supplier companies. I want to thank Profs. Adnan Aswad, Munna Kachhal and Armen Zakarian for giving me opportunities to develop and teach many courses in the Industrial and Manufacturing Systems Engineering, and Dean Subrata Sengupta for supporting the creation of the Vehicle Ergonomics Laboratory in the new Institute for Advanced Vehicle Systems Building. Roger Schulze, Director of the Institute for Advanced Vehicle Systems got me interested in working on a number of multidisciplinary programs in vehicle design. Together, we developed a number of vehicle concepts such as the Low Mass Vehicle, a new Model "T" concept for Ford's 100th anniversary, and a Reconfigurable Electric Vehicle. We also created a number of design projects by forming teams of our engineering students with students from the College for Creative Studies in Detroit, Michigan. My special thanks also go to James Dowd from Collins and Aikman and the Advanced Cockpit Enablers (ACE) team members for sponsoring a number of research projects on various automotive interior components and creation of a driving simulator to evaluate a number of advanced concepts in vehicle interiors.

Over the past forty-plus years, I was also fortunate to meet and discuss many automotive design issues with members of many committees of the Society of Automotive Engineers, Inc., the Motor Vehicle Manufacturers Association, the Transportation Research Board and the Human Factors and Ergonomics Society.

I would like to also thank Cindy Carelli from CRC Press, Taylor & Francis Group, for encouragement in preparing the proposal for this book, and Amy Blalock and her production group for turning the manuscript into this book. My thanks also go to Louis Tijerina, Anjan Vincent and Calvin Matle who spent hours reviewing the manuscript and providing valuable suggestions to improve this book.

Finally, I want to thank my wife, Rekha, for her constant encouragement and her patience while I spent many hours working on my computers writing the manuscript and creating figures included in this book.

Vivek D. Bhise
Ann Arbor, Michigan
December 26, 2010

1 Introduction to Automotive Ergonomics

ERGONOMICS IN VEHICLE DESIGN

Designing an automotive product such as a car or truck involves the integration of inputs from many disciplines (e.g., designers, body engineers, chassis engineers, powertrain engineers, electronics/electrical engineers, manufacturing engineers, product planners, market researchers, and ergonomics engineers). The design activities are driven by the intricate coordination and simultaneous consideration of many requirements (e.g., customer requirements, engineering functional requirements, business requirements, government regulatory requirements, and manufacturing and assembly requirements) and trade-offs between the requirements of different systems in the vehicle. The systems should not only function well, but they must satisfy the customers who purchase and use the products. The field of ergonomics or human factors engineering in the automotive product development involves working with many different vehicle design teams (e.g., exterior design team, interior design teams, package engineering teams, driver interface/instrument panel team, seat design team and management teams) to ensure that all important ergonomic requirements and issues are considered at the earliest time and resolved to accommodate the needs of the users (i.e., the drivers, passengers, personnel involved in assembly, maintenance, service, and so forth) while using (or working on) the vehicle.

Thus, the ergonomics engineers assigned to a vehicle program must understand vehicle systems, teams and team members assigned to design the vehicle systems, responsibilities and tasks performed by the teams and their timings to ensure that ergonomics support (e.g., ergonomic design guidelines, scorecards, recommendations and data) is provided to different teams at the right times during the vehicle development process.

Systems engineering is an engineering specialty that helps coordinate the work of many disciplines and design teams involved in the vehicle development process. It also ensures that important events in the vehicle program such as milestones (or gateways) are scheduled, and design reviews and management approvals are obtained before the vehicle development process moves to the next design phase. Systems engineers perform their functions by creating and following a Systems Engineering Management Plan (SEMP). The SEMP shows how, when and who, from various specialty disciplines – including ergonomics – would provide the necessary technical support.

DOI: 10.1201/9781003485582-1

OBJECTIVES

The objective of this book is to provide the reader a thorough understanding of ergonomic issues, design guidelines, models, methods to measure user performance and preference, and the various analysis procedures used in designing and evaluating automotive products. The second key objective is to provide sufficient background into how and when an ergonomics engineer should work with the specialists from other design disciplines (e.g., package engineering, body engineering, lighting engineering, climate control engineering, driver information and entertainment systems design, and so forth) to ensure that ergonomically superior vehicles are designed with a comprehensive understanding of their functional needs and trade-offs between different vehicle systems and their characteristics.

This book is organized in two volumes to develop an understanding of ergonomic issues and provide full ergonomics support to the vehicle programs,. The first volume provides the reader with important ergonomic principles, theories and models, design considerations, design practices, research studies and reference data related to the design of different chunks and systems within the vehicle. The second volume provides information on topics faced by the ergonomic engineers to better understand and provide support to the vehicle program in advanced topics (e.g., modeling driver vision to evaluate visibility and legibility), future vehicle features (e.g., driver assistance systems, electric and autonomous vehicles), detailed engineering, evaluation methods, cost-benefit analysis and product liability.

ERGONOMICS: WHAT IS IT?

Ergonomics is a multi-disciplinary science involving fields that have information about people (e.g., psychology, anthropometry, biomechanics, anatomy, physiology and psychophysics). It involves studying human characteristics, capabilities and limitations and applying this information to design and evaluate the equipment and systems that people use.

The basic goal of ergonomics is to design equipment that will achieve the best possible fit between the users (drivers) and the equipment (vehicle) such that the users' safety (freedom from harm, injury and loss), comfort, convenience, performance, preferences and efficiency (productivity or increasing output/input) are improved.

The field of ergonomics is also called Human Engineering, Human Factors Engineering, Engineering Psychology, Man-Machine Systems, or Human-Machine Interface (HMI) Design, and Human Centered Design. After the Second World War, the field of Human Factors emerged in the United States, mainly among the psychologists, to study the equipment and process design problems primarily from the human information-processing viewpoint. The field of ergonomics emerged in the European countries around 1949 to improve workplaces and jobs in the industries with an emphasis on biomechanical applications. The word, "ergonomics", the science of work laws (or the science of applying natural laws to design work), was coined by joining two Greek words "ergon" (work) and "nomos" (laws) (Jastrzebowski, 1857; Murrell, 1958). Over the past 50 years, the field covers both the information processing and physical work aspects and is more commonly known as "Human Factors

Engineering" or "Ergonomics" with about equal preference in the use of either name for the field. After the fuel economy crisis of the 1970s, the U.S. automobile industry began placing more emphasis on both the "Aerodynamics" and "Ergonomics" fields to satisfy the customers with their energy-saving and comfort/convenience needs. The use of both the somewhat similar sounding terms, aerodynamics and ergonomics, also was perceived to be somewhat appealing in marketing the products.

ERGONOMICS APPROACH

FITTING THE EQUIPMENT TO THE USERS

Ergonomics involves "Fitting the equipment to the people (or users)". This means that equipment should be designed such that people (population of users) can fit comfortably (naturally) within the equipment, and they can use the equipment without any awkward body postures, movements or errors.

It should be noted that ergonomics is NOT about fitting people to equipment (i.e., equipment should not be designed first and then people are simply asked to somehow adapt or force-fit to use it). In some cases, the equipment is designed such that only people with certain characteristics can fit or use them (which normally involves a personnel selection strategy, that is, placing restrictions on the "type" of people who can use the equipment).

DESIGNING FOR THE MOST

Ergonomics involves "Designing for the MOST" (i.e., to ensure that most users within the intended population of the users of the product can fit within the product). It should be noted that if we use other design strategies, like "designing for the average" or "designing for the extreme", only a few individuals within the user population will find the product to be "just right" (or fit very well) for them. Thus, designing for the most will involve making sure that the designer knows what the user population is and knows the distributions of characteristics, capabilities and limitations of the individuals in that population and uses the information to accommodate most (e.g., 95% of the users to design the product).

SYSTEMS APPROACH

Another important consideration involves "Human as a Systems Component". This means that the designer must treat the human as a component of the system that is being designed. The process for designing a vehicle should thus involve the considerations of the following major components: a) the driver/user, b) the vehicle and c) the environment (see Figure 1.1). The characteristics of all the components in the system must be considered in designing the vehicle. The vehicle design should not only involve designing all the physical components that fit and function well, but also make sure that the user is considered as a component and the user's characteristics are measured and used in designing the car or the truck – to ensure that the vehicle would meet the users' needs related to comfort, convenience, performance, preferences and

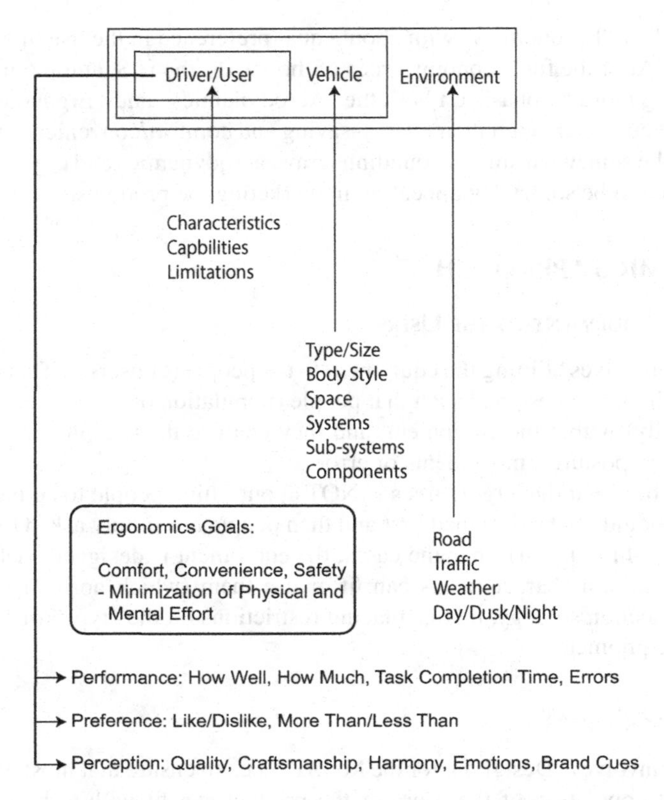

FIGURE 1.1 Ergonomics engineers' considerations are related to characteristics of the driver, the vehicle and the environment and their relationship to driver performance, preference and perception.

safety. It should be noted that during the design process of a physical product, the engineer designs each part of the product by paying attention to all its properties (e.g., dimensions, material, hardness, color, surface, how it fits/works with other components, etc.). Similarly, when the human is involved as an operator or the user of the product (e.g., a car or a truck), all relevant human (customer/user) characteristics must be studied and used in designing the product.

Thus, in designing a vehicle, a thorough understanding of the intended user population and the operating environment (which consists of the roadway, the traffic, the weather and operating conditions such as the dawn, day, dusk and night) of the vehicle must be considered. Figure 1.1 shows that when a driver/user operates a vehicle in a driving environment, the ergonomics engineer must consider the characteristics of all the components in the system and evaluate: a) how the driver/user will perform various tasks, b) the driver/user's preferences in using the product, and c) the pleasing perceptions created by experiencing the product such as the quality, craftsmanship, emotions evoked by the product and its resulting brand image. Bailey (1996) has proposed another but somewhat similar approach to conceptualize ergonomic problems. Bailey's approach involves considering the user, the activity (type of operation or usage) and the context (usage situation) in designing a system.

The word "systems" in systems engineering is also used to cover the following aspects of different systems in an automotive product:

1. An automobile product is a system containing a number of other vehicle systems (e.g., body system, powertrain system, chassis system, electrical system, fuel system, climate control system and entertainment system).
2. The design of the whole automobile thus will involve designing all the systems within the automobile such that the systems work together (i.e., the systems are interfaced or connected with other systems, and each system performs its respective functions) to create a fully functional vehicle and meet its customers' needs.
3. Professionals from many different disciplines (e.g., industrial design, mechanical engineering, electrical engineering, physics, manufacturing engineering, product planning, finance, business and marketing) are required to design (i.e., to make decisions related to the design) all the systems in the vehicle.
4. The vehicle has many different attributes (e.g., characteristics that its customers expect such as performance, fuel economy, safety, comfort, styling, package, weight and costs). And simultaneous inputs from professionals from many disciplines and specialists with deep knowledge in each of the vehicle systems are required to make decisions about proper consideration of levels of all the attributes and trade-offs between the attributes in designing all the systems within the vehicle.
5. The automotive product is a component of other larger systems (e.g., vehicle platform [which may be shared with other vehicle brands and models], highway transportation system, petroleum refining and fuel distribution system, electrical power generation and distribution system, financial system, and so forth).
6. The automobile works within different environmental, topographic and situational conditions (e.g., driving on a winding road at night in a thunderstorm).
7. All phases of the life cycle, from conceptualization of a new automotive product to its discontinuation (i.e., its disposal, scrappage, recycling, replacement, plant dismantling or retooling, etc.) must be considered during its design.

Thus, simultaneous considerations of many systems, many attributes, trade-offs between the attributes, life cycle, disciplines, other systems and working environments in solving problems (i.e., decision-making) constitute what is called the "Systems Approach". The systems approach is thus the primary and necessary part of systems engineering.

PROBLEM-SOLVING METHODOLOGIES

To solve different problems encountered during the development of a new automotive product, the ergonomics engineer relies on a number of different approaches.

Figure 1.2 shows three pure or basic approaches in solving a problem. The least time-consuming approach shown is the middle branch of Figure 1.2. It uses the "guess", or the guessed answer made by the decision maker. The "guessed" answer is generally not regarded as an objective answer as compared to the other two approaches, namely, using a model or performing an experiment shown in the outer branches of Figure 1.2. The modeling approach shown in the left branch of Figure 1.2 assumes that a well developed and validated model exists, and it can be exercised by inputting different combinations of the values of the input (or independent) variables to arrive at the best solution. The experimental approach is used where an ergonomics engineer designs and conducts an experiment to determine the best combination of independent variables needed to obtain the required level of output.

It may seem that the "guess" approach is a dangerous approach. However, in many cases, when a design recommendation is needed immediately (which occurs quite frequently in the auto industry – where your boss wanted an answer yesterday), the ergonomics engineer needs to provide his best "guessed" answer. Of course, it is assumed that the ergonomics engineer does his homework, which generally involves steps such as: a) reviewing the problem along with available sketches, drawings or hardware related to the problem, b) reviewing past studies, c) consulting other ergonomics experts, d) applying available models, and e) providing the best "guessed" answer with clearly specified assumptions. The "soundness" of the guessed answer will depend upon the level of expertise possessed by the decision maker and the amount of information available while making the guess. If more time is available, then additional literature or a data search benchmarking similar products developed by leading competitors, conducting analyses using pre-developed models, or even running a "quick-react" experiment could help in proposing a solution.

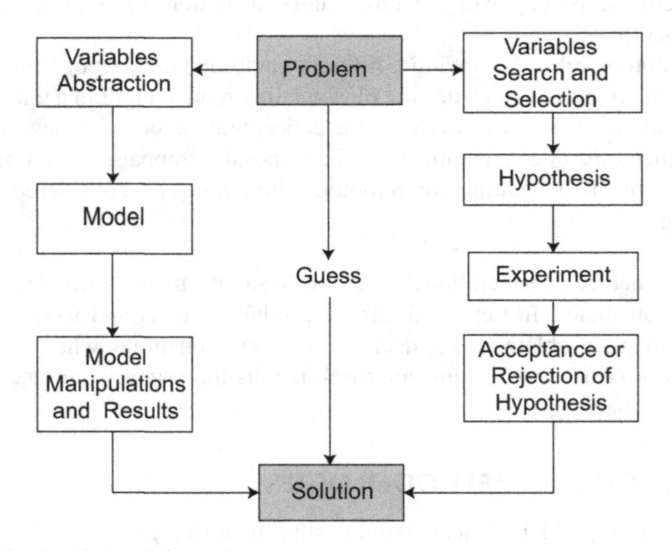

FIGURE 1.2 Problem-solving approaches.

Embedded within the thought processes involved in providing a solution is really the good old scientific method. This method is applied in all fields of science to develop solutions. The scientific method has been found to be very effective in solving ergonomic problems. The scientific method involves the following steps (remember the acronym DAMES suggested by Konz and Johnson, 2004):

D = *D*efine the problem broadly
A = *A*nalyze all variables that can affect the performance of people
M = *M*ake a search of solution space
E = *E*valuate alternatives – determine the best solution by applying the systems knowledge, models or by experimental research
S = *S*pecify and sell the solution

Since each new problem faced by an ergonomics engineer is generally "new" for which a model to predict a solution is not available, the experimental approach may be the best. The experimental approach, however, is generally very time-consuming and costly. In reality, since no experiment or model is complete (because they involve the use of only a few key variables and have limited validity and applicability), a proficient ergonomics engineer will use a unique combination of all three approaches shown in Figure 1.2 to solve each problem. Further, it should be realized that to resolve many design issues a number of iterations using the combinations of the above three approaches and other variations – such as brainstorming within and between teams, design reviews or structured walkthroughs with a team of experts – may be very useful.

ERGONOMICS RESEARCH STUDIES

Most research studies available in the ergonomics field can be categorized into the following three types or their combinations:

1. *Descriptive Research:* This type of research generally provides data describing human characteristics of different populations (e.g., anthropometric measurements such as stature, sitting height, shoulder width and elbow-to-elbow width and their distributions).
2. *Experimental Research:* This type of research generally involves experiments conducted to determine the effects of different combinations of independent variables on certain response variables under carefully manipulated and controlled experimental situations (e.g., determining the effects of different vehicle parameters such as seat height, rocker width, and rocker height on comfort ratings or performance measures during an entry into a vehicle).
3. *Evaluative Research:* This type of research generally involves comparisons of user performance (and/or preference) in using different designs (vehicle systems or features) (e.g., determining which of the four proposed entertainment system designs would be most convenient to use and/or preferred by the drivers by using 10-point rating scales).

ERGONOMICS ENGINEER'S RESPONSIBILITIES IN VEHICLE DESIGN

The inclusion of ergonomics engineers as a part of the vehicle development team and its many sub-teams is now an accepted practice in the automobile industry. The ergonomics engineers work from the earliest stages of new vehicle concept creation to the periods when the customer uses the vehicle, disposes it, and is ready to purchase his/her next vehicle.

The ergonomics engineer's major tasks during the life cycle of the vehicle are:

1. Provide the vehicle design teams with needed ergonomics design guidelines, information, data, analyses, scorecards and recommendations for product decisions at the right times (during team meetings, design reviews and at gateways, or milestones, in the vehicle program timing chart).
2. Make presentations of the ergonomic studies, scorecards and recommendations in front of the right level of decision makers (involving design teams, program managers, chief engineers, and senior management).
3. Apply available methods, models, standards and procedures (e.g., Society of Automotive Engineers [SAE, 2010] and company practices) to address issues raised during the vehicle development process.
4. Conduct quick-react studies (i.e., experiments) to answer questions raised during the vehicle development process.
5. Evaluate product/program assumptions, concepts, sketches, drawings, physical models/mock-ups/bucks, CAD models, mules, prototypes, and production vehicles made by the manufacturer and its competitors.
6. Participate in the design and data-collection phases of drive clinics and market research clinics involving concept vehicles and existing leading products as comparators (or controls) and in validation studies.
7. Obtain, review and act on customer feedback data from complaints, warranty, customer satisfaction surveys, market research data (e.g., J. D. Powers survey data [J. D. Powers, 2024]), vehicle inspection surveys with owners, automotive magazines, press, and so forth.
8. Provide ergonomics consultations to members of the vehicle development and attribute engineering teams.
9. Long Term: Conduct research, translate research results into design guidelines and develop design tools.

The ergonomics engineer's roles and responsibilities are discussed in more detail in Chapter 13.

HISTORY: ORIGINS OF ERGONOMICS AND HUMAN FACTORS ENGINEERING

Prehistoric Times and Functional Changes in Products

The history of ergonomics goes back to the designing of tools used by prehistoric man. These tools developed in the prehistoric era and evolved over many years. Many generations of families mastered certain trades and became "craftsmen". Certain designs that were functional lived on. Useless tools were not remade. Changes were made for functional improvements. Thus, when considering future product improvements, we should not introduce any design change for the sake of change. Changes should only be made if functional improvements can be achieved.

Air Force Research

After the Second World War, the U.S. Air Force systematically studied many problems experienced by the pilots. They found that pilot errors in using aircraft displays and operating controls were caused mainly by improper design of equipment and, thus, the equipment could be better designed to reduce pilot errors (Fitts and Jones, 1947a and 1947b). (Note: This topic is discussed in more detail in Chapter 7. Other approaches to improve human performance: Motivation, Training, Selection and Testing).

"Ergonomics" Coined

The word ergonomics was coined in England in 1949 by K. F. H. Murrell by joining two Greek words: "Ergon" = work, "nomos" = natural laws, to convey the philosophy of applying natural laws in designing for people. Thus, to design products for people, we should think about what comes naturally to people. In general, we have to think about the designs that naturally conform to people. The natural "fit" improves performance, that is, reduces operation/task completion time and errors, reduces learning and training times, and makes products more enjoyable, less difficult, and less boring.

HISTORY OF ERGONOMICS IN AUTOMOTIVE PRODUCT DESIGN

Some key historic events related to the applications of ergonomics to automotive design issues are presented below.

1918: SAE issued J585 standard on Tail Lamps and J587 standard on License Plate Illumination Devices.
1927: SAE issued J588 standard on Stop Lamps.
1956: Ford Motor Company established Human Factors Engineering Department.
1965: SAE publishes J941: Motor Vehicle Driver's Eye Locations (Eyellipse) recommended practice.
1966: Congress passes Safety Acts: National Traffic Safety and Motor Vehicle Safety Act and the Highway Safety Act.
1969: NHTSA (National Highway Traffic Safety Administration, Department of Transportation) publishes notices of proposed rulemaking in Crash Avoidance area (e.g., Vehicle Lighting – luminous intensity requirements on headlamps and signal lamps).
1976: SAE publishes J287: Driver Hand Control Reach recommended practice
1978: NHTSA enacted FMVSS 128 (Field of View Requirements on Motor Vehicles), rescinded later in 1979.
1984: Touch CRT introduced in cars (GM's Buick Riviera).

1986: Center High Mounted Stop Lamp (CHMSL) required on all vehicles by amending FMVSS 108 (See Chapter 9 on research related to requiring a CHMSL).
1997: Toyota launched "Prius" (Hybrid electric vehicle with state-of-power display).
1997: NHTSA published a report on Investigation of the Safety Implications of Wireless Communications in Vehicles (Goodman et al., 1997).
2000: Adjustable Pedals introduced in the U.S. market on light truck products.
2000: NHTSA hosted the first Internet Forum on Driver Distraction.
2001: Smart headlamps introduced in luxury vehicles.
2007: Ford introduced "Sync" to connect cell phones and i-pods and other USB-based systems.
2007: Rear seat entertainment systems with display screens available for rear passengers.
2010: Capacitive touch-screen technology introduced in vehicles.
2011: Plug-in Electric Vehicles sold in the U.S. market by major automotive manufacturers with power consumption gauges (eco-gauges).
2017 (circa): A combination of driver assistance features such as lane departure warning, forward collision warning, blind spot monitoring and warning, adaptive cruise control, driver alertness monitoring, and tire pressure monitoring have been offered by many auto manufacturers as standard and optional equipment. Many autonomous vehicles developed by different manufacturers began testing of the vehicles in controlled field tests and on limited public roads.
2022: Electric pickup trucks introduced in the U.S. market (e.g., Ford F-150 Lightning).

Presented in Appendix 1 are additional historic points in tracing the progress of Human Factors Engineering.

IMPORTANCE OF ERGONOMICS

Characteristics of Ergonomically Designed Products, Systems, and Processes

If a product (or a system) is designed well (i.e., meets ergonomics requirements), the following effects or outcomes are expected:

1. An ergonomically designed product should "fit" people well (like a well-fitting suit). (Note: One would use a well-fitting suit much more often than a poorly fitting one). Thus, when it is time to replace an old product, a customer will most likely purchase a newer version of the same product that fits him/her well. This suggests that ergonomically designed products will more likely be repurchased.
2. An ergonomically designed product can be used with minimal mental and/or physical work. Thus, as product usage increases, the customer will realize

the ease, comfort and convenience features and the absence of problems (e.g., difficulties, misuses and operating errors) while using the product.

3. An ergonomically designed product is "easy-to-learn". (Note: Owner's manuals of easy-to-learn products are seldom used. Easy-to-learn products work in an "expected" manner.)
4. A product with usability problems (i.e., the absence of ergonomics) can be quickly noticed – usually after use. Thus, the ergonomic characteristics of many products are not generally noticed in showrooms where the customer does not have an opportunity to use them.
5. Ergonomically designed products are generally more efficient (productive) and safer (less injurious).

Why Apply Ergonomics?

1. Create functionally superior products/processes/systems.
2. Costly and time-consuming redesigns can be avoided (with early incorporation of ergonomic inputs in the design process; superior products or systems can be developed without additional design iterations).
3. There are thousands of ways to design a product, but only a few designs are truly outstanding. (You want to find those "outstanding" designs quickly).

Ergonomics Is Not Common Sense

1. Common sense ideas/solutions are often wrong. For example, a designer wanted to create an instrument panel illuminated with "deep red" lighting for a new hot sports car. The ergonomics engineer reminded him that about 8% of U.S. males have a color deficiency in perceiving "the color red". The designer said, "but the air force used red-colored instrument panels in airplanes". The ergonomics engineer reminded the designer again saying "color-deficient persons cannot get a pilot's license, but a car is a consumer product, and you don't want to annoy these color-deficient males in using your vehicle. If you want red then we should add some yellow in it and make it orangish-red so that the red-color deficient people can still read the instruments".
2. Knowledge-based decisions are superior as they minimize usability problems. The ergonomist brings his/her specialized knowledge and data about the users during early phases in the design process.

ERGONOMICS AS AN IMPORTANT SPECIALTY DISCIPLINE IN AUTOMOTIVE DESIGN

Inputs from Other Disciplines

Designing a new automotive product requires input from many people with specialized backgrounds in many disciplines and vehicle systems. For example, mechanical engineers working in body engineering as compared to powertrain engineering require different knowledge (e.g., variables and considerations associated during design, manufacturing and assembly) to design their respective systems and lower-level systems. The body design work requires background in sheet metal parts design (e.g., structural analysis involving strengths, deflections, bending and tortional stiffnesses) and manufacturing and assembly processes (e.g., sheet metal parts bending, forming, shearing and spot-welding). Whereas the powertrain design engineer will require background in specialized areas such as heat management in internal combustion engines (or electric motors), durability of gears, shafts and bearings, and manufacturing processes such as casting, forging, machining, grinding, and assembly processes involving fasteners such as nuts and bolts. Other systems that are interfaced (or attached) to the vehicle body such as powertrain and climate control systems are also functionally very different. For example, wiring harnesses, tubes and hoses for providing fluids (fuel, oil, coolant) are connected to the engine and ducts carrying cold and hot air are connected to the climate control system.

Ergonomic considerations in designing body systems such as instrument panel, doors, seats and exterior lamps require evaluations of different issues (e.g., anthropometric dimensions and postures) as compared to powertrain design considerations such as packaging space, visibility over the hood and around pillars, vehicle acceleration, noise, vibrations, and so forth.

In designing many of the automotive systems and their lower-level sub-systems and components involve study of human interfaces (human sensory inputs, their perception and level of comfort/discomfort and driver response). Ergonomics engineers are often consulted to reduce the driver's discomfort, workload and errors in operating the equipment.

INTERFACES WITH MANY AUTOMOTIVE SYSTEMS

Different vehicle systems are also dependent upon the operation of other vehicle systems. For example, for the engine to provide power to its drive wheels, the engine needs to be physically and functionally connected to the transmission. Further, the engine and transmission need to be physically connected to the vehicle body with special mounts to reduce vibrations sensed by the driver and the passengers in the vehicle. To understand the effect of such interfaces on the comfort, safety and performance of driver and the passengers, the ergonomics engineers are also consulted.

An automobile is a complex product, and it has many systems. And many of the systems and their lower-level entities are interfaced with several other systems. Chapter 20 in Volume 2 provides more information on the types of interfaces and development of interface requirements.

VEHICLE ATTRIBUTES RELATED TO CUSTOMER SATISFACTION

The vehicle systems also perform many functions needed to satisfy many customer/user needs. The vehicle characteristics and features required to meet these customer needs are defined and called vehicle attributes. The vehicle attributes are managed by special attribute engineering departments such as vehicle package engineering, ergonomics, vehicle handling and dynamics, noise, vibrations and harshness (NVH), thermal and aerodynamics, performance and drivablity, emissions, weight, and so forth. The members of this attribute engineering department are constantly in communication with each other in their team meetings to review and solve many design problems. Systems engineering department has the primary responsibilities to ensure that the required attribute engineering personnel are present and perform their necessary tasks (e.g., consulting and evaluations involving testing, data collection and data analysis) to provide the required information at the required time to the different design teams.

APPLYING ERGONOMICS AS A VEHICLE ATTRIBUTE

Figure 1.3 shows how customer needs are used to develop vehicle systems through use of vehicle attributes. The first box on the left shows that the customer needs are determined by interviewing a representative sample of customers (from the population of drivers and users in the market segment of the vehicle being developed) and from feedback from owners (after use of their new vehicles). The customer needs are then used to define attributes (or characteristics) that the vehicle must possess.

The second box from the left shows a list of attributes that a vehicle must have to sell well. (The number of vehicle attributes, their descriptions and titles may vary between different vehicle manufacturers.) The goal here is to develop a comprehensive and complete list of vehicle attributes so that when all vehicle requirements are defined no customer needs are missed. Each vehicle attribute is further decomposed into sub-attributes, sub-sub-attributes and sub-sub-sub-attributes and so on. The attribute decomposition ensures that all lower-level customer needs are satisfied by lower levels of attributes.

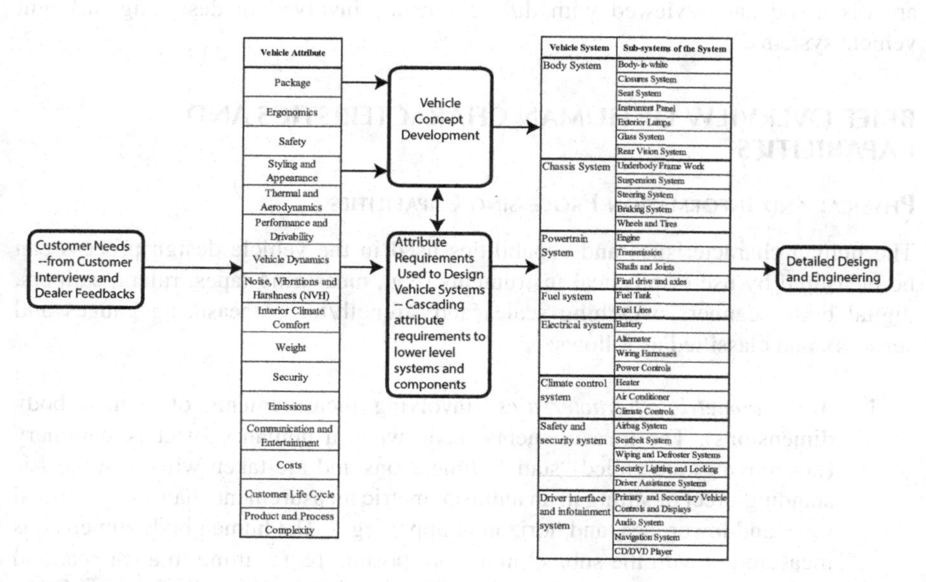

FIGURE 1.3 This flow diagram showing steps from customer needs determination to detailed design and engineering with listings of vehicle attributes and vehicle systems.

The third column from the left in Figure 1.3 shows that the list of vehicle attributes and their requirements are used to develop different vehicle concepts. The listings of vehicle systems and their sub-systems that need to be packaged (configured, allocated space and placed within the vehicle) are shown in the fourth column from the left.

The detailed vehicle design (the last box in Figure 1.3) involves creating designs of all vehicle systems including all of their lower-level entities to ensure that each of the entities will perform their allocated functions, and they work together by using interfaces (e.g., physical fasteners, electrical wires, tubes, and hoses. See Chapter 20) between the entities within the vehicle.

To facilitate this process, the vehicle-level attribute requirements are cascaded down (i.e., assigned) to vehicle systems and their lower-level entities (i.e., sub-systems, sub-sub-systems and so on until the lowest component level entity is defined in detailed engineering phase [Chapter 21]). This process is described in more detail in the next chapter.

Ergonomics is one of the vehicle attributes. Meeting all requirements of the ergonomics attribute and its sub-attributes should meet the overall goal of designing an ergonomically superior vehicle. The sub-attributes of the ergonomics attribute can be defined as: (a) controls and displays; (b) occupant package; (c) field of view and driver vision; (d) entry/exit; (e) exterior interfaces, and so forth. The ergonomics attribute is typically shared with occupant packaging and mechanical packaging sub-attributes (within package attribute). Similarly other attributes that require careful integration with ergonomics are safety, performance and drivability, vehicle dynamics, NVH, interior climate comfort, communications and entertainment and customer life cycle. Ergonomics considerations and requirements related to location and operation of all vehicle systems (including their low-level entities) and their interface requirements are discussed and reviewed with different teams involved in designing different vehicle systems.

BRIEF OVERVIEW OF HUMAN CHARACTERISTICS AND CAPABILITIES

Physical and Information Processing Capabilities

The human characteristics and capabilities used in the vehicle design process can be measured by use of physical instruments (e.g., measuring tapes, rulers, calipers, digital body scanners, weighing scales and strength/force measuring gauges and sensors). and classified as follows:

1. *Anthropometric characteristics* (involving measurements of human body dimensions). The measurements made when a human subject is stationary (not moving) are called "static" dimensions and are taken when a subject is standing-erect or sitting in an anthropometric measurement chair (with vertical torso and lower legs, and horizontal upper legs). The human body dimensions measured is with the subject in a work posture (e.g., sitting in a car seat and performing a task) are called "functional" anthropometric dimensions. Other measurements of the human body (and body segments) such as surface areas,

volumes, center of gravity, and weights are also considered to be part of anthropometry (science of human body dimensions) (See Chapter 4 for more details).
2. *Biomechanical characteristics* (e.g., ability to produce forces/strength and motion) (see Chapter 4 for more details).
3. *Information Processing Capabilities.* These are mental (cognitive) capabilities involving the acquisition of information through various sensors (in the human eyes, ears, joints, vestibular tissues, etc.), transmitting this sensed information to the brain, recalling information stored in the memory, processing the information to make decisions (detecting, recognizing, comparing, selecting, etc.) and making responses (e.g. motor action – generating a body movement, activation a control, or making a verbal response). (See Chapter 6 for more details).

In general, many human abilities degrade as people get older. The degradation in most human abilities is about 5–10% per decade after about 25 years of age. With practice, humans can perform complex tasks with very little or no conscious effort. However, humans are not consistent or precise in performing tasks like machines. Thus, human performance in most tasks varies considerably. The variability in the same subject performing the same task is called the "within subject variability", whereas "between subject variability" is the difference in performance when a different subject performs the same task. When designing a vehicle, the ergonomics engineers must make sure that most users in the population can perform the tasks associated with the vehicle.

IMPLEMENTING ERGONOMICS – CONCLUDING COMMENTS

The following chapters of this book covers ergonomic concepts, issues and methods used in designing different automotive products and their features. Volume 1 of these two volumes deals with how the anthropometric, biomechanical and information processing considerations are used in designing aspects of occupant package, seats, controls, displays, instrument panels, window openings, entry/exit, vehicle service related issues, craftsmanship and descriptions on how the ergonomics engineers work with the teams involved in the vehicle design process. Volume 2 covers more advanced and research-oriented issues such as modeling of human vision to predict visibility and legibility, performance measurement, evaluation of driver workload, dealing with the special-user populations, and research needs to enable implementation of new technologies in future vehicles.

REFERENCES

Bailey, R. W. 1996. *Human Performance Engineering.* Upper Saddle River, New Jersey: Prentice Hall PTR.

Fitts, P. M. and R. E. Jones.1947a. Analysis of Factors Contributing to 460 "Pilot Errors" Experiences in Operating Aircraft Controls, Memorandum Report TSEAA-694-12, Aero Medical Laboratory, Air Materiel Command, Wright-Patterson Air Force Base, Dayton, Ohio. In *Selected Papers on Human Factors in the Design and Use of Control Systems,* ed. H. Wallace Sanaiko, 332–358, New York: Dover Publications, Inc.

Fitts, P. M. and R. E. Jones.1947b. Psychological Aspects of Instrument Display. I: Analysis of Factors Contributing to 270 "Pilot Errors" Experiences in Reading and Interpreting Aircraft Instruments, Memorandum Report TSEAA-694-12A, Aero Medical Laboratory, Air Materiel Command, Wright-Patterson Air Force Base, Dayton, Ohio. In *Selected Papers on Human Factors in the Design and Use of Control Systems*, ed. H. Wallace Sanaiko, 359–396, New York: Dover Publications, Inc.

Goodman. M. J., F. Bents, L. Tijerina, W. W. Wierwille, N. A. Lerner, and D. Benel. 1997. An Investigation of the Safety Implications of Wireless Communications in Vehicles. Report no. DOT HS 808 635. Washington, D.C.: U. S. Department of Transportation, the National Highway Traffic Safety Administration.

J. D. Powers. 2024. Customer Surveys on Initial Quality, In-Service and Product Appeal. www.jdpower.com/business/us-initial-quality-study-iqs (accessed March 8, 2024)

Jastrzebowski, B. W. 1857. *An Outline of Ergonomics, or the Science of Work Based upon the Truth Drawn from the Science of Nature*. Przyrodo i Premyst, 29–32, Proznan. Warszawa: Central Institute for Labour Protection. (Reprinted as Commemorative edition published on the occasion of the *XIVth Triennial Congress of the International Ergonomic Association and the 44th Annual Meeting of the Human Factors and Ergonomics Society* [San Diego, CA], 2000.)

Konz, S. and S. Johnson. 2004. *Work Design – Occupational Ergonomics*. Sixth Edition, IBSN# 1-890871-48-6. Scottsdale, AZ: Holcomb Hathaway, Publishers, Inc.

Murrell, K. F. H. 1958. The Term "Ergonomics". *American Psychologist, 13, Issue 10*, p. 602.

Society of Automotive Engineers, Inc. 2010. *The SAE Handbook*. Warrendale, PA: SAE.

2 Ergonomics in the Systems Engineering Process

INTRODUCTION

The emphasis in this book is on designing ergonomically superior automotive products that are designed and made by people with specialized skills and disciplines and used by people. Thus, these products must be developed to satisfy the needs of customers who want vehicles with certain characteristics (e.g., body style, size, level of luxury and features).

Automotive products are very complex. The complexity in any product can be directly related to a number of systems contained in the product: number of subsystems in each of the systems, number of components in each of the subsystems, and number of interfaces between different components, subsystems, and systems. The development process of complex products can be made more effective through a systematic implementation of the systems engineering (SE) processes, application of a number of its techniques, and by coordinating activities of many different teams involved in designing many systems within the vehicle.

OBJECTIVES

The objectives of this chapter are to provide a deeper understanding of the SE, its approach, processes, issues, techniques, advantages, and disadvantages. In this chapter, we develop the basic understanding in the SE and the product development processes by defining many terms and considerations used in the implementation of the processes. We also study how ergonomics as a specialized discipline and a vehicle attribute should be managed to provide support to different co-located design teams and coordinated with the systems engineering management plan.

VEHICLE ATTRIBUTES

Vehicle attributes are the characteristics that a vehicle must have to satisfy its customers. The attributes are derived from the customer needs, business needs and the government requirements (or standards) that the vehicle must meet. Ergonomics is also one of the vehicle attributes, and the ergonomic engineers assigned to the vehicle program must work with many teams involved in designing the vehicle.

DOI: 10.1201/9781003485582-2

SYSTEMS ENGINEERING FUNDAMENTALS

What Is a System?

A system consists of a set of components (or elements) that work together to perform one or more functions. The components of a system generally consist of people, hardware (e.g., parts, tools, machines, computers, and facilities), or software (i.e., codes, instructions, programs, databases) and the environment within which it operates. The system also requires operating procedures (or methods) and organization policies (e.g., documents with goals, requirements and rules) to implement its processes to get its work done. The system also works under a specified range of environmental and situational conditions (e.g., temperature and humidity conditions, vibrations, magnetic fields, and power/traffic flow patterns). The system must be clearly defined in terms of its purpose, functions, and performance capability (i.e., abilities to perform or produce output at a specified level under specified operating environment).
Some definitions of a system:

1) A system is a set of functional elements organized to satisfy specified objectives. The elements include hardware, software, people, facilities and data.
2) A system is a set of interrelated components working together toward some common objective(s) or purpose(s) (Blanchard and Fabrcky, 2011).
3) A system is a set of different elements so connected or related as to perform a unique function not performable by the elements alone (Rechtin, 1991).
4) A system is a set of objects with relationships between the objects and between their attributes (Hall, 1962).

The set of components have the following properties (Blanchard and Fabrycky, 2011):

a) Each component has an effect on the whole system.
b) Each component depends upon other components.
c) The components cannot be divided into independent subsystems.

WHAT IS SYSTEMS ENGINEERING?

Systems Engineering is a multidisciplinary engineering decision-making process involved in designing and using products and systems. The SE activities involve both the technical and management activities from an early design stage to the end of the life cycle of a product or a system. The SE is also a multidisciplinary approach, that is, it involves specialized professionals from many different disciplines to work simultaneously together and consider many design and operational issues and trade-offs between product characteristics to enable the realization of a successful product or a system. SE focuses on defining customer needs and required functionality early in the development cycle, documenting requirements, and proceeding with the design synthesis and system (product) validation while considering the complete problem (INCOSE, 2006).

Objective of SE

The objective of the SE is to ensure that the product (or the system) is designed, built, and operated so that it accomplishes its purpose of satisfying its customers in the most cost-effective way possible by considering performance, safety, costs, schedule, and risks.

Characteristics of SE Approach

The basic characteristics of the SE approach are as follows:

1. *Multidisciplinary*: The SE is an activity that knows no disciplinary bounds. It involves professionals from a collection of disciplines throughout the design and development process. The professionals work together (simultaneously and co-located under one roof) constantly communicating and helping each other with all aspects of the product. The types of disciplines to be included depend on the type and characteristics of the product and the scope of the product program.

   For example, the SE application for an automotive product will require disciplines such as engineering (e.g., mechanical, electrical/electronics, computer and information science, chemical, manufacturing, industrial, human factors/ergonomics, quality, environmental, and safety engineering), sciences (e.g., physics, chemistry, and life sciences), industrial designers (who define the sensory form of the product, i.e., the look, feel, sound, and smell of interior and exterior of the product, such as styling and appearance of surfaces of the product, touch feel of the surface and material characteristics, sounds of operating equipment, and smell of materials), market researchers (who define the product needs, its market segment, customers, market price, and sales volumes), management (e.g., program and project management personnel involving product planners, accountants, controllers, and managers), plant personnel involved in its manufacturing and assembly, insurers, distributors, and dealers.

   It is important to get inputs from all the disciplines that affect the characteristics and uses of the product at the early stages of the product development. This ensures that their needs, concerns, and trade-offs between different multidisciplinary issues are considered and resolved early during product development, and costly changes or redesigns are avoided.
2. *Customer focused*: The SE places continuous focus on the customers, that is, the product design should not deviate from the needs of the customers. The customers should be involved in defining the product needs and in subsequent evaluations to ensure that the product being designed will meet their needs.
3. *Product-level requirements first*: The SE places concentrated effort on initial definition of the requirements at the overall or the whole product level. For example, at the product level, the requirements on an automotive product will be based on all basic major attributes (that are derived from the needs of its external and internal customers) of the vehicles such as safety, fuel economy,

drivability (ability to maneuver, accelerate, decelerate, and cornering or turning), seating comfort, thermal comfort, styling, and costs.

It is important to realize that the customer buys the "integrated whole" product for his/her use and not as a mere collection of independent components that form the product. Thus, the requirements on the systems, subsystems, and components of the product should be derived only after the product-level requirements are clearly understood and defined.

4. *Product life cycle considerations*: The SE includes considerations of the entire life cycle of the product being designed – throughout all stages from "Concept Development to Disposal of the Product" (from lust to dust). Thus, it is the application of all relevant scientific and engineering disciplines in all the phases of the product such as designing, manufacturing, assembly, testing and evaluation, uses under all possible operating conditions, service and maintenance, disposal (or recycling) and retirement from service – which the product encounters throughout its life cycle.
5. *Top-down orientation*: The SE considers "top-down" approach that views the product (or the entire system) as a whole and then sequentially breaks down (or decomposes) the product into its lower levels such as systems, subsystems, and components.
6. *Technical and management involvement*: The SE is both a technical and a management process. It involves making all the technical decisions related to the product and its life cycle as well as the management of all the tasks to be completed in a timely manner to implement the SE process and apply the necessary techniques.
7. *Technical process*: The technical process of the SE is an analytic effort necessary to transform the operational needs of the customers into a design of the product (or a system) with proper size, capacity, and configuration. It creates documentation of the product requirements and drives the entire technical effort to evolve and verify an integrated and life-cycle balanced set of solutions involving the users and the product in its usage situations.
8. *Management process*: The management process of the SE involves assessing costs and risks, integrating the engineering specialties and design groups, maintaining configuration control, and continuously auditing the effort to ensure that the cost, schedule, and technical performance objectives are satisfied to meet the original operational need of the product and the product program.
9. *Product-specific orientation*: The SE implementation (i.e., its details such as steps, methods, procedures, team structure, tasks, and responsibilities) depends on the product being produced (i.e., its characteristics) and the company producing it (i.e., different companies generally have somewhat different processes, timings, organizational responsibilities, and brand-specific requirements).

The *NASA Systems Engineering Handbook* (NASA, 2016) describes the SE as both the art and science of developing an operable system (or a product) capable of meeting requirements within often-opposed constraints. It further describes the SE as a holistic, integrative discipline, wherein the contributions of structural engineers, electrical engineers, mechanism designers, power engineers, human factors engineers, and many more disciplines are evaluated and balanced, one against another, to produce a coherent whole that is not dominated by the perspective of a single discipline (NASA, 2016).

AUTOMOTIVE PRODUCT AS A SYSTEM

An automotive product is considered as a system that involves a number of lower- (or second) level systems such as body system, chassis system, powertrain system, fuel system, electrical system, climate control system, braking system, and so on. Each of the systems within the automotive product can be further decomposed into subsystems, sub-subsystems, sub-sub-subsystems and so on, until the lowest level components are identified. For example, the body system includes body frame subsystem, body panels subsystem, closure subsystem (which includes hood sub-subsystem, doors sub-subsystem, trunk or lift-gate sub-subsystem), exterior lamps subsystem (headlamps, tail lamps and side-marker lamps), seats subsystem, instrument panel subsystem, interior trim components subsystem, and so forth.

SYSTEMS, SUBSYSTEMS, AND COMPONENTS IN AN AUTOMOTIVE PRODUCT

Table 2.1 illustrates major systems, subsystems and sub-subsystems or components within a typical automotive product with an internal combustion engine. The definitions and contents of various vehicle systems illustrated in this table can vary

TABLE 2.1
Major Systems and Their Sub-systems in a Typical Automotive Product

Vehicle System	Sub-systems of the System	Ergonomic Considerations during Design Phases
Body System	Body-in-white	Binocular obscurations caused by pillars; Occupant space; View over hood/beltline; Mirror locations; Heat shields/protection
	Closures System	Door handle location, shape, opening/closing forces; Opening dimensions and angles
	Seat System	SgRP locations; Seating comfort; Entry/exit;
	Instrument Panel	Types of controls and displays; Layout; Operability; Veiling glare; Warning signals and messages, On-board diagnostics.
	Exterior Lamps	Luminous intensity; Visibility; Glare;
	Glass System	Fields of view; Transmissivity; Glare; Wiped and defrosting areas
	Rear Vision System	Mirror size, location; Blind area sensors and fields

(continued)

TABLE 2.1 (Continued)
Major Systems and Their Sub-systems in a Typical Automotive Product

Vehicle System	Sub-systems of the System	Ergonomic Considerations during Design Phases
Chassis System	Underbody Frame Work	Ground clearance
	Suspension System	Floor height; Entry/exit
	Steering System	Steering wheel diameter, location, rim and spokes shape/size, materials; Obscurations
	Braking System	Stopping distances and decelerations; Brake pedal location and forces;
	Wheels and Tires	SgRP height; Entry/exit
Powertrain System	Engine	Engine temp, RPM, sound
	Transmission	Shifter type location, feedback; Shift feel; Sound
	Shafts and Joints	Rear occupant space and tunnel size
	Final drive and axles	Wheelbase and occupant compartment length
Fuel system	Fuel Tank	Fuel display location. Type; Fuel filler location
	Fuel Lines	Hand/finger clearances in engine compartment
Electrical system	Battery	Location/reach/clearances for service
	Alternator	Output display; Malfunction warning
	Wiring Harnesses	Malfunction warning
	Power Controls	Gains, rates, sounds, feedbacks of power control operations
Climate control system	Heater	Locations and size of air registers
	Air Conditioner	Locations and size of air registers
	Climate Controls	Types of controls and displays; Layout; Operability
Safety and security system	Airbag System	Location, size, warning messages
	Seatbelt System	Belt-fit comfort, anchor point locations, strength
	Wiping and Defroster Systems	Wiper pattern, field of view; wiper speeds; defrosting time to clear snow
	Security Lighting and Locking Systems	Lamp locations, beam pattern, timings and operational features
	Driver Assistance Systems	Controls, displays and operability; driver workload and distractions
Driver interface and infotainment system	Primary and Secondary Vehicle Controls and Displays	Locations of foot pedals, steering wheel and gear shifter; operating forces and feedback
	Audio System	Understanding/Operability; sound quality
	Navigation System	Understanding operability; content and format of displayed information
	CD/DVD Player	Understanding/Operability; sound quality

somewhat between different vehicle makes and models. Further, implementation of different technologies used in performing different vehicle functions can have a major effect on the design of any vehicle system. In fact, one of the challenges the vehicle engineering groups face is – how to divide the entire vehicle into different systems, subsystems, sub-subsystems, and so on and to assign design responsibilities to various engineering teams. This issue of division or decomposition of an automotive product for management of various product development activities and their interfaces is covered in Chapters 7, 8, 12 and 20.

The key tasks of systems designers are to ensure that each system performs its functions and systems through their interfaces with other systems work harmoniously to meet the customer needs of the whole product. Thus, the task of designing the vehicle requires a lot of understanding of systems, their functions, interactions between systems, and trade-offs between vehicle attributes to come up with a balanced vehicle design. Details related to these issues are covered throughout this book.

MANAGING A COMPLEX PRODUCT

The need for the SE arose with the increase in complexity of the products. (Note: Designing very simple products can be accomplished with a very small number of people.) The increased product complexity in turn increases the number of interactions (i.e., relationships) between many components and also increases the challenges in designing for high levels of reliability. The complexity is not only due to the engineering aspects of the systems but also due to the organization and management of people from multiple disciplines, data and making a multitude of decisions that affect many systems.

Therefore, a complex product can be divided (or decomposed) into a number of manageable entities. This decomposition is an important task in the management of the product development process. A complex product can be decomposed into several hierarchical levels such as systems, systems into subsystems, and subsystems into components. The number of levels of divisions depend on many factors such as the functions of the product, the past design experience and problems encountered in developing similar products, ability of the design team to deal with many design issues simultaneously, level of the design and engineering details that need to be analyzed and evaluated, stringency in meeting requirements, program schedule, and so forth.

Figure 2.1 shows an illustration of a product decomposition tree. This figure presents a tree diagram (an upside-down tree) showing a top-down progressive decomposition of a product (P) into its systems (S1 to S4), each system into its subsystems (e.g., SS11, SS12, SS21, …, SS42), and each subsystem into its components (C111, C112, …, C425).

Table 2.2 provides an example of systems and components of a laptop computer (a complex product but less complex than an automotive product). The first column of the table provides a brief description of the function of each system. The second and third columns provide the system name and its components, respectively.

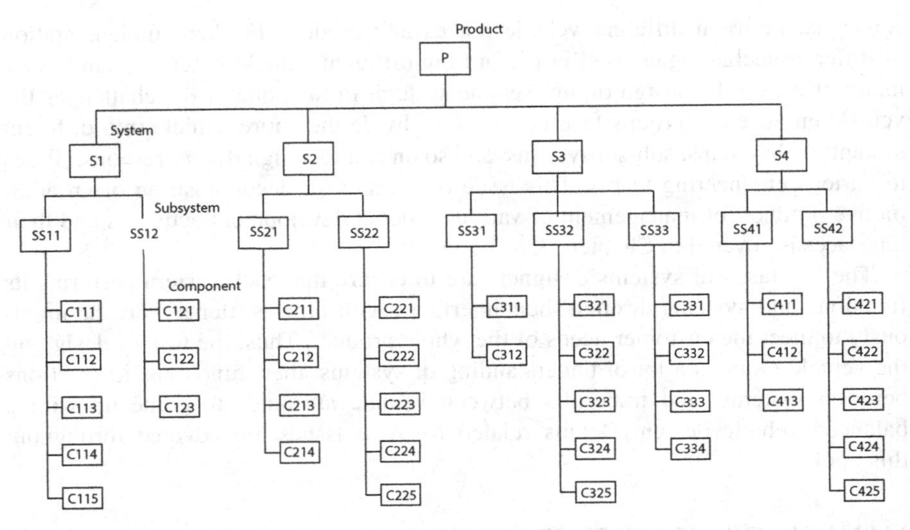

FIGURE 2.1 Illustration of top-down decomposition of a product into systems, subsystems, and components.

TABLE 2.2
Functions, Systems, and Components of a Laptop Computer

Function	System	Components
Hold all parts together	Chassis system	Top cover, hinges, magnesium chassis, palm rest, expansion cover, bottom cover
Display visual information	Display system	14-inch display, protective glass, magnesium bezel
Present audio data	Audio system	Speakers, audio cable, stereo audio system, display connector
Input data	Input system	Keyboard, touch pad and track point, microphone
Process data	Electronic processing system	Printed circuit board (system board with a multi-core processor)
Store data	Memory system	DVD Drive, 64 GB solid-state drive, memory chips
Provide electrical power	Power system	AC to DC converter, replaceable battery, connectors, cables, internal battery
Communicate with the Internet	Wireless communication system	Wireless local network, wireless wide network
Protect components from overheat	Cooling system	Cooling fan, fan motor, heat sink, and shield

It should be noted that each system exists to serve at least one or more functions required for the product to meet its requirements. It is also important to understand and keep track of the functions of each system, because a design team involved in designing each system must make sure that the system performs its functions.

AUTOMOTIVE PRODUCT DEVELOPMENT PROCESS

WHAT IS AUTOMOTIVE PRODUCT DEVELOPMENT?

The automotive product development process involves designing and engineering of a future automotive product. The automotive product (i.e., a vehicle) can be a car or a truck or a variant such as a pickup truck, a station wagon, an SUV (i.e., sports utility vehicle), or a van. The manufacturing and assembly operations are generally assigned to a different group (i.e., generally not a part of the product development). However, selected representatives from manufacturing and assembly operations should actively participate in the teamwork during the product development process.

The automotive products are generally produced in large quantities (about 10,000–700,000 vehicles per year per model and at rate typically of about 40–70 vehicles/hour on an assembly line), then shipped to dealers in many locations and sold to customers to meet their transportation needs. The vehicles must be safe, efficient, economical, dependable, "fun to drive and use" and "pleasing" to the customers. The vehicles also must have the necessary characteristics such as performance (i.e., operating capabilities), styling/appearance (form), quality (customer satisfaction) and craftsmanship (perception of being well-made). The customers must "enjoy owning the vehicles" – that is, the vehicles must have all the necessary attributes and the right features to meet their lifestyles.

Flow Diagram of Automotive Product Development

The vehicle development process generally begins with the members of the design team understanding the customers and their needs and ends up with the customers providing their feedback after using the vehicle. Figure 2.2 shows the major phases in the vehicle-development process along with the production, marketing, sales and vehicle usage phases. Based on an understanding of the customer needs, government requirements and business needs of the company, the design team consisting of members from different disciplines (e.g., industrial design, product architects, engineering, manufacturing, product planners and market researchers) generally develops attribute requirements at the vehicle level and creates the vehicle specifications. The information is used by the team to develop one or more vehicle concepts (in the form of sketches, drawings, CAD models, mock-ups or vehicle bucks). The vehicle concepts are iteratively improved by using customer feedback and suggestions by different team members and are market researched along with the vehicle's leading competitors to determine if a leading concept could be selected for subsequent detailed design and engineering work. During the detailed engineering phase, CAD models of all entities from components to the whole vehicle are prepared along with materials selection and many supporting engineering analyses and tests to ensure that the design will meet the attribute requirements of the vehicle. Based on the selected product design, manufacturing processes and suppliers are chosen. The production equipment and plants are designed and built or modified for their manufacturing and assembly. Marketing, sales and distribution plans are developed. The early production parts and systems are assembled into prototype vehicles. All entities from component to major vehicle systems are tested to verify that they meet their respective requirements. The assembled systems are installed into vehicle bodies and prototype vehicles are created. These prototyped vehicles are further tested to verify and validate vehicle level requirements. Next, the results of the evaluation tests are presented to the company management. After the final approval to produce the vehicle is given by senior management, the vehicle is "launched" (i.e., production begins). The produced vehicles are shipped to the dealerships for sale. As the purchased vehicles are used by the customers, feedback from the customer experience (i.e., data from field-operating performance, customer likes/dislikes, vehicle repairs and warranty work) are continuously collected and provided for improving existing products and designing future products.

To support the entire vehicle development process, resources (e.g., dollars, people, equipment, and facilities) are needed. Budgets and schedules are created to manage the entire product-development process. The costs of all the product-development activities are borne by the company, its investors or partially supported through low interest loans or tax incentives from local, state or federal government. The organization begins to make money only after the vehicles are sold, that is, revenues are generated from the vehicle sales.

Ergonomics engineers are involved in providing necessary data on driver and passenger related ergonomic issues and considerations in understanding most of the phases shown in the flow diagram. Table 2.3 provides examples of ergonomic

TABLE 2.3
Examples of Ergonomic Considerations in Vehicle Systems and Subsystems

Vehicle System	Sub-systems of the System	Ergonomic Considerations during Design Phases	Chapters providing additional information
Body System	Body-in-white	Binocular obscurations caused by pillars; Occupant space; View over hood/beltline; Mirror locations; Heat shields/ protection	4.5, 8, 10
	Closures System	Door handle location, shape, opening/closing forces; Opening dimensions and angles	4, 5, 8, 10
	Seat System	SgRP locations; Seating comfort; Entry/exit;	4,5

TABLE 2.3 (Continued)
Examples of Ergonomic Considerations in Vehicle Systems and Subsystems

Vehicle System	Sub-systems of the System	Ergonomic Considerations during Design Phases	Chapters providing additional information
	Instrument Panel	Types of controls and displays; Layout; Operability; Veiling glare; Warning signals and messages, On-board diagnostics.	4,5, 6,7, 8
	Exterior Lamps	Luminous intensity; Visibility; Glare;	8, 9
	Glass System	Fields of view; Transmissivity; Glare; Wiped and defrosting areas	8
	Rear Vision System	Mirror size, location; Blind area sensons and fields	8
Chassis System	Underbody Frame Work	Ground clerance	5
	Suspension System	Floor height; Entry/exit	5
	Steering System	Steering wheel diameter, location, rim and spokes shape/size, materials; Obscurations	4, 5, 7, 8, 10
	Braking System	Stopping distances and decelrations; Brake pedal location and forces;	4, 5
	Wheels and Tires	SgRP height; Entry/exit	5
Powertrain System	Engine	Engine temp, RPM, sound	5, 6, 7
	Transmission	Shifter type location, feedback; Shift feel; Sound	5, 6, 7
	Shafts and Joints	Rear occupant space and tunnel size	5
	Final drive and axles	Wheelbase and occupant compartment length	5
Fuel system	Fuel Tank	Fuel display location. Type; Fuel filler location	5
	Fuel Lines	Hand/finger clearances in engine compartment	5
Electrical system	Battery	Location/reach/clearances for service	11

(continued)

TABLE 2.3 (Continued)
Examples of Ergonomic Considerations in Vehicle Systems and Subsystems

Vehicle System	Sub-systems of the System	Ergonomic Considerations during Design Phases	Chapters providing additional information
	Alternator	Output display; Malfunction warning	7, 11
	Wiring Harnesses	Malfunction warning	7
	Power Controls	Gains, rates, sounds, feedbacks of power control operations	7
Climate control system	Heater	Locations and size of air registers	7
	Air Conditioner	Locations and size of air registers	7
	Climate Controls	Types of controls and displays; Layout; Operability	7
Safety and security system	Airbag System	Location, size, warning messages	7
	Seatbelt System	Belt-fit comfort, anchor point locations, strength	7
	Wiping and Defroster Systems	Wiper pattern, field of view; wiperspeeds; defrosting time to clear snow	7, 8
	Security Lighting and Locking Systems	Lamp locations, beam pattern, timings and operational features	7, 8
	Driver Assistance Systems	Controls, displays and operability; driver workload and distractions	7, 8
Driver interface and infotainment system	Primary and Secondary Vehicle Controls and Displays	Locations of foot pedals, steering wheel and gear shifter; operating forces nd feebbacks	5, 7
	Audio System	Understanding/Operability; sound quality	7
	Navigation System	Understanding operability; content and format of displayed informatin	6, 7
	CD/DVD Player	Understanding/Operability; sound quality	6, 7

considerations and chapters in this book that provide more details about the issues and their evaluations.

SYSTEMS ENGINEERING PROCESSES IN PRODUCT DEVELOPMENT

SYSTEMS ENGINEERING PROCESS

The SE Process during product development includes the following basic tasks:

1. Define the objectives of the product and the product program
2. Establish product performance requirements (requirement analysis)
3. Establish the functionality of the product (functional analysis)
4. Develop alternate design concepts for the product (architectural synthesis)
5. Select a baseline product design (selection of a balanced product design)
6. Verify that the baseline product design meets requirements (verification)
7. Validate that the baseline product design satisfies its users (validation)
8. Iterate the above process through lower levels (cascading product requirements to lower [decomposed] levels of entities through allocation of functions and design synthesis)

It is important to realize that product development typically involves many activities and tasks. The tasks can be organized and described differently depending on objectives or viewpoints of the describer (e.g., a product design engineer versus a systems engineer). A product design engineer may describe the product development and manufacturing process shown in Figure 2.2, with the following major phases: (1) understanding the customer needs and product planning; (2) product concept development; (3) detailed design and engineering (see Chapter 21 for details); (4) production planning process (includes production equipment and facilities design process); (5) production (manufacturing and assembly); (6) product distribution and sales; and (7) collection of customer feedback, whereas a systems engineer will describe the product development process differently as shown in Figure 2.3.

Figure 2.3 shows the following major SE tasks in the product development process: (1) defining customer needs; (2) conducting requirements analysis (defining the requirements of the product); (3) conducting functional analysis (i.e., determining what functions need to be performed by the product, its systems, subsystems, and components) and allocating functions (i.e., assigning the functions to each system, subsystem, and component of the product); (4) conducting design synthesis (i.e., integrating, evaluating, and refining the whole product configuration and its architecture); and (5) formulating a balanced product design by determining trade-offs between requirements and different vehicle features or issues related to performance, safety, quality (includes perceived quality/craftsmanship, reliability, manufacturing and assembly), costs, and timing schedules.

These tasks are performed iteratively (in the sense that the process is applied again in each iteration) as more design issues and product details (e.g., packaging,

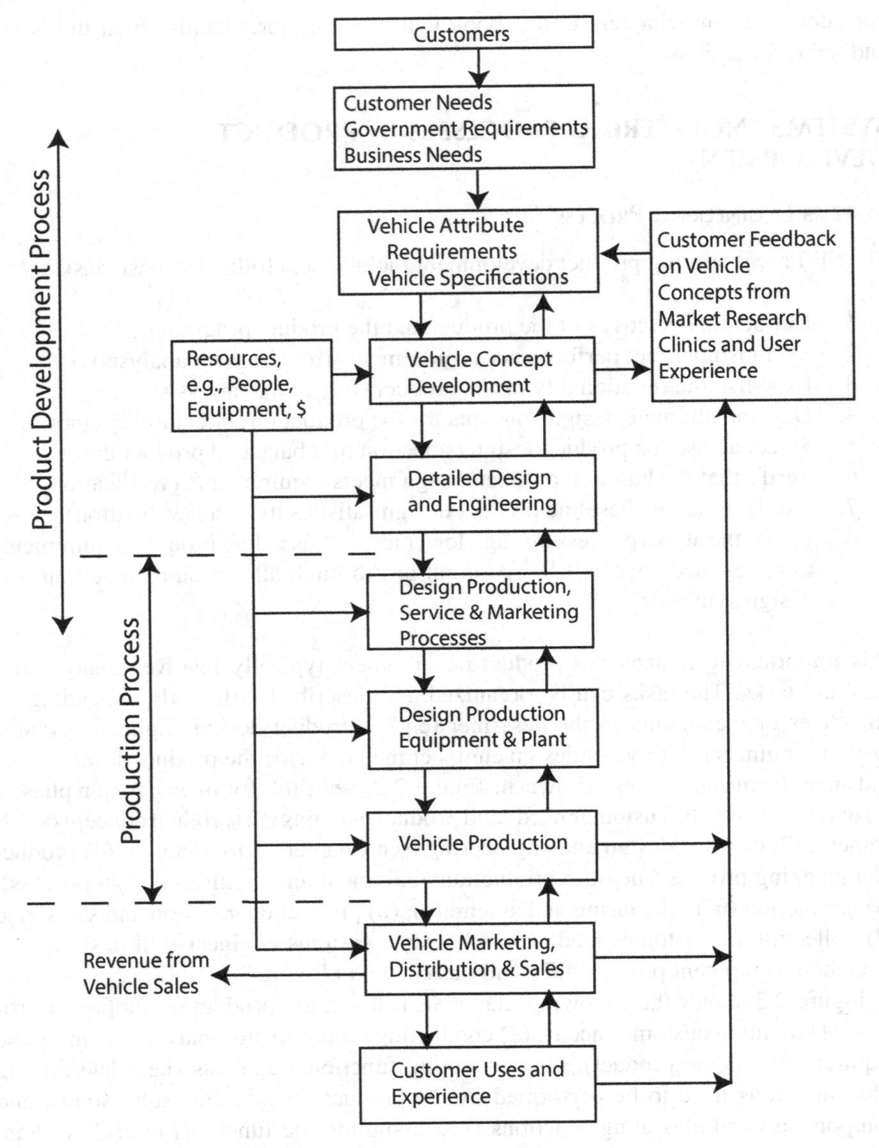

FIGURE 2.2 Vehicle development and production process.

functionality, assembly and manufacturing, usability and styling) are analyzed during each successive iteration. The overall product design with its systems is evaluated and redesigned (reconfigured, adjusted, or refined) in each iteration until an acceptable product is realized. Multifunctional co-located design teams with constant communication and design reviews between team members and simultaneous (concurrent)

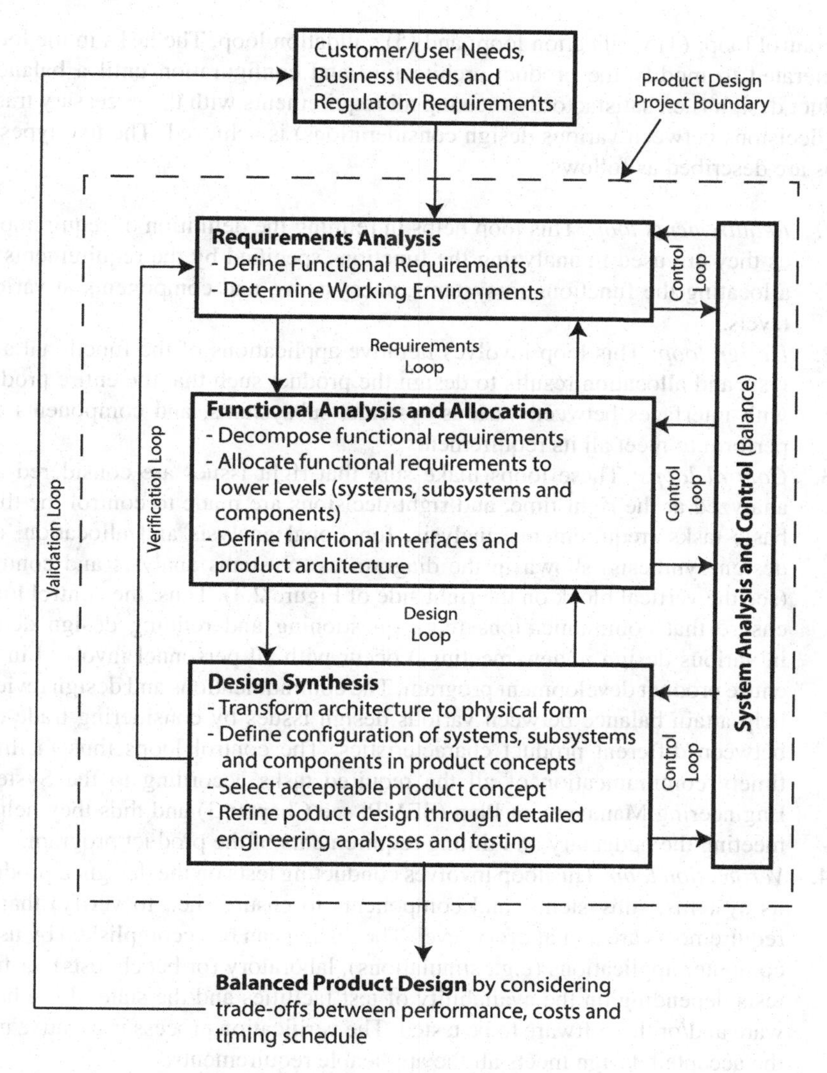

FIGURE 2.3 Systems engineering process with major tasks and loops.

engineering approach in performing the above activities is important for successful completion of the product design activities.

Five Loops in the Systems Engineering Process

It is important to realize that the SE process shown in Figure 2.3 contains the following five types of loops (feed forward and back): (1) requirements loop; (2) design loop;

(3) control loop; (4) verification loop; and (5) validation loop. The tasks in the loops are iterated to modify the product architecture and configuration until a balanced product design (i.e., satisfactorily meeting all requirements with the necessary trade-offs decisions between various design considerations) is achieved. The five types of loops are described as follows:

1. *Requirements loop*: This loop helps in refining the definition of requirements as they are used in analyzing the functions specified by the requirements by allocating the functions to systems, subsystems, and components at various levels.
2. *Design loop*: This loop involves iterative applications of the functional analysis and allocation results to design the product such that the entire product with interfaces between various systems, subsystems, and components can perform to meet all its requirements.
3. *Control loops*: These loops make sure that right issues are considered and analyzed at the right time, and right decisions are made to control the three basic tasks (requirements analysis, functional analysis and allocation, and design synthesis) shown in the diagram as "Systems analysis and control" (see the vertical block on the right side of Figure 2.3). Thus, the control loops ensure that communications (e.g., questioning and refining design details in various design review meetings) occur with all personnel involved in the entire product development program. The communications and design reviews help attain balance between various design issues by considering trade-offs between different product characteristics. The control loops thus facilitate timely communication of all the required tasks according to the Systems Engineering Management Plan (SEMP; see Chapter 3) and thus they help in meeting the budgetary and timing requirements of the product program.
4. *Verification loop*: This loop involves conducting tests on the designed product, its systems, subsystems, and components to ensure (i.e., to verify) that all requirements are met at every level. The testing can be accomplished by using computer applications (e.g., simulations), laboratory (or bench tests), or field tests depending on the availability of test facilities and the state of the hardware and/or the software to be tested. The verification process is iterative until the accepted design meets all the applicable requirements.
5. *Validation loop*: This loop involves tests and evaluations that are conducted to ensure that the product will meet all its stated customer needs, that is, the product will be acceptable to the customers. The validation tests are generally performed at the whole product (i.e., vehicle) level. However, system – or subsystem – level tests using customers are also useful in validating designs at the lower levels. The iterations of the validation loops and any changes resulting from the failures or deficiencies in the validation testing are generally expensive. If the product is designed right with the proper analyses in all the preceding tasks and their iterations, then the validation tests can be performed with minimal number of iterations.

Major Tasks in the Systems Engineering Process

The major tasks shown in the three middle blocks and verification and validation loops of Figure 2.3 are described in the following section with more details.

Requirements Analysis

Requirements analysis is critical to the success of a product. The requirements should be documented, actionable, measurable, testable, traceable, related to identified customer and business needs (including meeting government rules and requirements), and defined to a level of detail sufficient to account for all the product attributes (i.e., product characteristics desired by the customers). This analysis either verifies that the existing requirements are appropriate or develops new requirements that are more appropriate for the mission/operation of the product.

The analysis includes the following: (1) development of measures suitable for ranking alternative designs in a consistent and objective manner and (2) evaluations of the impact of environmental factors and operational characteristics of the performance of the product and minimaly acceptable functional requirements. These measures and evaluations should also consider the impact of the design on costs and schedule. Each requirement should be periodically examined for validity, consistency, desirability, and attainability.

Functional Analysis and Allocation

The purpose of the functional analysis is to determine the functions of each system, subsystem, or component of the product. The functions are determined from the performance requirements. For a given system, subsystem, or a component to meet its specified performance requirement, the systems engineer must progressively identify and analyze system functions and subfunctions in order to identify alternatives to meet the system requirements. The function identification task is performed in conjunction with function allocation and design synthesis activities that generally involve making assumptions regarding possible design configurations of the product and its systems. All specified modes of operation and situations (e.g., normal, unusual, or emergency) are considered in this analysis, and functions and subfunctions of all systems, subsystems, and components are identified. The process of allocation of function involves the following: (1) assignment of a requirement to a function; (2) assignment of a system element to a requirement; and (3) division of a requirement among entities and assignment of each component to its higher level entity (e.g., subsystem or system). For example, a requirement on the total weight of the system can be considered by allocating target weight to each component (or percentage of the requirement satisfied or supported by each component).

During the allocation process, the SE activity allocates performance and design requirements to each system function and subfunction. These derived requirements are stated in sufficient detail to permit allocation of one or more functions to entities such as hardware, software, procedural data, or personnel. The SE also identifies any special personnel skills or peculiar design requirements that must be considered to perform the allocated functions. Allocation activities are performed in conjunction with the functional analysis and synthesis activities. Traceability of the allocated system requirements should be maintained to every function allocated to each entity.

The functions assigned to a system, subsystem, and component will be dependent on the type of technologies selected for the product. For example, for an automotive product, selection of technology for its power train system will affect functions of many other systems, their subsystems and components. If a gasoline engine is selected, then the fuel system and the entities (systems, subsystems, and components) of the engine will be very different than if an electric powertrain is selected. Thus, during the early iterations of the functional analysis, many different technologies and the level of readiness of the technologies should be considered to determine how various different requirements can be met.

Design Synthesis

During the synthesis, the SE together with the representatives of the hardware, software, and other appropriate engineering specialties (e.g., vehicle assembly) develops a product architectural design (i.e., overall configuration of the product) that is sufficient to meet the performance and design requirements that are allocated in the detailed design (see Chapter 21). Design of the product architecture occurs simultaneously with the allocation of requirements and analysis of the product and system functions.

The design is generally documented by using block diagrams, flow diagrams, and/or package engineering drawings (or computer-aided design [CAD] models). These diagrams will (1) portray the arrangement of items that make up the baseline design; (2) identify each element along with techniques for its test, support, and operation; (3) identify the internal and external interfaces; (4) permit traceability to source requirements through each decomposition level; and (5) provide procedures for comprehensive change control. Several interface and flow diagramming techniques that can be used in the design synthesis are covered in Chapter 20 and 21. Various block diagramming techniques used for system breakdown and functional analyses are also described by Blanchard and Fabrycky (2011) and Bhise (2017).

The documentation of work performed in this phase becomes the primary source of data for developing, updating, and completing the product, systems, and subsystems specifications, interface control documentation, specification trees, and test requirements. Interface control requirements and drawings should be established, coordinated, and maintained. Changes to these documents are maintained and disseminated to all appropriate participating engineering groups (Note: Interface diagrams and interface matrices are covered in Chapter 20).

During the final configuration and requirements definition, the SE uses the specifications as a mechanism to transfer information from the systems requirements analysis to system architecture design and system design tasks. Joint signoffs of specifications by the specification authors, the designers, and the design engineers pertaining to the SE and the design engineering disciplines ensure understanding and buy-in into the overall product design. The specifications should ensure that the requirements are testable and are stated at the appropriate specification level.

Specialty engineering functions (e.g., human factors/ergonomics engineering, safety engineering, quality engineering, and others; also called attribute engineering functions) participate in the SE process in all phases (see Table 2.5 showing organization matrix). They are responsible for reliability, maintainability, testability, manufacturing capability, quality management, human factors, safety, and design to meet their cost targets. Specialty engineering shall be involved in the issuing of design requirements and the monitoring of the progress of the design and performance analysis to ensure that the design requirements are met, and the trade-offs made between various product attributes are acceptable.

Verification

During requirements verification, the SE and test engineering conduct tests to verify that the completed product and its systems meet all the requirements contained in the requirements specifications. Tests conducted to verify requirements are performed using hardware and software (or acceptable computer simulations of both hardware and software) configured to the final design.

Validation

The whole product is tested under all types of situations expected during the operating life of the product. The testing should involve a representative sample of customers to ensure validity. In addition, industry experts (i.e., subject matter experts) who are very knowledgeable (and even more critical than the customers) about the product attribute issues should perform the validation evaluations. The outcomes of the tests are used to determine whether the product will be acceptable to its customers.

Verification versus Validation

The difference between verification and validation can be better understood by studying the locations of the verification and validation blocks in the flow diagram presented in Figure 2.4. It shows the validation loop on the right side and the verification loop on the left side of the diagram. The validation is generally performed at the whole product level (see the right side loop beginning from the "Balanced product" in Figure 2.4). The verification loop generally evaluates systems, subsystems, or components that are lower level functions (as compared with the product level evaluated for validation purposes). Volume 2 (Chapters 19 and 22) provides more information on verification and validation procedures and methods.

Subsystems and Components Development

Cascading requirements from a higher level to a lower level (e.g., product to systems, systems to subsystems, and subsystems to components) require considerable knowledge about possible configurations, technologies, and trade-offs between product characteristics in developing basic architecture of the products and its systems. The term "cascading requirements" means transfer or allocation of a higher level requirement to a lower level entity. This means that a lower level entity must be designed to meet a function allocated to a higher level entity. The cascading of requirements from

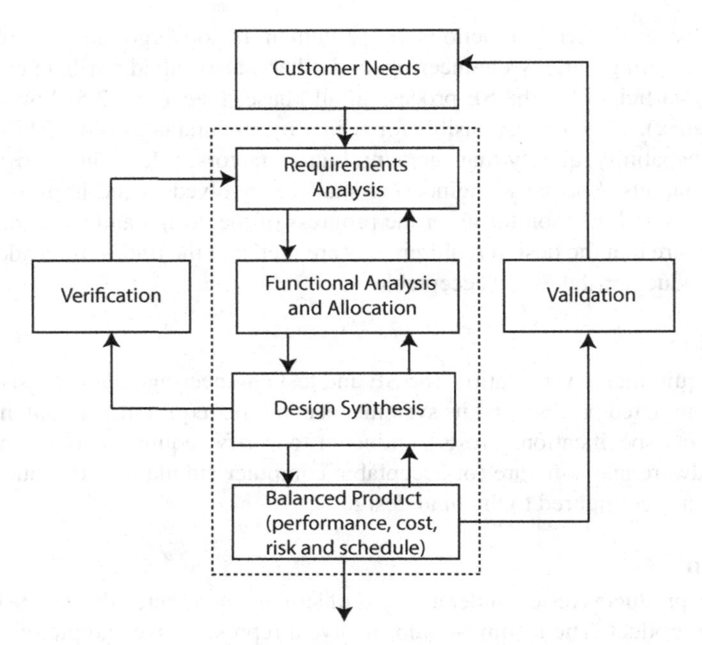

FIGURE 2.4 Verification and validation processes in product design.

a higher to a lower level involves the following: (1) careful use of functional analysis and allocation to ensure that no requirements are dropped or misallocated (i.e., the requirement is allocated to an entity as compared with another entity that can perform the required functions better or when designed in a different configuration); (2) specifying subsystems with minimal interfaces with other systems and subsystems to satisfy the requirement with lower costs and reduced variability; and (3) specifying each subsystem requirement with clearly defined targets, test procedures, evaluation measures, and criteria for acceptable performance.

Design reviews with multifunctional team of experts are generally very useful to evaluate alternate configurations with different allocations of functions needed to fulfill the cascading of the product- or higher-level requirements. The reviews should be conducted openly with the clear intention of working together to improve the customer acceptability of the product. The following issues should be discussed during the design reviews: (1) results of functional analysis and function allocation issues; (2) interfaces between systems and subsystems; (3) trade-offs to be considered between functions and features of different systems and subsystems; (4) effects on costs, reliability, and major product attributes (e.g., safety or weight); and (5) feasibility analysis results, design challenges, and problems. The feasibility analyses should consider issues such as (1) manufacturing and assembly feasibility and meeting or exceeding the targets specified in all applicable requirements; (2) feasibility of meeting time schedules and budgets; and (3) technical and personnel support, and development challenges.

Example of Cascading a Requirement from the Product Level to a Component Level

Let us assume that an automotive climate control engineer is asked to design a climate control for a passenger car. The requirement on the climate control system would be that the occupants of the vehicle should be comfortable inside the passenger compartment by controlling the temperature to a customer preferred level inside the compartment within 20°–29°C (68°–85°F) when the outside air temperature is anywhere between -7°C and 49°C (-20°F in the winter to 120°F in the summer).

The subsystems of the climate control system are as follows: (1) hot air exchanger; (2) cold air exchanger; (3) air blower with a variable speed motor; (4) duct system; (5) hot coolant fluid (water) delivery from the engine cooling system; (6) compressed refrigerant from the air-conditioning pump; (7) hoses and tubes connecting the heat exchangers of the refrigerant and hot coolant fluid; (8) thermostat; (9) air blower speed controller; and (10) electrical system. Other vehicle systems that affect the occupant thermal comfort are vehicle body (window glass areas and body insulation) and seating subsystem (i.e., seating geometry and materials).

The climate control system requirements can be cascaded down to (1) the heating system involving hot air exchanger, air blower, blower speed controller, thermostat, and hot coolant supply system; (2) the cold air supply system involving the cold air exchanger, air blower, blower speed controller, thermostat, and the compressed refrigerant supply system; (3) windows, glass, and vehicle body that conducts heat from outside air temperature and radiant heat from the sun load through the glass areas; and (4) heat transfer characteristics of the seat upholstery.

The cascaded requirements are shown in Table 2.4. The vehicle (product)-level requirement is stated in the second column from the left. The cascaded system-, subsystem-, and component-level requirements are stated in fourth, sixth, and eighth columns from the left, respectively.

The requirements and the requirements-setting processes associated with various ergonomic analyses are presented in subsequent chapters.

Iterative Nature of the Loops within the Systems Engineering Process

The SE process is based on an iterative, top-down, hierarchical decomposition of the product-level requirements to the component-level requirements. The decomposition process is supported by studies and analyses performed by specialists from different disciplines that analyze the basis for significant decisions and the options considered. The iterative, top-down, hierarchical decomposition methodology includes the parallel activities of the functional analysis, allocation, and synthesis. The iterative process begins with product-level decomposition and proceeds through the major system level, the functional subsystem level, and the hardware/software configurations of entities at the lower levels. As each level is developed, the activities of functional analysis, allocation, and synthesis should be completed before proceeding to the next lower level.

When carryover components are used, bottom-up iterative processes also need to be implemented to ensure that the subsystems and systems are designed to meet the

TABLE 2.4
An Illustration of Requirements Cascade from Product Level to Component Level

Product Level	Vehicle Level Requirement	System Level	System Level Requirement	Subsystem Level	Subsystem Level Requirement	Components	Component Level Requirement
Product = Passenger Car	Comfortable thermal environment inside the passenger compartment	Climate Control System	Provide hot or cold air according to thermostat and blower speed in accordance with outside heat load	Heating System	Provide hot air according to thermostat and blower speed in accordance with outside heat load	Hot air exchanger	Allow hot water to circulate
						Air Blower	Blow air at set speed
						Hot water supply	Send hot water to hot air exchanger
						Thermostat	Control hot water inflow
						Air blower speed controller	Control blower speed to speed setting
						Heater air ducts	Pass hot air through floor vents

		Hot water Hoses and pipes	Maintain hot water flow
Cooling System	Provide cold air according to thermostat and blower speed in accordance with outside heat load	Cold air exchanger	Allow compressed refrigerant to pass through expansion valve in the cold air exchanger
		Air Blower	Blow air at set speed
		Compressed refrigerant supply	Send refrigerant to expansion valve
		Thermostat	Control refrigerant flow
		Air blower speed controller	Control blower speed to speed setting
		A/C air ducts	Pass cold air through air registers in the instrument panel

(*continued*)

TABLE 2.4 (Continued)
An Illustration of Requirements Cascade from Product Level to Component Level

Product Level	Vehicle Level Requirement	System Level	System Level Requirement	Subsystem Level	Subsystem Level Requirement	Components	Component Level Requirement
						Refrigerant hoses and pipes	Maintain refrigerant flow
		Vehicle Body	Protect occupants from exterior heat	Body-in-white	Hold positions of climate control, seats, window openings and window glass.	Window openings	Reduce incident radiant heat load
						Window glass	
					Secure insulation materials in body cavities	Insulation	Reduce conductive heat load
				Seat	Position occupants in the vehicle interior space	Seat upholstery material	Reduce heat conductance of upholstery material
		Electrical system	Provide electrical energy to the blower and thermostat	Control power flow through climate controller unit	Control circuits to blower, thermostat and hot water and refrigerant fluid flow		

needs to interface the carryover components. However, the bottoms-up iterations can reduce design flexibility in creating the upper-level systems. The constraints placed by the carryover components can reduce the performance and efficiencies of the upper-level systems and the product.

Incremental and Iterative Development Approach

In some cases, the product development process is started with incomplete information before customer needs are fully known or understood. This could be because of time pressures or other urgent demands (e.g., political, security, competition) to begin the product development process quickly. As more and more information on customer needs and competitive products is available, customer requirements are refined, and the product being developed is modified. The product development process under such situations can become very chaotic, volatile, and inefficient. However, providing flexibility in responding to new information is also important. INCOSE (2006) refers to such process as "evolutionary development" where speed, responsiveness, flexibility, and adaptability are important. Despite the downside of an unstable, chaotic project, this approach can allow learning through continuous changes and investigations of a number of possible product features and product architectures. And thus, it avoids the loss of large investments and may generate short-term or localized solution optimizations.

SYSTEMS ENGINEERING "V" MODEL

All the above discussed steps and considerations can be described along with a horizontal time axis in a model. The model is presented in Figure 2.5, and it is known as the SE "V" model (refer to Blanchard and Fabrycky [2011] for more information). The model is described in the context of development of a new automotive product. The model shows basic phases of the entire product program on a horizontal time axis which represents time (*t*) in months before Job#1. In the automotive industry, "Job#1" is defined as the event when the first production product (vehicle) is shipped out of the assembly plant. The product program generally begins many months prior to Job#1. The beginning time of the program depends on the program's complexity (i.e., the changes in the new product as compared with the out-going product) and the management's approval to begin the product development process. In the early stages prior to the official start of a product program, an advanced product planning activity (usually the advanced product planning department or a special product planning team) determines the product characteristics and preliminary architecture (e.g., vehicle type, size, and type of powertrain), performance characteristics, the intended market (i.e., countries where the product will be sold), and a list of reference products (used for benchmarking) that the new product may replace or compete with. A small group of engineers and designers from the advanced design group are selected and asked to generate a few early product concepts to understand design and engineering challenges. A business plan including the projected sales volumes, the planned life of the product, the program timing plan, facilities and tooling plan, manpower plan,

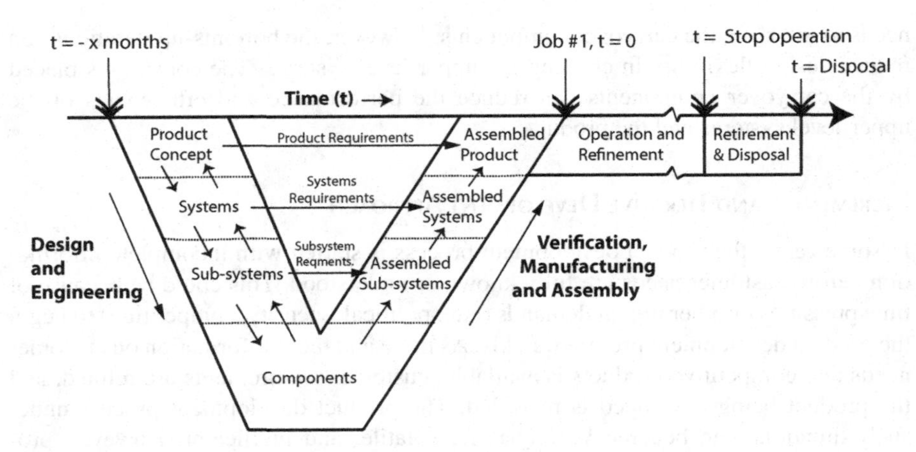

FIGURE 2.5 Systems engineering "V" model showing life cycle activities from the concept design to the disposal of a complex product.

and financial plan (including estimates of costs, capital needed, revenue stream, and projected profits) are developed and presented to the senior management along with all other product programs planned by the company (to illustrate how the proposed program will fit in the overall corporate product plan). The product program typically begins officially after the approval of the business plan by the company management. This program approval event is considered to occur at *x*-months prior to Job#1 in Figure 2.5.

At minus *x*-months, the chief product program manager is selected, and each functional group (such as design, body engineering, chassis engineering, power train engineering, electrical engineering, aerodynamics engineering, packaging and ergonomics/human factors engineering, and manufacturing engineering) within the product development and other related activities is asked to provide personnel to support the product development work. The personnel are grouped into teams, and the teams are organized to design and engineer various systems and subsystems of the product.

Figure 2.5 shows that the first major phase after the team formation is to create an overall product concept. During this phase, the designers (industrial designers) and the package engineers work with different teams to create the product concept that involves creating (1) early drawings or CAD models of the proposed product; (2) computer-generated 3D life-like images or videos of the proposed vehicle (fully rendered with color, shading, reflections, and textural effects); and (3) physical mock-ups (foam core, clay, wooden, or fiberglass bucks to represent the exterior and interior surfaces of the proposed vehicle). The images and/or models of the proposed vehicle are shown to prospective customers in market research clinics and to the management. Their feedback is used to further refine the product concept.

As the product concept is being developed, each engineering team decides on how each of the systems can be configured to fit within the product space (defined by the product concept) and how the various systems can be interfaced with other systems to work together to meet all the functional, ergonomic, quality, safety, and other requirements of the product. This is shown as the "Systems" phase in Figure 2.5. If problems about any of the systems are discovered during this phase, the designs are sent back to refine or modify the product concept (The feedback is represented as the up arrow from the systems to product concept shown in Figure 2.5).

As the systems are being designed, the next phases involve a more detailed design, that is, design of subsystems of each system and component within each of the subsystems. These subsequent phases, straddled in time to the right, are shown as "Subsystems" and "Components". The above phases forming the left half of the "V" represent the time and activities involved in "Design and engineering". The up arrows in the left half of the "V" indicate the iterative nature of the SE loops covered earlier (see Figures 2.2 and 2.3).

The right half of the "V", moving from the bottom to the top, involves testing, assembly, and verification where the components are produced and tested to ensure that they meet their functional characteristics and requirements (developed during the left half of the "V"). The components are assembled to form subsystems that are tested to ensure that they meet their functional requirements. Similarly, the subsystems are assembled into systems and tested; and, finally, the systems are assembled to create the whole product. At each phase, the corresponding assemblies are tested to ensure that they meet the requirements considered during their respective design phases (i.e., the assemblies are verified). These requirements are shown as the horizontal arrows between the left and the right sides of the "V" in Figure 2.5. It should be noted that down arrows between various assembly steps in the right half of "V" are not shown in Figure 2.5. The down arrows will indicate failures in the verification steps. When failures occur, the information is transmitted to the respective design team to incorporate design changes to avoid repetition of such failures.

The engineers and technical experts assigned to various teams in the product program work throughout all the above phases and continuously evaluate the product design to verify that the product users can be accommodated, and they will be able to use the product under all foreseeable usage situations. Early production vehicles developed just before Job#1 are usually used for additional whole vehicle evaluations for the product validation purposes (see Volume 2, 9 for additional information on verification and validation testing).

After Job#1, the product is transported to the dealerships and sold to the customers. The model in Figure 2.5 also shows a time period (on the right side of the V) called "Operation and refinement". During this time period, the product is being purchased, used, and maintained by the customers and serviced by dealers (or other repair shops). The product may also be refined (revised with minor changes during the existing model cycle or updated as refreshed or new models every few years) with some changes during its operational time. As the product becomes old and outdated, the product is pulled from the market. This marks the end of the life cycle of the product. At that point, the assembly plant and its equipment are recycled or retooled for the next product (as a next model year product), a totally new product is produced in the plant, or the plant is closed. As the products reach the end of their useful life, the products are sent to scrapyards where many of the components may be disassembled.

The disassembled components are either recycled for extraction of the materials or sent to junkyards.

The website of this book contains an Excel spreadsheet to create a financial plan of a vehicle program with various activities and monthly cost estimates of the activities (see Chapter 23).

NASA DESCRIPTION OF THE SYSTEMS ENGINEERING PROCESS

The *NASA Systems Engineering Handbook* describes the SE as including 17 processes that are grouped into the following three sets of processes: (1) system design processes; (2) product realization processes; and (3) technical management processes (see Figure 2.5). The descriptions of each set of three processes as provided in the *NASA Systems Engineering Handbook* (2007) are provided below.

System design processes: These NASA defined system design processes (see left side of Figure 2.5) for a complex product design problem are really the part of the "Design and Engineering" activities that take place in the left side of the SE "V" model shown in Figure 2.5 (These processes are numbered as 1 to 4 in Figure 2.6). The four-system design processes define stakeholder (customer) expectations, generate technical requirements, logically decompose the problem and the product, and convert the technical requirements into a technical solution (i.e., a product concept) that will satisfy stakeholder expectations. These processes are applied iteratively to each level of the product, which is product to systems, systems to subsystems, and

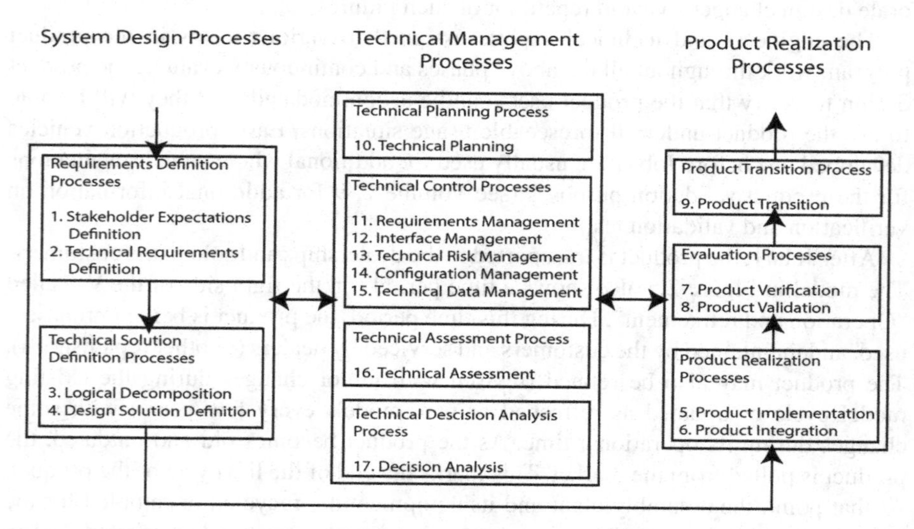

FIGURE 2.6 Systems engineering process described in the NASA handbook.

Source: Redrawn from National Aeronautics and Space Administration (NASA), *NASA Systems Engineering Handbook*, Report no. NASA/SP-2007-6105 Rev1, NASA Headquarters, Washington, DC, 2007.

subsystems to component level until the lowest level components are defined to the point where they can be built, bought, or reused (from carryover components). The outputs of the system design processes are fully defined technical requirements on all levels of the product and the design solution defining the product.

Product realization processes: Product realization processes are shown on the right side of Figure 2.5. These processes are numbered as 6 to 9 in Figure 2.6. These processes are applied to each operation/mission of the product in the system structure, starting from the lowest level product (component level) and working up to higher level integrated entities. These processes are used to create the design solution for each entity (e.g., by assembling components into subsystems, subsystems into systems, and systems into product – referred to here as the product implementation or product integration processes, [numbered 5 and 6]). The processes numbered 7, 8, and 9 are for verification, validation, and transitioning up to the next hierarchical level of the product. The realized product design solutions thus meet the stakeholder expectations (Note that these processes take place on the right side of the SE "V" model shown in Figure 2.5).

Technical management processes: The technical management processes (shown in the middle section of Figure 2.6 and numbered 10 to 17) are used to establish and evolve technical plans for the project; to manage communication across interfaces; to assess progress against the plans and requirements for the systems, products, or services; to control technical execution of the project to completion; and to aid in the decision-making process.

The above SE processes are used both iteratively and recursively. It should be noted that the horizontal two-way arrows between the three sections of Figure 2.6 indicate the iterative and recursive nature of the entire process. The "iterative" is the "application of a process to the entities at different levels with the same product or set of products to correct a discovered discrepancy or other variation from requirements", whereas "recursive" is defined as adding value to the product (or system) "by the repeated application of processes to design next lower level system or to realize next upper level of the products within the product (or system) structure". This also applies to repeating application of the same processes to the system structure in the next life cycle phase to mature the system definition and satisfy phase success criteria. The technical processes are applied recursively and iteratively to break down the initializing concepts of the product (or system) to a level of detail concrete enough that the technical team can implement a product from the information. Then, the processes are applied recursively and iteratively to integrate the smallest product entity into greater and larger systems until the whole of the product (or system) has been assembled, verified, validated, and transitioned.

The above-described system design processes, product realization processes, and technical management processes can thus be considered as the SE implementation in the traditional product development process. The performance and effectiveness of the product development process can thus be managed by the technical and management processes within the SE process. This issue is covered in the next section.

MANAGING THE SYSTEMS ENGINEERING PROCESS

The Systems Engineering Management Plan (SEMP) is a documentation created to plan the SE activities in terms of what needs to be done (e.g., types of analyses, methods, and procedures), when the activities need to be done (e.g., sequence of activities in the process), and who performs the activities (i.e., certain specified engineering departments or disciplines).

The SEMP is basically a planning document, and it should be prepared by the SE activity after the concept design phase. The SEMP is covered in detail in Chapter 3 and Bhise (2017).

RELATIONSHIP BETWEEN SYSTEMS ENGINEERING AND PROGRAM MANAGEMENT

The SE emphasizes technical issues by analyzing, measuring, and evaluating risk reduction, whereas the program (or project) management deals with the management issues. The main aim of the systems engineer is to create a successful product, whereas that of the program manager is to successfully complete the program on time and within the cost constraints. Both the disciplines, the SE, and the program management, overlap in terms of understanding the importance of the various tasks that need to be completed in the process of designing and producing the product. Figure 2.7 presents a Venn diagram showing responsibilities of the SE and the Project (or program) Management in a product program (Note: A program may include a

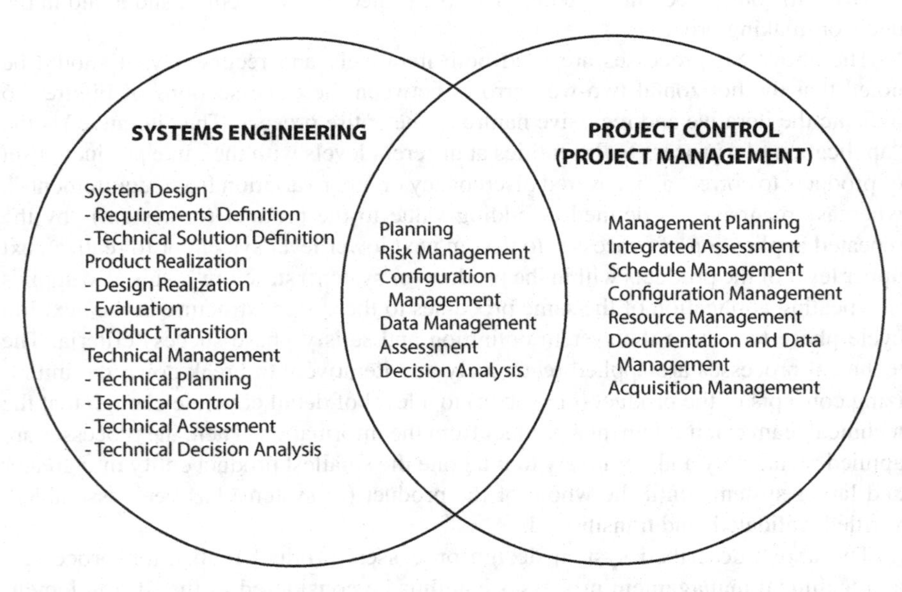

FIGURE 2.7 Relationship between systems engineering and project management.

Source: Redrawn from National Aeronautics and Space Administration (NASA), *NASA Systems Engineering Handbook*, Report no. NASA/SP-2007-6105 Rev1, NASA Headquarters, Washington, DC, 2007.

number of projects; thus, program management can include management of a number of projects). The SE covers the technical tasks and issues, and the project (or program) management covers management of timings of planned tasks, resources, schedules, documentation, and data. The tasks that are common to both are shown in the overlapping parts of the Venn diagram in Figure 2.7. These overlapping areas contain common or shared data needed by both disciplines in areas such as overall product planning (with schedule and budget), product configuration, internal and external risks, decisions and analyses that affect the program, data, and so forth.

The systems engineer, as compared with program management, has to concentrate more on the early stages of the conceptual and preliminary design to ensure that technical aspects such as the requirements are properly considered, and systems are being designed by proper consideration of trade-offs and allocation of functions within the packaging space of the product. These early steps have a major influence in ensuring that the right product is being designed.

Program management is the function of planning, overseeing, and directing the numerous activities required to achieve the requirements, goals, and objectives of the customer and other stakeholders within specified cost, quality, and schedule constraints.

ROLE OF SYSTEMS ENGINEERS

The systems engineers will usually play the key role in leading the development of the product and/or system architecture, defining and allocating requirements, evaluating design trade-offs, balancing technical risks between systems, defining and assessing interfaces, and providing oversight on verification and validation activities. The systems engineers will usually have the prime responsibility in developing many of the project documents (including the SEMP), requirements/specification documents, verification and validation documents, certification packages, and other technical documentation (NASA, 2016).

The SE is about trade-offs and compromises, about generalists rather than specialists. The SE is about looking at the "big picture" and not only ensuring that they get the design right (i.e., meets requirements), but that they get the right design. Thus, a system engineer needs to perform the following tasks:

1. Understand customer and program needs
2. Obtain required data
3. Develop SEMP
4. Communicate the SEMP to program teams
5. Provide recommendations to program teams on SE tasks
6. Assist teams in conducting necessary trade-off analyses
7. Continuously communicate with program teams to perform above tasks

Integrating Engineering Specialties into the Systems Engineering Process

As part of the technical effort, specialty engineers (who have specialized in an engineering discipline, for example, mechanical engineering) in cooperation with the SE and subsystem designers often perform tasks that are common across disciplines. Foremost, they apply specialized analytical techniques to create information needed by the project/program manager and the systems engineer. They also help define and write system requirements in their areas of expertise, and they review data packages, engineering change requests (ECRs), test results, and documentation for major program reviews. The project/program manager and/or systems engineer needs to ensure that the information and products generated add value to the project commensurate with their cost.

The specialty engineering technical effort should be well integrated into the vehicle program. The roles and responsibilities of the specialty engineering disciplines should be summarized in the SEMP. The specialty engineering disciplines to be included in a product program will vary depending on the product needs; however, the following disciplines are generally included:

1. Quality Engineering
2. Mechanical Engineering
3. Electrical/Electronics/Computer Engineering
4. Industrial Engineering/Operations Research
5. Human Factors Engineering
6. Safety Engineering
7. Reliability Engineering
8. Manufacturing, Processing, Assembly and Plant Engineering

In addition to the above engineering disciplines, the product design team generally includes members from the following disciplines:

1. Styling/Industrial Design
2. Environmental/Sustainability Engineering
3. Energy Systems Engineering
4. Marketing, Sales, and Dealerships
5. Finance, Product Planning, and Program/Project Management
6. Legal Support on Regulations, Patents, and Liabilities

Matrix Management of Vehicle Lines with Corporate Management Organizations

The integration of various specialty and attribute engineering into the vehicle programs involves balancing of available expertise between the vehicle program (line) organizations and corporate staffs (or offices; also called core engineering staffs in product development and manufacturing and assembly) within an automotive company. It is accomplished primarily by creating a matrix management structure. Table 2.5 shows an example of a matrix management structure where the line organizations responsible for the development of each vehicle are shown as columns, and corporate staffs are shown as rows. At the beginning of each vehicle program, a chief program manager is selected. The first task of the chief program manager generally is to secure and transfer the expertise and manpower needed to carry out the program from each corporate staff. The corporate staff have the main responsibilities to develop expertise (e.g., experts and trained engineers who can be loaned to each line organization; necessary software/analysis tools, test facilities and laboratories; develop new technology applications to implementation ready status) to ensure that all vehicle programs get the necessary technical support to develop their vehicles.

ROLE OF COMPUTER-ASSISTED TECHNOLOGIES IN PRODUCT DESIGN

CAD AND CAE

The way we use computers in the life cycle of a product has changed dramatically over the past 15–30 years. Computer-assisted technologies form the backbone of product visualization, configuration, and engineering evaluations. CAD systems are used to create three-dimensional solid models of parts, assemblies, and movements of parts. The CAD helps enormously in visualization of the product and in communicating product configuration, features, and details. CAD also helps in early phases of packaging of systems and components in the product space. Many different product configurations with different allocations of spaces for different systems, subsystems, and components can be created in a short time to enable visualization of different possible arrangements (concepts and layouts). The CAD models also help in communicating and comparing product information (e.g., drawings, dimensions, tolerances for assembly, fits, clearances, or interferences between parts) between different product design, engineering, and testing activities of the equipment producers and their suppliers. Ergonomics engineers frequently refer to and use the CAD models to conduct additional analyses to evaluate issues related to vehicle seating package, locations of pedals, steering wheel and gear shifter, instrument panel layouts, visibility of controls and displays, maximum and minimum hand reach zones, field of view, entry/exit, and so forth.

Computer-Assisted Engineering (CAE) applications can be used to analyze structural, thermal, fluid/air flows, and many functional aspects of the products. Prototype parts and systems can be created for evaluations. Many computer simulations can be conducted to evaluate different product uses over a large number of use cycles under different usage conditions before costly field testing is undertaken.

The computer-assisted/integrated technologies can save time, increase speed and efficiency, reduce costly errors (e.g., avoiding misreading of dimensional values), and rework in accomplishing many tasks during the life cycle of complex products.

FORMATION OF TEAM STRUCTURE AND TEAMS

Development of an automotive product requires many people from different disciplines and specializations. The number of people and teams required will depend upon the scope of the vehicle development program and the automotive company. However, about 400 to 1,200 engineering personnel from different specializations such as body engineering, chassis engineering, electrical engineering, powertrain engineering, and

TABLE 2.5
Matrix Management of Vehicle Programs with Corporate Staffs

Corporate Management Organization							
Vehicle Centers -->			Cars				
Vehicle Nameplate -->			Car #1	Car #2	Car #3	Car #4	Car #5
Corporate Staffs							
	Product Planning						
	Finance and Accounting						
	Marketing and Sales						
	Legal Affairs						
	Quality Engineering						
Product Development							
	Systems Engineering						
	Attribute Engineering Departments						
		Package					
		Ergonomics					
		Safety					
		Styling and Appearance					
		Thermal and Aerodynamics					
		Performance and Drivability					
		Vehicle Dynamics					
		Noise, Vibrations and Harshness (NVH)					
		Interior Climate Comfort					
		Weight					
		Security					
		Emissions					
		Communication and Entertainment					
		Costs					
		Customer Life Cycle					
		Product and Process Complexity					
	Vehicle Systems Engineering						
		Body System					
		Chassis System					
		Powertrain System					
		Fuel System					
		Electrical System					
		Climate Control System					
		Safety and Security System					
		Driver Interface and Infotainment System					
Manufacturing and Assembly							
	Manufacturing Plants Management & Eng.						
	Vehicle Assembly Plants Mgnt & Eng.						

Product Programs Management by Vehicle Lines

SUVs					Pickup Trucks				Vans/MPVs			Commercial Trucks		
SUV #1	SUV #2	SUV #3	SUV #4	SUV #5	Pickup #1	Pickup #2	Pickup #3	Pickup #4	Van/ MPV #1	Van/ MPV # 2	Van # 3	Comm #1	Comm #2	Comm #3

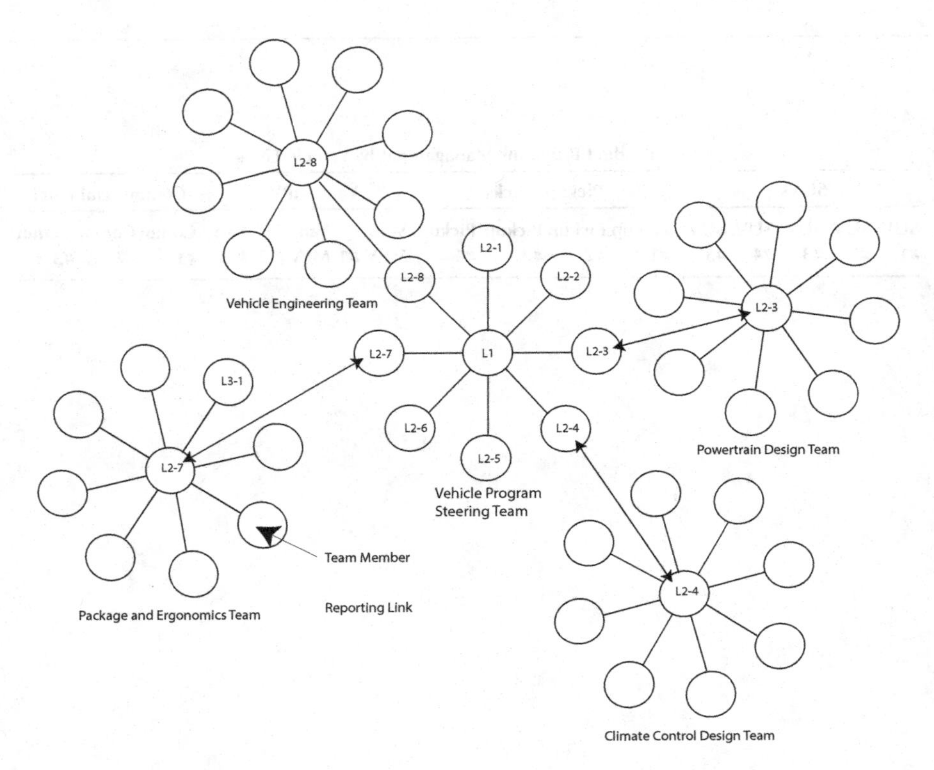

FIGURE 2.8 Team structure in an automotive product development program.

so on, are needed in a typical vehicle program in Western automotive companies. The entire design project is usually organized by using many teams, each undertaking design of certain portion or systems or subsystems of the vehicle. The structure of each team with team leader and number of team members, technical qualifications of each team member, responsibilities of each team members, progress reporting, problem resolution and communication methods are strictly enforced to ensure that all vehicle systems and interfaces between the systems can be designed so that all identified engineering requirements are met.

The highest level team in a vehicle program are typically headed by the chief vehicle program manager, and the membership of the team consists of high-level managers and chief engineers of major engineering offices (see Figure 2.8). This team in some auto companies is called the "Vehicle Program Steering Team".

The organization structure of the vehicle program steering team with top level (Level 1) and next level (Level 2) is illustrated below:

Vehicle Program Steering Team

L1 = Vehicle Program Manager (Level 1)
  L2-0 = Program Management Manager (Level 2)
  L2-1 = Body Engineering Chief Engineer (Level 2)

L2-2 = Chassis Engineering Chief Engineer (Level 2)
L2-3 = Powertrain Chief Engineer (Level 2)
L2-4 = Climate Control Chief Engineer (Level 2)
L2-5 = Electrical Engineering Chief Engineer (Level 2)
L2-6 = Fuel System Chief Engineer (Level 2)
L2-7 = Package and Ergonomics Engineering Chief Engineer (Level 2)
L2-8 = Vehicle Engineering Chief Engineer (Level 2)
L2-9 = Manufacturing Engineering Chief Engineer (Level 2)
L2-10 = Chief Designer (Level 2)
L2-11= Vehicle Attribute Engineering Chief Engineer (Level 2)

The next level teams, headed by each Level 2 chief engineer with membership of Level 3 manages can be illustrated as follows:

Body Engineering Team

L2-1 = Body Engineering Chief Engineer (Level 2)
L21-1= Body Structural Engineering Manager (Level 3)
L21-2= Body Closures Engineering Manager (Level 3)
L21-3= Body Safety Systems Manager (Level 3)
L21-4= Body Electrical Engineering Manager (Level 3)
L21-5= Body Lighting Engineering Manager (Level 4)
L21-6= Instrument Panel Engineering Manager (Level 3)
L21-7= Seating Systems Engineering Manager (Level 3)
L21-8= Body Trim Components Engineering Manager (Level 3)

Vehicle Attribute Engineering Team

L2-11 = Vehicle Attribute Engineering Chief Engineer (Level 2)
L211-1= Vehicle Dynamics Engineering Manager (Level 3)
L211-2= Aerodynamics Engineering Manager (Level 3)
L211-3= Thermal Management Engineering Manager (Level 3)
L211-4= Noise, Vibrations and Harshness Engineering Manager (Level 3)
L211-5= Craftsmanship Engineering Manager (Level 3)

Similarly, the next level teams headed by each Level 3 chief manager with membership of Level 4 supervisors are illustrated as follows:

Body Closures Team

L21-2= Body Closures Engineering Manager (Level 3)

L212-1= Hood Engineering Supervisor (Level 4)
L212-2= Front Doors Engineering Supervisor (Level 4)
L212-3= Rear Doors Engineering Supervisor (Level 4)
L212-4= Trunk/Liftgate Engineering Supervisor (Level 4)

Body Exterior Lighting Team

L21-5= Body Lighting Engineering Manager (Level 3)
L215-1= Front Lamps Engineering Supervisor (Level 4)
L215-2= Rear Lamps Engineering Supervisor (Level 4)
L215-3= Side Marker and Courtesy Lamps Supervisor (Level 4)

TREATING SUPPLIERS AS PARTNERS

It is important to realize that, depending upon the automotive company, about 35–75 percent of the content in automotive products are produced and supplied by supplier companies. Thus, the quality of the vehicle depends upon the quality of the entities supplied by the suppliers and how these entities interface and work together with entities provided by different suppliers and the automotive company. Many of the suppliers are selected early and are asked to participate in the product development process and given the tasks of designing the entities that they would provide. The suppliers, thus, should be treated as partners during the entire product development, production and the automotive assembly processes.

It is thus very important to select the right set of suppliers. Supplier selection criteria typically include: (a) expertise in systems engineering and specialized disciplines needed to develop the entities; (b) production capability in terms of required levels of quantities with specified quality and price; (c) demonstrated flexibility to change during early design stages; (d) must be very dedicated and responsive in meeting key product requirements (e.g., high fuel economy); (e) have the ability to incorporate innovative methods and technologies; and (f) the ability to support globally (on products marketed in many countries).

Ergonomics engineers thus will often find themselves working with supplier personnel involved in designing components and systems and providing them ergonomics support.

IMPORTANCE OF SYSTEMS ENGINEERING

The SE with the assistance from the other engineering disciplines establishes the baseline product (or system) design, allocates system requirements, establishes measures of effectiveness for ranking alternative designs, and integrates the design among the design disciplines. The SE is responsible for verifying that the developed product (or system) meets all the requirements defined in the product specifications. The SE also provides the analyses to ensure that all the requirements will be met. Thus, the products developed by application of SE principles, processes, and techniques will benefit from the following:

1. The right products will be developed because the SE will make sure that (a) the customer needs are obtained and translated into requirements; (b) the requirements are used by multidisciplinary teams for product development;

(c) best product configurations are selected because of the iterative and recursive applications; (d) all product entities are verified to assure compliance with their requirements and, finally; (e) the product is validated. Thus, the customers will like the products and will be very satisfied.
2. Product development time can be reduced by avoiding costly delays.
3. Costly redesign and rework of problems will be reduced.
4. The product will remain on the market for a longer time.

ADVANTAGES AND DISADVANTAGES OF THE SYSTEMS ENGINEERING PROCESS

The major advantages of the implementation of the SE process in the development of a complex product program are as follows:

1. It will help in reducing costs and time overruns.
2. It will help in creating products that the users want (i.e., ensure customer satisfaction).

The disadvantages of incorporation of SE functions in a product development program are as follows:

1. It adds people to the payroll and thus increases costs to the program.
2. It creates an additional documentation burden by creating the SEMP.
3. It creates more work for team members in communicating with the SE activities.

CHALLENGES IN COMPLEX PRODUCT DEVELOPMENT

Implementation of the SE process in a major product program faces many challenges. There are many basic needs in successful implementation of the SE. Some important considerations in the SE implementation include the following:

1. Commitment of everyone in the organization from the top management to the team members to understand and follow the SE process.
2. Availability of resources to recruit experienced systems engineers with formal training in the SE process and its techniques.
3. Organization culture to work in teams and focus on satisfying customer needs.
4. Commitment in systematically following the iterative and recursive process from requirements analysis to development of balanced product concept.
5. Availability of detailed product engineering capability with test facilities and resources to perform the verification and validation tests.
6. Top-notch program management and communication capability.
7. Availability of the state-of-art computer-assisted design, analysis, and engineering capabilities.

Some challenges and issues in the SE implementation in the product development programs are provided below.

1. Reducing the time required in implementation, integration, and execution of the SE activities (This in turn can reduce program costs).
2. Reducing chances of long delays in reworking and cost overruns or even cancellation of programs. Some reasons for such problems are as follows:
   a. Customer needs may change with time due to technological, political, economic factors – many of which cannot be predicted well during the early stages of the program.
   b. Technology readiness may be overestimated (sufficient time not allocated to develop new systems).
   c. Impossible or overly ambitious requirements are given to the product development teams, for example, during the development of the F-111 aircraft, the Air Force and the Navy provided very different requirements for the aircraft (Kamrani and Azimi, 2011).
   d. Changes in priorities during the product program, for example, new information during the Desert Storm operation required the Department of Defense to change priorities of the basic needs for a new airplane (Kamrani and Azimi, 2011).
   e. Late inclusion of new customers with different requirements (e.g., shifting the market strategy of selling a vehicle designed for one market to another, such as from the US market to the European or Asian market or vice versa) can have a major impact on the success of the product.
   f. Unexpected results or surprises in the product evaluations (tests) requiring a lot of reanalysis and redesign of the product.
   g. Personnel with the required expertise and training not available for the program (New team members must be trained in the SE processes and techniques).
3. Initial estimates of program time and budget may be overly optimistic or unrealistic (This would apply more to new products with no historical data on similar products).
4. Difficulties in resolving trade-offs between different product characteristics or product attributes. For example, the Systems Engineer is faced with decisions such as:
   a. To reduce costs at a constant level of risk, performance must be reduced.
   b. To reduce risk at a constant cost, performance must be reduced.
   c. To reduce costs at a constant performance level, higher risks must be accepted.
   d. To reduce risk to a constant performance level, higher costs must be accepted.

MODEL-BASED SYSTEMS ENGINEERING

Model-based systems engineering (MBSE) is now rapidly gaining interest among the SE and product-design community. It is a systems engineering methodology that focuses on creating and using predeveloped models as the primary means of information exchange between engineers, rather than on document-based information exchange. It helps to support system requirements, design, analysis, verification and validation activities from the beginning in the conceptual design phase and continuing throughout development and later life-cycle phases. The MBSE methodology is used in many industries, such as aerospace, defense, rail, and automotive.

MBSE is the formalized application of modeling to support system or product development activities involving system requirements, design, analysis, verification, validation and all life-cycle phases of the system (or product). It is, thus, an implementation of the SE process that focuses on use of a software application (i.e., a model) as the primary means of information exchange, rather than on document-based information exchange. It helps in developing and managing requirements and interfaces between systems with complex functionality and capabilities.

The model-based approach as compared to the traditional document-based approach is considered superior because of the ability of the software to obtain inputs from many team members and to maintain a detailed database on relationships and traceability created by the software applications. The model-based approach is also found to reduce time required to develop product/system specifications, and it is also more accurate and error-free.

CONCLUDING REMARKS

The SE process is powerful in supporting complex decision-making tasks involved in multidisciplinary considerations. The process is very useful in designing complex automotive products and systems. It is, however, difficult to master as there is no unique implementation procedure. In fact, it is a very flexible process in the sense that it can be applied in different levels of detail depending on resources available and the implementer's knowledge of the process and its techniques. The implementation of the SE process can vary between organizations and also within different products of the same organization. Not all programs require a very detailed and formal SE implementation. Smaller product programs cannot generally afford to have a separate SE department or systems engineers dedicated to the program.

The SE process is especially useful in designing very large and complex products (or systems) that have hundreds or thousands of components. Beginning the process with customer needs and performing requirements analysis, iteratively and recursively, provides a tremendous advantage in defining what needs to be done. Conducting functional analysis and design synthesis iteratively helps in creating and evaluating a number of possible alternate configurations. The processes of verification and validation ensure that the product meets its requirements and the product itself is right for its customers.

REFERENCES

Bhise, V. D. 2017. *Automotive Product Development: A Systems Engineering Implementation*. ISBN: 978-1-4987-0681-0. Boca Raton, FL: CRC Press.

Blanchard, B. S. and W. J. Fabrycky. 2011. *Systems Engineering and Analysis*. Upper Saddle River, NJ: Prentice Hall PTR.

Hall, A. D. 1962. *A Methodology for Systems Engineering*. New York, NY: D. Van Nostrand Company, Inc.

International Council on Systems Engineering (INCOSE). 2006. *Systems Engineering Handbook – A Guide for System Life Cycle Processes and Activities*. Report no. INCOSE-TP-2003-003-03, Version 3.

Kamrani, A. K. and M. Azimi (Eds.). 2011. *Systems Engineering Tools and Methods*. Boca Raton, FL: CRC Press.

National Aeronautics and Space Administration (NASA). 2016. *NASA Systems Engineering Handbook*. Report no. NASA/SP-2016-6105 Rev2. NASA Headquarters, Washington, DC. 20546. www.nasa.gov/wp-content/uploads/2018/09/nasa_systems_engineering_handbook_0. (accessed March 9, 2024).

Rechtin, E. 1991. *Systems Architecting, Creating and Building Complex Systems*. Englewood Cliffs, NJ: Prentice Hall.

3 Decision Making and Risks in Automotive Product Programs

INTRODUCTION

Decisions are made throughout the life cycle of every product program. Decisions are made whenever alternatives exist and the most desired alternative needs to be selected. The selected alternative should result in reducing risks and increasing benefits. Many different criteria can be used in selecting an alternative. The early decisions are related to the type of product to be designed (e.g., an automotive manufacturer needs to decide on the type of vehicle to be designed), requirements on the product, and its characteristics (e.g., how far should an electric vehicle travel on a fully charged battery). Later, the decisions are related to the number of systems and/or features in the product, their functions, and how the systems should be configured and packaged within the product space.

Decisions are also made during each phase and at each milestone of a product development program where the management decides whether to proceed to the next phase, make changes to the product being designed or even to scrap the program (if the program is found to be not competitive in its market segment). Early decisions have a major impact on the overall costs and timings of the program – because the later decisions depend on the design-specific parameters and their values selected in the earlier phases of the program. For example, powertrain type, its dimensions, its location in the vehicle space (e.g., front-wheel drive or rear-wheel drive), and the technologies to be implemented in a new automotive product will affect decisions related to the design of other vehicle systems (e.g., fuel system, cooling system, and space available to package suspensions).

After the product is introduced on the market, customer feedback is received. The manufacturer needs to decide whether to make any changes and what to change if negative feedback is received. The reasons for customer dissatisfaction need to be understood, and decisions need to be made on how and when to fix any defects in the product. Furthermore, after the product is marketed for an extended time, decisions need to be made as to what product characteristics should be revised, how to revise, and when to revise.

DOI: 10.1201/9781003485582-3

All the above described decisions involve risks. For example, adding more features (or capabilities) than what customers need, and over-designing will waste resources. Conversely, failures in incorporating any customer-desired major changes and under-designing the product will result in loss of sales or even degrade the reputation of the product and its brand in the marketplace.

The decision-making models and methods covered in this chapter include a) decision matrix involving combinations of alternatives and outcomes with decision-making principles such as maximizing expected value, achieving an aspiration level, most probable principle, Laplace principle, Maximin principle, Maxmax principle and Hurwicz principle; b) analytical hierarchical method to obtain weights of alternatives; c) Pugh diagram with weighted multiple product attributes to compare and select a product concept with other benchmarked products; d) benchmarking and breakthrough; e) quality function deployment to help identify important function specifications of products related to their customer needs; and f) failure modes and effects analysis. This chapter also provides examples of types of risks such as technical risks, schedule risks and financial risks, methods to assess risks, for example, risk matrix, and trade-offs between product attributes considered during product development.

PROBLEM-SOLVING APPROACHES

Many different approaches are used to solve problems in various scientific and engineering fields. The approaches are generally similar, and there are some variations in their applications. The scientific method is an efficient method of doing research (learning by doing) (Konz, 1990). Table 3.1 compares the steps in the scientific method with those in the engineering design method and the six-sigma quality engineering approaches. DMAIC and IDOV are the acronyms of the two six-sigma approaches. The acronym, DAMES, was suggested by Konz (1990) to remember the steps in the engineering design method (Note: The acronyms can be found in the third, fifth, and seventh columns of Table 3.1 by reading vertically down). All the four approaches presented in Table 3.1 are generally used in an iterative manner until a satisfactory solution is found. Each step of any of these problem-solving approaches requires decision making. Decisions are made during each step to decide on what to do, how to do it, when to do it, what alternative to choose, and so forth.

Early steps in all problem-solving methods involve data collection and analysis of the available data to understand the problem and to develop possible alternatives. The alternatives are evaluated by conducting experiments or exercising available models (which are generally verified by earlier research). The results of these experiments and model applications are used to make the decisions to select acceptable solutions.

DECISION MATRIX

ALTERNATIVES, OUTCOMES, PAYOFFS, AND RISKS

Systems Engineering (SE) involves decision making such as, what needs to be done, when, how, and how much, and considering the trade-offs between possible design product attributes and their requirements. In a decision-making situation, the

TABLE 3.1
Problem-Solving Approaches

Step no.	Scientific Method	Engineering Design Method	Six-Sigma Improvement Method	Design for Six-Sigma (DFSS)
1	State the problem	D *Define Problem*	D *Define Problem*	I *Identify Requirements*
2	Construct hypothesis or model	A *Analyze – Determine relevant variables*	M *Measure – Gather data on issues and variables*	D *Design – Characterize it by invention and innovation*
3	Apply analysis to the model. Use the model to predict what can happen	M *Make Search for Possible Solutions*	A *Analyze – Analyze gathered information*	O *Optimize the Design – Conductanalyses and/or experiments*
4	Design and conduct an experiment under real situations	E Evaluate possible solutions to determine best solution	I Improve–Develop improvements	V Verify the Design-Conduct evaluations/ tests to verify and validate the design
5	Compare model predictions with experimental results	S Specify Solution	C Control–Establish process controls	

decision maker (e.g., engineer, designer, program manager or a management team) is faced with the task of deciding on an acceptable alternative among several possible alternatives. The decision maker also needs to consider possible outcomes (i.e., what will happen in the future), and the costs or benefits (called the payoffs) associated with each combination of an alternative and an outcome. Further, each possible outcome may or may not occur in the future. There are many different decision models to help determine a desired or an acceptable alternative (Blanchard and Fabrycky, 2011). Let us assume the following:

A_i = ith alternative, where $i = 1, 2, \ldots, m$
O_j = jth outcome, where $j = 1, 2, \ldots, n$
P_j = the probability that jth outcome will occur, where $j = 1, 2, \ldots, n$
E_{ij} = evaluation measure (payoff – positive for benefit [profit] and negative for cost [loss]) associated with ith alternative and jth outcome

The decision evaluation matrix associated for the above problem is presented in Table 3.2. Many principles can be used to select a desired alternative. The principles are described here.

TABLE 3.2
Decision Evaluation Matrix

	Probabilities of Outcomes					
	P_1	P_2	P_3	.	.	P_n
Alternative			**Outcomes**			
	O_1	O_2	O_3	.	.	O_n
A_1	E_{11}	E_{12}	E_{13}	.	.	E_{1n}
A_2	E_{21}	E_{22}	E_{23}	.	.	E_{2n}
A_3	E_{31}	E_{32}	E_{33}	.	.	E_{3n}
.	.	.	.	.	.	.
.	.	.	.	.	.	.
A_m	E_{m1}	E_{m2}	E_{m3}			E_{mn}

Let us assume that an automotive manufacturer wants to select a design for its future mid-size SUV. The manufacturer is considering the following five alternatives:

A_1 = Develop a new mid-size SUV with a 250 HP 2.0 L ICE and 2-row seating
A_2 = Develop a new mid-size SUV with a hybrid powertrain with 300 HP 2.0 turbo-boost plus an electric motor and 2-row seating
A_3 = Develop a new mid-size SUV with a 320 HP 2.3 L turbo-boost plus an electric motor (hybrid) and 3-row seating
A_4 = Develop a new mid-size SUV with a 335 HP electric powertrain with twin motors (full-electric) with 3-row seating
A_5 = Improve an existing 280 HP 2.3L turbo-boost (ICE) SUV with 2-row seating

Six possible outcomes assumed by the manufacturer are as follows:

O_1 = Economy does not change – oil prices remain low, and the battery technology does not improve
O_2 = Economy improves by 5%, oil prices remain low and battery technology does not improve
O_3 = Economy degrades by 5%, oil prices increase by 30% and the battery technology does not improve
O_4 = Economy does not change; oil prices remain low and battery technology improves by 50%
O_5 = Economy improves by 5%, oil prices remain low and battery technology improves by 50%
O_6 = Economy degrades by 5%, oil prices increase by 30% and the battery technology improves by 50%

The evaluation measure is the present value of revenues minus costs in billions of dollars during the vehicle program from -40 to +60 months from Job#1. The values

TABLE 3.3
Decision Matrix for an Auto Manufacturer's Problem

	Probability of Outcome					
	0.5	0.1	0.1	0.1	0.1	0.1
	Outcomes					
Alternatives	**O1**	**O2**	**O3**	**O4**	**O5**	**O6**
A1	$2.40	$2.62	$2.19	$2.84	$4.04	$3.93
A2	$2.95	$3.19	$1.66	$3.44	$4.80	$4.67
A3	$1.85	$2.05	$1.67	$2.24	$3.29	$3.19
A4	$1.43	$1.60	$1.26	$1.78	$2.71	$2.63
A5	$2.86	$3.10	$2.62	$3.35	$4.68	$4.55

Note: The numbers in the matrix are in billion dollars.

of the evaluation measure for each combination of the above five alternatives and the six outcomes are provided in Table 3.3. The values of the evaluation measure were obtained by exercising a cost model for the vehicle program for all combination of alternatives and outcomes. The model is presented in Volume 2, Chapter 23. Table 3.3 also provides probabilities for each of the outcomes assumed by the manufacturer. The assumed values of monthly vehicle sales, manufacturer's suggested vehicle price (MSRP) and cost of purchased price of parts and systems per vehicle produced under the 30 combinations of alternatives and outcomes are provided in Table 23.2 of Volume 2, Chapter 23.

Maximum Expected Value Principle

One commonly used principle to select an alternative is based on the maximum expected value. The expected value of $A_i = \{E_i\}$ can be computed as: $\sum_j \left[P_j \times E_{ij}\right]$.

Thus, under this principle, the decision maker will select the alternative with the maximum expected value, which is defined as: max $\{E_i\}$ for $i = 1, 2, \ldots, m$. The selection of the alternative and application of the above principle are illustrated in the following example.

The following computation illustrates the computation of the expected value of A_1:

$$\begin{aligned}\text{Expected value of } A_1 = \{E_1\} &= (0.5\times 2.4)+(0.1\times 2.62)\\ &+(0.1\times 2.19)+(0.1\times 2.84)\\ &+(0.1\times 4.04)+(0.1\times 3.93)\\ &= 2.76\end{aligned}$$

The expected values of A_2, A_3, A_4, and A_5 are 3.25, 2.17, 1.71, and 3.25 billion dollars, respectively. Thus, the alternatives A_2 and A_5 have the maximum expected value of 3.25 billion dollars among the five alternatives and A_2 or A_5 will be selected under

TABLE 3.4
Alternatives Selected by Five Principles

	Probability of Outcome										
	0.5	0.1	0.1	0.1	0.1	0.1					
	Outcomes										
Alternatives	**O1**	**O2**	**O3**	**O4**	**O5**	**O6**	**Expected Value Principle**	**Laplace Principle (Average Value)**	**Maximin Principle (Min Values)**	**Maxmax Principle**	**Hurwicz Principle (with $\alpha = 0.5$)**
A1	$2.40	$2.62	$2.19	$2.84	$4.04	$3.93	$2.76	$3.00	$2.19	$4.04	$3.12
A2	$2.95	$3.19	$1.66	$3.44	$4.80	$4.67	$3.25	$3.45	$1.66	$4.80	$3.23
A3	$1.85	$2.05	$1.67	$2.24	$3.29	$3.19	$2.17	$2.38	$1.67	$3.29	$2.48
A4	$1.43	$1.60	$1.26	$1.78	$2.71	$2.63	$1.71	$1.90	$1.26	$2.71	$1.99
A5	$2.86	$3.10	$2.62	$3.35	$4.68	$4.55	$3.26	$3.53	$2.62	$4.68	$3.65

Note: The selected alternatives are shown by highlighting (in gray) the most desired payoff under each of the five principles shown in the last five columns of this table.

the maximum expected value principle (see the column labeled as "Expected Value Principle" in Table 3.4).

Other Principles

Six additional principles that can be used to select an alternative are described below.

1. *Aspiration level*: The principle of aspiration level assumes that the decision maker needs to meet a certain aspirational (or desired) level such as a minimum acceptable profit level or a maximum amount of tolerable loss. If we assume that the decision maker in the above example (Table 3.3) wants to make at least 4 billion dollars profit, then he would consider alternatives A_1, A_2 and A_5 (because these three alternatives include the outcomes with payoff of at least 4 billion dollars). On the other hand, if he does not want to incur a profit below 3 billion dollars, he would not consider alternative A_4.
2. *Most probable future*: The decision maker may decide based on the most likely (or most probable) outcome (which has the highest probability of occurrence). In our above example (Table 3.3), the outcome O_1 has the highest probability (0.5) of occurrence. Under this situation (outcome O_1), selection of alternative A_2 will ensure the maximum profit of 2.95 billion dollars.
3. *Laplace Principle*: The Laplace Principle assumes that the decision maker does not have any information on the probability of occurrences of any of the outcomes and, thus, he assumes that all the outcomes are equally likely. In our above example, under this principle, all the occurrence probabilities will be equal to 1/6. Thus, the decision maker can simply take the average value of all E_{ij} for each alternative (i.e., over each i) and select the alternative with the maximum profit. In our above example, under this principle, the decision maker would select alternative A_5 with the maximum average profit of 3.53 billion dollars (see the column labeled as "Laplace Principle" in Table 3.4).
4. *Maximin Principle*: This principle is based on the "extremely pessimistic view" of the decision maker (i.e., nature will do its worst under every alternative). Therefore, the decision maker will select the alternative that maximizes the value of the proceed (profit) among the minimum values of all alternatives (i.e., the decision maker will reduce his loss by selecting the alternative with the least loss [or select the alternative with the highest profit among the minimum values]). The profit (P_i) in ith alternative can be defined as follows:

$$P_i = \max_i \left\{ \min_j E_{ij} \right\}$$

Table 3.4 shows that, under this principle, the decision maker will select alternative A_5, which has the highest value (billion \$2.62) among the lowest possible values of the evaluation measure among all the alternatives (see the column labeled as the "Maximin Principle" in Table 3.4).

5. *Maximax Principle*: This principle is based on the "extremely optimistic view" (think about the best possible) of the decision maker. The decision maker will select the alternative that maximizes the maximum values in each alternative, that is, to take the maximum of the maximum values in each alternative. The profit (P_i) in ith alternative can be defined as follows:

$$P_i = \max_i \left\{ \max_j E_{ij} \right\}$$

Table 3.4 shows that under this principle, the decision maker will select alternative A_2, which has the highest value (billion $ 4.80) among the highest possible values of the evaluation measure among all the alternatives (see the column labeled as the "Maximax Principle" in Table 3.4).

6. *Hurwicz Principle*: This principle is based on a compromise between optimism (Maximax principle) and pessimism (Maximin principle). The profit (P_i) in ith alternative is computed based on the selection value of index of optimism (α) as follows:

$$P_i = \alpha \left[\max_i \left(\max_j E_{ij} \right) \right] + (1-\alpha) \left[\max_i \left(\min_j E_{ij} \right) \right]$$

where α = index of optimism. α can vary within $0 \leq \alpha \leq 1$

$\alpha = 1$ indicates that the decision maker is extremely optimistic

$\alpha = 0$ indicates that the decision maker is extremely pessimistic

The value of P_i should be computed for each alternative using the above formula and the alternative with the maximum value of P_i should be selected. The last column of Table 3.4 illustrates that for $\alpha = 0.5$, alternative A_5 will be selected because it has the highest value (3.65 billion dollars) in the last column when the values were computed using the above expression for P_i.

EVALUATING CHANGES IN VEHICLE DESIGNS

Application of the above described decision matrix to changes made in the vehicle design (e.g., features or options added or modified) require redefinitions of alternatives, and re-estimation of values of associated parameters and evaluation measures. For example, whether to add a navigation system in a new vehicle, the decision maker will have to estimate costs and benefits of each alternative. The outcomes need to include different percentages of vehicles that can be sold with the navigation system. The costs of purchasing a navigation system from a supplier, assembling it in a vehicle, and testing to ensure that it works properly, and the costs of updating, repair and maintenance of the navigation units can be estimated from internal data. The benefits gained from each navigation system can be determined by developing a benefits estimation model that can include variables such as cost of time and fuel saved in reaching destination, costs reduced due to reduction in number of accidents in reaching the destination, revenue gained due to increase in vehicle sales, and customer's willingness to pay increased price.

Minor changes made in the vehicle design for ergonomic improvements are difficult to evaluate by using the decision matrix. However, the costs and benefits of each use can be made based on measures such as time saved per use, frequency of use (number of uses/year), costs of incorporating the change (fixed costs for tooling and equipment) and variable costs per unit with the design change, increase in vehicle sales due to the design change, and increase on vehicle price due to the ergonomic change, and so forth. Table 23.4 presents results of a sensitivity analysis exercise conducted using the spreadsheet model by combination of adding the costs of ergonomic design changes of 0, 10, 50, 100 and 1000 dollars per vehicle and increasing the vehicle price by 0, 10, 50, 100 and 1000 dollars.

CARE AND PRECAUTIONS IN APPLYING DECISION MATRIX

Decision makers need information to help decide on all the basic parameters (e.g., the variables covered in the earlier section, such as number of alternatives, possible outcomes, probabilities of the outcomes, costs or benefits associated with each alternative in each outcome) of the problem. Without the availability of reliable information, the decisions made by the decision maker may not be very useful or could even be very misleading. Further, care must be taken in selecting the decision maker to ensure that he or she is not biased and does not have any misconceptions or preconceived notions related to the product concepts, technologies considered in the concepts, customer expectations, and so forth.

OTHER TECHNIQUES USED IN DECISION MAKING

Many other techniques are also available to gather information and display the information in a format that will help the decision maker to understand the problem (see Bhise (2023) for more information on the tools. The formats of many of the tools are visual in nature and thus they promote better understanding of the magnitudes and relationships between variables or events. Some examples of the tools and their applications are: (1) Fish diagram is used commonly to illustrate causes of a problem; (2) Pareto chart helps in identifying the top few issues that contribute to most of the problems; (3) the quality function deployment (QFD) chart provides a list of engineering specifications and their relative importance ratings; (4) Pugh chart helps in selecting a product concept by considering multiple attributes; and (5) a risk analysis helps assess the level of risk in a given situation.

OBJECTIVE MEASURES AND DATA ANALYSIS METHODS

Depending upon the task used to evaluate a product, task performance measurement capabilities and instrumentation available, the ergonomics engineer would design an experiment and procedure to measure dependent measures. The objective measures can be based on physical measure such as time (taken or elapsed), distance (position or movements in lateral, longitudinal or vertical directions), velocities, accelerations, events (occurrences of predefined events, e.g., stop lamp activation on a lead vehicle, detection distance to a target on the roadway), and measures of user's physiological state (e.g., heart rate). The recorded data are reduced to obtain the values of the dependent measures and their statistics such as means, standard deviations, minimum, maximum, percentages above and/or below certain preselected levels. The measured values of the dependent measures are then used for statistical analyses based on the experiment design selected for the study. Examples of applications involving objective measures are provided in Chapter 15.

SUBJECTIVE METHODS AND DATA ANALYSIS

The ergonomics engineers use the subjective methods because in many situations: a) the subjects are better able to perceive characteristics and issues with the product, and thus, they can be used as the measurement instruments, b) suitable objective measures do not exist, and c) the subjective measures are easier to obtain.

The two most commonly used subjective measurement methods used during the vehicle development process are: a) rating on a scale and b) paired comparison based methods (Satty, 1980). These methods are presented in Chapter 19. Other subjective methods based on assessments provided by team members included in this chapter are (1) Pugh diagram; (2) Weighted Pugh analysis; (3) Weighted total score for concept Selection; (4) Quality function deployment; (5) Task analysis; and (6) Failure modes and effects analysis. These methods are described below.

PUGH DIAGRAM

The Pugh diagram is a simple but effective tool in understanding how many attributes can be used to study comparisons between different products (or product concepts) to select the best product. The tool can also help in improving the selected product in additional iterations of comparisons.

The Pugh diagram is a tabular formatted tool consisting of a matrix of product attributes (or characteristics) and alternate product concepts along with a benchmark (a reference) product called the "*datum*". The diagram helps to undertake a structured concept selection process and is generally applied by a multi-disciplinary team to converge on a superior product concept. The process involves creation of the matrix using inputs from all the team members. The rows of the matrix consist of product attributes based on the customer needs and the columns represent different alternate product concepts.

The evaluations of each product concept on each attribute are made with respect to the *datum*. The process uses classification metrics of "same as the datum" (S), "better than the datum" (+), or "worse than the datum" (-). The scores for each product concept are obtained by simply adding the number of plus and minus signs in each column. The product concept with the highest total score ("sum of plus" signs minus "sum of minus" signs) is considered to be the preferred product concept. Several iterations are employed to improve the preferred product concept by combining the best features of highly ranked concepts for each attribute until an acceptable concept emerges and becomes the new benchmark.

Table 3.5 presents an example of a Pugh diagram created to develop a new concept (a target vehicle) for a 2020 model of Jeep Cherokee (mid-size SUV) by comparing with three 2015 model year SUVs. The table shows how each vehicle compares with the 2015 Jeep Cherokee vehicle (used as the datum) for each vehicle attribute. (Note that "+", "S" and "-" symbols respectively indicate that the vehicle in the column has better, same as, and worse than the datum. The sum of number of attributes receiving "+" sign minus the sum of attributes receiving "-" sign provides a total score, which is a measure of improvement in each vehicle over the datum vehicle. Thus, the 2020 Jeep Cherokee received a total score of 5, which is higher than the corresponding scores of the other two benchmarked vehicles.

TABLE 3.5
Pugh Diagram for Evaluating a New Mid-size SUV Concept

Vehicle Attribute	2020 Jeep Cherokee (Target)	2015 Jeep Cherokee (Trailhawk) (DATUM)	2015 Ford Escape (4WD Titanium)	2015 Honda CRV (Touring AWD)
Durability	+		S	+
Off Road Capability	+		-	-
Fuel Economy	+		+	+
NVH	S		+	+
Handling & Dynamics	S		S	+
Towing Capacity	S		S	-
Ride Comfort (On Road)	-		S	S
Ease of maintenance & modification	+		S	S
Cost	S		-	-
Weight	+		+	+
Safety	S		S	+
Aesthetics	+		S	S
Aero/Thermal	S		+	+
Sum of +	6		3	7
Sum of -	1		2	3
Sum of S	6		8	3
Total score	5		1	4

It should be noted that the above Pugh diagram was created with the assumption that all vehicle attributes are equally important to the customer. And hence scores of sum of plus signs and sum of minus signs were simply obtained by adding number of attributes in which the vehicle corresponding to a column was better or worse than the datum. However, in reality, some attributes may be more important to customers than other attributes. Thus, the analysis can be modified to consider different importance weights to each attribute. The weighted Pugh analysis is covered in the next section.

WEIGHTED PUGH ANALYSIS

There are many different methods to include importance weights to each attribute. Table 3.6 shows a modified Pugh diagram for the above problem. Here an additional column of importance rating is included in the Pugh diagram. The importance weighting is based on a 10-point scale where a rating of 10 is assigned as most important and rating of 1 is assigned as least important. The "+", "S" and "-" signs in each column (except the datum) are replaced by 1, 0 and -1 weights, respectively. The final score called the "weighted sum" is computed for each column by adding the sums of importance rating multiplied by the 1, 0 or -1 score in the column for each vehicle.

A comparison of weighted sum values for the three vehicles shows that the 2020 Jeep Cherokee received the highest score of 37 and the 2015 Ford Escape received the lowest score of 10. The design team will study the numbers and find further opportunities to improve the weighted score for the new vehicle.

TABLE 3.6
Weighted Pugh Diagram

Vehicle Attribute	Importance Rating	2020 Jeep Cherokee (Target)	2015 Jeep Cherokee (Trailhawk) (DATUM)	2015 Ford Escape (4WD Titanium)	2015 Honda CRV (Touring AWD)
Durability	10	1		0	1
Off Road Capability	7	1		-1	-1
Fuel Economy	8	1		1	1
NVH	8	0		1	1
Handling & Dynamics	8	0		0	1
Towing Capacity	3	0		0	-1
Ride Comfort (On Road)	6	-1		0	0
Ease of maintenance & modification	6	1		0	0
Cost	8	0		-1	-1
Weight	5	1		1	1
Safety	9	0		0	1
Aesthetics	7	1		0	0
Aero/Thermal	4	0		1	1
Weighted Sum		37		10	34

WEIGHTED TOTAL SCORE FOR CONCEPT SELECTION

During the product development process, the decision makers (e.g., usually top management) are faced with the decision to select a concept and proceed with its detailed design and engineering work. The selection is complicated because the product concepts need to be evaluated by considering many attributes of the product. The attributes are generally developed from the customer needs obtained from extensive interactions with the customers (e.g., by conducting market surveys and/or from customer feedback). The customers can also be asked to provide importance ratings (or weights) of each of the attributes. The weights can also be developed by using the Analytical Hierarchical Process covered in Chapter 19. The customers (or the design team members) can also be asked to rate each product concept on each attribute. All the above information can then be used to determine total weighted score of each product concept. The product concepts can be compared based on the total weighted score and the concept with the highest total score can be selected.

The computation of the total weighted score (T_j) of the jth product concept is described by the following mathematical expression.

$$T_j = \sum_{i=1}^{n} w_i R_{ij}$$

Where, T_j = Total weighted score of jth product concept by considering all the n attributes. Note: $i = 1, 2, \ldots, n$

w_i = weight of ith product attribute

R_{ij} = Ratings of concept C_j on i^{th} attribute

Table 3.7 provides an example of the above weighting scheme. Each product concept is defined as C_j, where $j = 1$ *to* 4; and each product attribute is defined as A_i, where $i = 1$ *to* 5. The ratings (R_{ij}s) are provided using a 10-point scale, where 1 = poor and 10= excellent. The attribute weights (w_i) were obtained by using a 5-point scale, where 1= not important and 5= very important. The total weight score ($T_1 = 119$) of

TABLE 3.7
Illustration of Total Weighted Score of Product Concepts Based on Attribute Weights and Ratings of Product Concepts by Attributes

Attribute	Attribute weight (wi)	Product Concepts			
		C1	C2	C3	C4
A1	5	10	8	5	7
A2	3	5	8	9	4
A3	5	7	4	5	7
A4	1	5	8	3	6
A5	2	7	9	6	8
Total weighted score		*T1*	*T2*	*T3*	*T4*
		119	110	92	104

concept C_1 was the highest; and the concept C_3 *with* $T_3 = 92$) scored the lowest. Thus, the product concept C_1 can be selected for further development, or the ratings data can be used to come up with further modifications of the product concepts. New ratings can be obtained after the modifications to iterate the above procedure until an acceptable product concept is achieved.

The above method is used in the QFD (to compute absolute importance scores of functional specifications), and it can be considered as a modified scoring method for the Pugh Diagram. The QFD tool is described later in this chapter.

Pugh Diagrams in Ergonomic Evaluations

Pugh diagrams and weighted scoring techniques presented above can also be used for evaluation of any of the vehicle entities (i.e., components and systems) based on selected vehicle attributes (or subattributes), customer needs or vehicle systems. For example for ergonomic evaluation of a vehicle component such as a gear shift lever, the following subattributes can be included in the Pugh diagram: a) location of the gear shifter from the steering wheel, b) reach distance to the gear shifter knob, c) clearance between the gear shifter knob and the steering wheel (if the shifter is steering column mounted), d) shifting force, e) shifting movement feel, f) feel of gear detents, g) visibility of shifter location, and h) legibility of shifter position labels.

BENCHMARKING AND BREAKTHROUGH

Benchmarking and Breakthrough methods are generally used during the very early stages of a product development program. From the information gathered during the benchmarking exercises, the product designers can realize the "gaps" between the characteristics and capabilities of the products of their competitors and their new product concepts. Whereas the breakthrough approach forces the design teams to look beyond the existing products and technologies and thus think about developing a totally new product or features and achieve huge improvements over the existing product designs.

Benchmarking

Benchmarking is a process of measuring products, services, or practices against the toughest competitors, or those companies recognized as the industry leaders. Thus, it is a search for the industry's best products or practices that can lead to superior performance.

A multi-functional team within the product development community is usually selected to perform the product benchmarking activities. The benchmarking exercise typically starts with identifying the toughest competitors (e.g., very successful and recognized brands as the industry leaders), and their products (models) that serve similar customer needs to the manufacturer's proposed product. The selected competitor products are used to compare against the target product. The target product is the product considered by the manufacturer to be its future product (or an existing model of the future product). The team gathers all important competitive products and information about the products and compares the competitors' products to their target product through a set of evaluations (e.g., measurements of product characteristics, tearing-down the products into their lower level entities for close examinations, evaluations by experts for study of performance, capabilities, unique features, materials and manufacturing processes used by the competitors, estimates of costs to produce the benchmarked products, and so forth.

The information gathered from the comparative evaluations is usually very detailed. However, the depth of evaluations included in benchmarking can vary between problem applications and companies. For example, a benchmarking of an automotive disc brake may involve comparisons based on part dimensions, weights, materials used, surface characteristics, strength characteristics, heat dissipation characteristics, processes needed for its production, estimated production costs, features that would be "liked very much" by the customers, features that can be "hated" by the customers, features that would create "Wow" reactions among the team members and potential customers, special performance tests such as part temperatures during severe braking torque applications, brake "squealing" sound, and so on. In addition, digital pictures and videos can be taken to help visualize differences in the benchmarked products and their components.

The gathered information is generally summarized in a tabular format with product characteristics listed as rows and different benchmarked products represented in columns. Bhise (2023) provides more information on the benchmarking technique.

Breakthrough

The breakthrough approach involves throwing away all the existing product designs (and processes) and brainstorming to develop a totally new design to obtain huge potential gains in terms of performance improvements, costs and added customer satisfaction. Breakthrough designs typically require radically new thought dimensions and lead to adoption of new technologies. Thus, implementation of a breakthrough design creates new problems in systems integration and management. Bhise (2023) provides more information on the breakthrough approach.

QUALITY FUNCTION DEPLOYMENT

Quality function deployment (QFD) is a technique used to understand customer needs (voice of the customer) and transform customer needs into engineering characteristics of a product (or a process) in terms of functional (or design) requirements (or specifications). It relates "what" (what are the customer needs) to "how" (how should engineer design to meet the customer needs), "how much" (magnitude of design variables, i.e., their target values) and also provides competitive benchmarking information and importance ratings on functional specifications – all in one diagram. Thus the information contained in the QFD diagram can be used by the product designers for decision making related to selection of important functional specifications and their target values. The QFD was originally developed by Dr. Yoji Akao of Japan

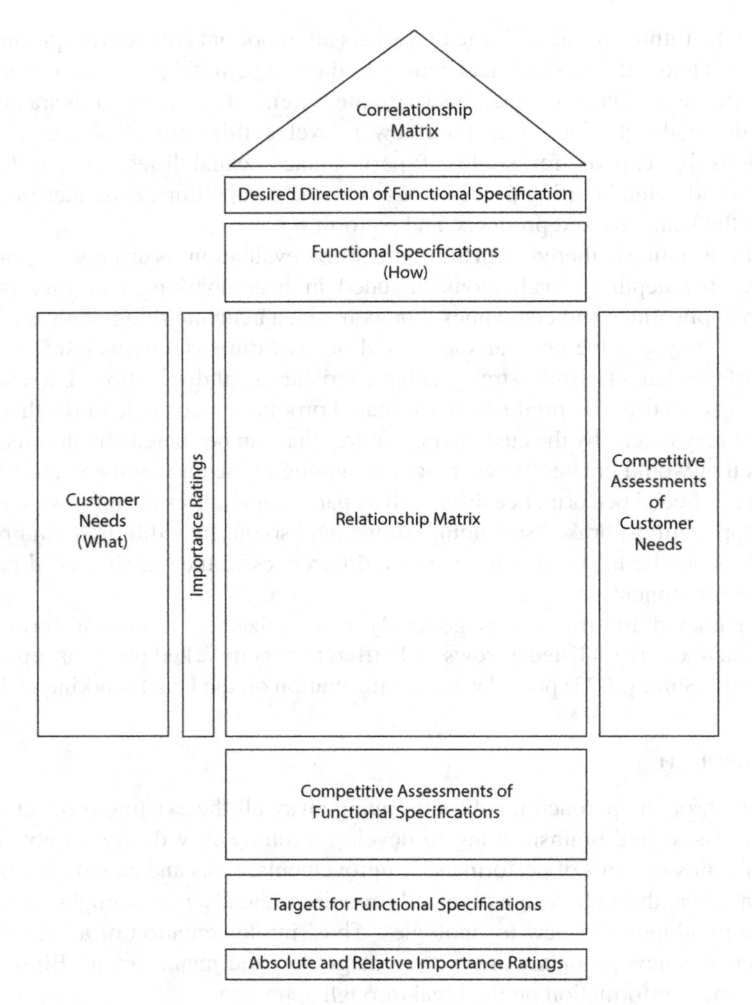

FIGURE 3.1 Structure of an QFD diagram showing its regions and contents.

in 1966 (Akao, 1991). The QFD is applied in a wide variety of applications and is considered a key tool in the Design for Six Sigma (DFSS) projects. It is also known as the "House of Quality" because the correlations matrix drawn on the top of the QFD matrix diagram resembles the roof of a house.

Figure 3.1 illustrates the basic structure (or regions) of the QFD. The contents of each region of the QFD are described below (Note: An example of a completed QFD chart is presented in the next section in Figure 3.3).

1. *Customer Needs (What)*: The needs of the customers as expressed by "what the customer wants in the product" are listed sequentially in rows of this left-most region. Each customer need should be described by using the customer's

words (as the customer would describe it in his/her own words, for example, give me a vehicle that would last for a long time, a vehicle that looks good, that works flawlessly, energy efficient, fun to drive, and so forth). The list of customer needs should be sorted into different categories and duplicated needs (even with different wordings) should be removed. The categorized needs should be organized in a hierarchical order such as primary needs, secondary needs (within each primary need) and tertiary needs (within each secondary need). The ith customer need (tertiary) is defined as C_i. (Note: The mathematical definitions of all the QFD variables are described later in this section).

2. *Importance Ratings*: This column provides importance ratings for each of the customer needs. The column of importance rating is provided to the right of the customer needs column. The importance ratings (or weights) can be obtained by using a number of different weighting techniques (e.g., rating scales, analytical hierarchical process). However, a 10-point rating scale is commonly used, where 10= extremely important and 1= not at all important. The importance rating for i^{th} customer need is defined as W_i.
3. *Functional Specifications (How)*: The function specifications are created by the engineers involved in the product development to define how the product (or the entity for which the QFD is prepared) "should function" or "should be designed to meet its customer requirements". The functional specifications describe "How" the engineers will address the customer needs. Thus, these specifications should be described using technical terms and variables that are used and selected by the engineers such as functions to be performed, types of mechanisms, materials, dimensions, strengths (capabilities), manufacturing processes, test requirements, and so forth. Here, the engineer should list each functional (or engineering) specification in a separate column. The functional specification columns are provided to the right of the importance ratings column. The jth functional specification is defined as F_j.

   The functional specifications should cover engineering considerations, methods or variables that need to be considered during product development. Some examples of the variables used for the functional specifications are: engineering requirements that must be met (e.g., max force in a specified direction and number of cycles of force applications), type of construction (e.g., welded versus assembled using fasteners, material to be considered to make the entity (e.g., steel, high-strength steel, aluminum, carbon fiber, etc.), type of production process to be used (e.g., extrusion versus cast), locations (e.g., installation locations expected by the customers for operation and service), physical space (e.g., envelope size, volume) needed for the specified entity, product characteristics (e.g., maximum achievable acceleration), capacity or capabilities of the entity, durability (e.g., works without a failure for 100,000 cycles under specified conditions), and so forth.
4. *Relationship Matrix*: The relationship matrix is formed by the customer needs as its rows and functional specifications as its columns. Each cell of the matrix represents the strength of relationship between the customer need

and the functional specification defining the cell. The weights of 9, 3 or 1 are commonly used to define strong, medium or weak relationship, respectively. The following coded symbols are used to illustrate the strengths of relationships: the two concentric circles (for 9=strong), one open circle (for 3=medium) and a triangle (for 1=weak). A cell is left blank when no relationship exists between the customer need and functional specifications defining the cell. The relationship in a cell of the relationship matrix is defined as R_{ij} (i.e., relationship between i^{th} customer need and j^{th} functional specification).

5. *Desired Direction of Functional Specification*: This row (placed above the functional specifications row) shows an up-arrow, a down-arrow or a "0" (zero) to indicate the desired direction of the value of each engineering specification (defined in the column). The up-arrow indicates that a higher value is desirable. The down-arrow indicates that a lower value is desirable. And "0" indicates that the functional specification is not dependent on either increase or decrease in its value. Thus, a quick visual scan of this row gives graphic information on whether the desired values of the functional specifications included in different columns need to be larger, smaller or not dependent on their values.
6. *Correlation Matrix (Roof)*: The correlation matrix is formed by relationship between combinations of any two engineering specifications defined by the cells of the matrix. The direction and strength of the relationship is indicated in the cell by a positive or a negative number (defined as I_{jk}) in the cell. Coded symbols are also used to indicate the direction and strength of the relationships. Only half of the matrix above the diagonal is shown as the roof of the QFD chart (see the roof in Figure 3.3).
7. *Absolute Importance Ratings of Functional Specifications*: The absolute importance rating of each functional specification (defined as A_j) is computed by summing weighted relationships between customer needs and the functional specification. The weighting of relationships is based on the importance rating (W_i) of each customer need and the strength of the relationships (R_{ij}'s). The expression for computation of values of A_j's is given later in this section. The absolute importance ratings are presented in a row just above the last row of the QFD chart.
8. *Relative Importance Ratings of Functional Specifications*: The relative importance rating (defined as V_j) of each functional specification is expressed as percentage of ratio of its contribution (A_j) to the sum of all A_j's. The expression for computation of values of V_j's is given later in this section. The relative importance ratings are presented in the last row of the QFD chart.
9. *Competitive Assessment of Customer Needs*: Each product used in benchmarking (along with the manufacturer's target product) is rated to determine how well each customer need (C_i) is satisfied by the product. The rating is usually given by the customers (or by the product development team members after they are very familiar with the customer needs and the products) by using a 5-point scale, where 1= poor (The product poorly meets the customer need) and 5 = excellent (The product meets the customer need

at a high excellence level). The ratings of each product are plotted on the right side of the relationship matrix.

10. *Competitive Assessment of Functional Specifications*: Each product used in benchmarking (along with the manufacturer's target product) is rated to determine how well each functional specification is satisfied by the product. The rating is usually provided by the technical experts or the product development team members (after they are very familiar with each functional specification and the products) by using a 5-point scale, where 1= poor (The product poorly meets the functional specification) and 5 = excellent (The product meets the functional specification at a high excellence level). The ratings of each product are plotted on the bottom side just below the relationship matrix.
11. *Targets for Functional Specification*: The target values for each of the functional specification (provided in the columns of QFD) are provided below the competitive assessment plot of the functional specifications. The target values are determined by the team after extensive discussions on all the collected data. The target should be specified precisely. Examples of target are: a) specification of material to be used, for example, aluminum, b) class level to be achieved (e.g., best-in-class, among the leaders, slightly above the average, average, or below average product among all the current products in its class); c) minimum value to be achieved, for example, minimum engine torque output should be 300 ft-lb at 3000 RPM, or d) minimum rating level to be achieved, for example, ratings value greater than or equal to 4 on a 5-point scale, where 5 = excellent and 1 = poor.

The following mathematical definitions will clarify the above-described variables and their relationships:

$C_i = i^{th}$ customer need, where $i = 1, 2, ..., m$

$F_j = j^{th}$ functional specification, where $j = 1, 2, ..., n$

$W_i =$ Importance rating of ith customer need, where $i = 1, 2, ..., m$

(The value of importance rating can range from 1 to 10, where 10= extremely important and 1 = not at all important.)

$R_{ij} =$ Relationship between i^{th} customer need and j^{th} functional specification.

(The values assigned to R_{ij} will be 9, 3 or 1 to define strong, medium and weak relationship, respectively. The ij^{th} cell in the relationship matrix will be left unfilled (i.e., blank) if there is no relationship between C_i and F_j.)

$I_{jk} =$ Relationship between j^{th} functional specification and k^{th} functional specification where j and $k = 1, 2, ..., n$ and $k \neq j$. (Thus, it is the interrelationship between two functional specifications shown in the roof of the QFD diagram.)

The values of I_{jk} can be as follows:

+9 = Strong positive relationship between j^{th} functional specification and k^{th} functional specification

+3 = Positive relationship between j^{th} functional specification and k^{th} functional specification

0 = No relationship between j^{th} functional specification and k^{th} functional specification
-3 = Negative relationship between j^{th} functional specification and k^{th} functional specification
-9 = Strong negative relationship between j^{th} functional specification and k^{th} functional specification

A_j = Absolute importance rating of jth functional specification

$$= \sum_i \left[W_i \times R_{ij} \right]$$

V_j = Relative importance rating of jth functional specification (%)

$$= 100 \times A_j / \left[\sum_j A_j \right]$$

An Example of the QFD Chart

A driver's side front door trim panel had to be designed for a new mid-sized 4-door sedan. The door trim panel covers the inner side (occupant side) of the steel doors and includes items such as inside door opening handle, door pull handle, armrest, door mounted switches (e.g., mirror, window and door lock switches), courtesy lights, speakers and map pocket storage (which typically stores bottles and an umbrella) (see Figure 3.2). The appearance of the door trim panel is important as it should match in color and materials with the instrument panel and other interior trim parts and seats. The manufacturer formed a team of interior designers, package engineers, body engineers, electrical engineers, ergonomics engineers, market researchers and engineers from the suppliers of the door trim panel, switches, speakers and window raising and lowering mechanism to create a QFD chart for the front door trim panel.

The team interviewed a number of customers (i.e., the owners of their current vehicle model and two of their best competitors) and asked them about their needs and expectations about the door trim panel of their future vehicle. The customers first started telling the team what they wanted a good door trim panel. The team members kept on asking a number of probing questions: What do you mean by a good door trim panel? What would you like in the door trim panel? The customers responded saying that a good trim panel means: a) it should look good, b) it should be easy to use, c) it should have plenty of features, d) it should be safe, e) it should have enough storage capacity, and finally, f) it should make the vehicle interior look spacious (not crammed with little clearances and its outer surfaces should not be too close to their hips and shoulders of the occupants). The above needs were considered as the secondary needs. The team then probed more into each of the secondary needs and created lists of tertiary needs. The primary, secondary and tertiary needs were listed on the left side of the QFD as "whats" (i.e., what the customers wanted) in the QFD shown in Figure 3.3.

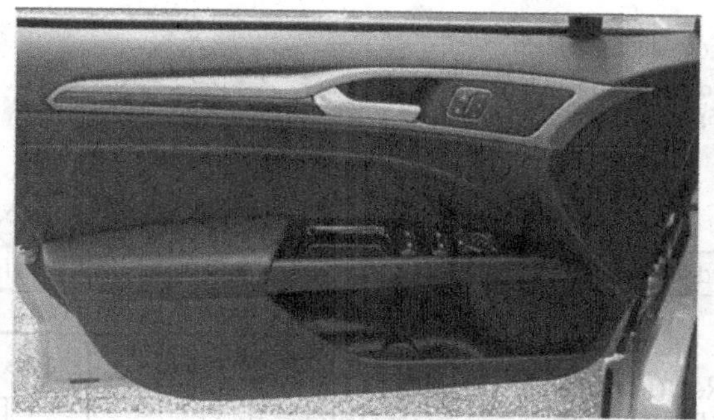

2015 Ford Fusion

2015 Toyota Camry

2015 Honda Accord

FIGURE 3.2 Door trim panels of three mid-sized vehicles.

The team members also asked customers to rate each of the tertiary needs on a 10-point importance scale, where 1= not at all important and 10 = extremely important. The importance ratings are provided to the right side of the customer requirements column in Figure 3.3.

The team brainstormed and created a list of "Hows", that is, how would they design the door trim panel? How should it function? What would be their technical

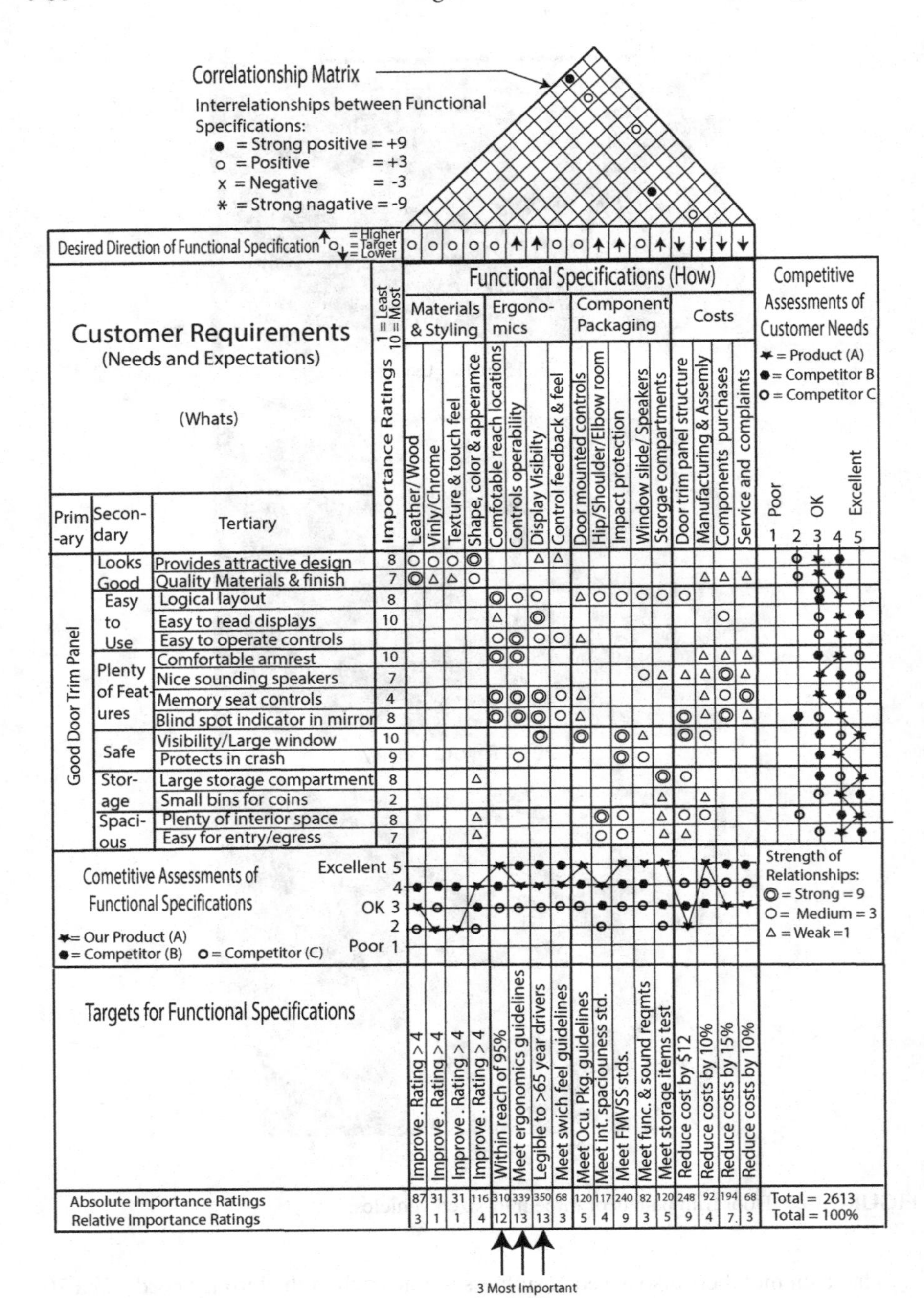

FIGURE 3.3 A QFD chart for an automotive door trim panel.

descriptors or functional specifications of the door trim panel? The functional specifications developed by the team are listed as column headings, in the QFD chart. The functional issues of importance to the engineers were categorized into the following groups: a) materials and styling, b) ergonomics, c) component packaging and d) costs. Functional considerations under each of the groups are listed under separate columns. The functional specification columns are placed immediately to the right of the importance ratings in Figure 3.3.

Next, the team discussed every combination of customer needs and functional specifications and provided strength of their relationships using the following scale: a) strong relation (weight=9), b) medium relation (weight=3), and c) weak relation (weight=1). The symbols corresponding to the weights are placed in the cells of the relationship matrix (see Figure 3.3). Similarly, the relationships between every pair of functional specifications were discussed by the team and symbols corresponding to very positive to very negative relationships are placed in the interrelationship (correlation) matrix shown on the top of the QFD chart as its roof.

Based on the information gathered during the customer interviews, the team members rated each of the three vehicles (their current product "A" and two competitors called Competitor "B" and "C") and plotted the ratings on each of the tertiary customer needs and functional specifications. The plot of the competitive assessment of customer needs is provided as the rightmost part of the QFD chart. The plot of competitive assessment of functional specifications is provided below the relationship matrix (i.e., the matrix of customer needs and functional specifications) (see Figure 3.3).

The team developed targets (i.e., what target values or rating levels, guidelines and/or requirements to use for designing their future instrument panel) for each of the functional specifications by discussing how their product compared with their two competitors and their marketing goals. The targets are shown in a section below the competitive assessments of the functional specifications in the QFD.

Finally, the absolute and relative importance ratings of each of the functional specifications were computed and entered in the last two rows of the QFD chart. The three functional specifications that received the highest importance ratings are: (1) display visibility (i.e., displays visible to the driver without any obscurations and legible labels and graphics); (2) control operability (i.e., how the controls are configured and their operating motions); and (3) comfortable reach locations (i.e., placement of the controls/items within the driver's hand reach zone). These three functional specifications must get very high priority in designing the door trim panel.

Advantages and Disadvantages of QFD

Developing a single QFD chart can be very time consuming as it takes many hours of teamwork involving meetings, discussions, customer visits, benchmarking of the competitive products, development of targets, and so on. The advantage is that it exposes the entire product design team to all aspects of the decisions to be taken during the product development process. The process of developing a QFD thus educates the team, documents the collected information, and prioritizes information needed during its development. Thus, when the team actually starts developing the product, the subsequent decisions generally would take much less time as the team members would have already discussed all the issues and they would be very aware of most of the interfacing and trade-off considerations. The product developed using QFD, therefore, would have a better chance of designing the right product and satisfying its customers.

TASK ANALYSIS

Task analysis is a simple but powerful method to determine user problems in product designs. It involves an ergonomics engineer to breakdown a tasks into a series of subtasks or steps and analyze the demands placed on the user in performing each step and compare against the capabilities and limitations of the users. Chapter 11 provides an example of a task analysis in Table 11.2 to predict possible errors in opening the hood and checking the engine oil level.

FAILURE MODES AND EFFECTS ANALYSIS

The failure modes and effects analysis (FMEA) method was used in the 1960s as a systems safety analysis tool. It was used in the early days in the defense and aero-space systems design to ensure that the product (e.g., an aircraft, spaceship or a missile) was designed to minimize probabilities of all major failures by brainstorming and evaluating all possible failures that could occur and acting on the resulting prioritized list of the corrective actions. Over the past twenty years, the method has been routinely used by the product design and process design engineers to reduce the risk of failures in the designs of products and processes used in the many industries (e.g., automotive, aviation, utilities, and construction). The FMEA conducted by product design engineers is typically referred to as DFMEA (the first letter "D" stands for design). And the FMEA conducted by the process designers is referred as PFMEA (the first letter "P" stands for process). In many automotive companies, product (or process) design release engineers are required to perform the task of creating the FMEA chart and to demonstrate that all possible failures with the risk priority number (RPN) over a certain value are prevented.

FMEA is a proactive and qualitative tool used by quality, safety and product/process engineers to improve reliability (i.e., eliminate failures – thus, improve quality and customer satisfaction). Development of a FMEA involves the following basic tasks:

a) Identify possible failure modes and failure mechanisms.
b) Determine the effects or consequences that the failures may have on the product and/or process performance.
c) Determine methods of detecting the identified failure modes.
d) Determine possible means for prevention of the failures.
e) Develop an action plan to reduce the risks due to the identified failures.

FMEA is very effective when performed early in the product or process development and conducted by experienced multi-functional team members as a team exercise.

The method involves creating a table with each row representing a possible failure mode of a given product (or a process) and providing information about the failure mode using the following columns of the FMEA table:

1. Description of a system, subsystem or component
2. Description of a potential failure mode of the system, subsystem or component
3. Description of potential effect(s) of the failure on the product/systems, its subsystems, components or other systems
4. Potential causes of the failure
5. Severity rating of the effect due to the failure
6. Occurrence rating of the failure
7. Detection rating of the failure or its causes
8. Risk priority number (RPN); it is the multiplication of the three ratings in items 5, 6 and 7 above
9. Recommended actions to eliminate or reduce the failures with higher RPNs
10. Responsibility of the persons or activities assigned to undertake the recommended actions and target completion date
11. Description of the actions taken
12. Resulting ratings (severity, occurrence and detection) and risk priority numbers (after the action is taken) of the identified failures in item 2 above.

Examples of rating scales used for severity, occurrence and detection are presented in Tables 3.8 to 3.10 respectively. The definitions of the scales generally vary between

TABLE 3.8
An Example of a Rating Scale for Severity

Rating	Effect	Criteria: Severity of Effect
10	Hazardous -- without warning.	Very high severity rating when potential failure mode affects safe product operation and/or involves non-compliance with government regulations without warning.
9	Hazardous -- with warning	Very high severity rating when potential failure mode affects safe product operation and /or involve non-compliance with government regulations with warning.
8	Very High	Product inoperable with loss of primary function
7	High	Product operable with reduced level of performance. Customer dissatisfied.
6	Moderate	Product operable but usage with reduced level of comfort or convenience. Customer experiences discomfort.
5	Low	Product operable but usage without comfort or convenience. Customer experiences discomfort.
4	Very Low	Minor product defect (e.g., noise, vibrations, poor surface finish) only noticed by most customers
3	Minor	Minor product defect only noticed by average customer.
2	Very Minor	Minor product defect only noticed by discriminating customer.
1	None	No effect.

TABLE 3.9
An Example of a Rating Scale for Occurrence

Rating	Probability of Failure	Possible Failure Rates
10	Very High: Failure is almost inevitable	≥ 1 in 2
9		1 in 3
8	High: Repeated failures	1 in 8
7		1 in 20
6	Moderate: Occasional failures	1 in 80
5		1 in 800
4	Low: Relatively low failures	1 in 2,000
3		1 in 15,000
2	Remote: Failure is unlikely	1 in 150,000
1		≤ 1 in 1500,000

TABLE 3.10
An Example of a Rating Scale for Detection

Rating	Detection	Criteria: Likelihood of Detection by Design Control
10	Absolutely Uncertain	Design Control will not and/or cannot detect a potential cause or mechanism for the failure mode; or there is no Design Control.
9	Very Remote	Very remote chance that the Design Control will detect a potential cause/mechanism for the failure mode.
8	Remote	Remote chance that the Design Control will detect a potential cause/mechanism for the failure mode.
7	Very Low	Very low chance that the Design Control will detect a potential cause/mechanism for the failure mode.
6	Low	Low chance that the Design Control will detect a potential cause/mechanism for the failure mode.
5	Moderate	Moderate chance that the Design Control will detect a potential cause/mechanism for the failure mode.
4	Moderately High	Moderately high chance that the Design Control will detect a potential cause/mechanism for the failure mode.
3	High	High chance that the Design Control will detect a potential cause/mechanism for the failure mode.
2	Very High	Very high chance that the Design Control will detect a potential cause/mechanism for the failure mode.
1	Almost Certain	Design Control will almost certainly detect a potential cause/mechanism for the failure mode.

different organizations depending upon the type of industry, product or process, nature of the failures, associated risks to humans and costs due to the failures.

An Example of an FMEA

An automatic transmission in an automobile will not operate properly if the transmission fluid leaks out. An engineer designing a transmission fluid hose conducted an FMEA (Failure Modes and Effects Analysis) to evaluate possible failures caused by the hose. The FMEA is presented in Table 3.11. The hose involved in this example consists of a nylon tube with connectors inserted into each end. Ferrules are crimped onto each end to help hold the connectors on. A conduit made of plastic covers the hose, to protect it from heat and moving parts near the engine. The hose carries transmission fluid from the reservoir to the clutch actuation system. The transmission fluid in the hose is pressurized during operation to about 6.5 bars (94.3 psi). The hose must also be able to withstand temperatures above 60° C (140° F). During the development phase, the design itself is proven in a series of tests referred to as the DVP&R (Design Verification Plan and Report). The DVP&R is conducted once for each design, so it does not consider all the sources of variability that the product and materials are exposed to during the life of the part. There were three failures with RPN over 50 in the FMEA. The actions taken by the engineer reduced the RPNs of all the three failure modes (see Table 3.11).

OTHER PRODUCT DEVELOPMENT TOOLS USED IN DECISION MAKING

During the product development process, tools from many areas such as systems engineering, specialty engineering areas, and program and project management are used to manage both the technical and business activities of the program. These tools provide important information in management and technical decision making activities during the product planning and development process. Business plan and Systems Engineering Management Plan (SEMP) are two important tools used in automotive product development and are presented below.

TABLE 3.11
An Example of a FMEA

FAILURE MODES AND EFFECTS ANALYSIS	
Part Number: 1008-92 Description: Powertrain Fluid Hose Assembly Model Year: 2017	Original Date: 1/03/2015 Date Revised: 3/03/2016 Prepared by: T. James

TABLE 3.11 (Continued)
An Example of a FMEA

No.	Item/ Function	Potential Failure Mode	Potential Effect(s) of Failure	Severity Rating (S)	Potential Causes of Failure	Occurrence Rating (O)	Current Design Controls
1	Hose-Allows transmission fluid to travel and exert pressure to point of use	Leak in the hose	A slow leak would not be noticed immediately. The driver would notice gear shifting delays over time and finally the vehicle will stop shifting gears.	8	Hole in the hose. Defective hose material. Hose material degrades.	3	100% leak test with air.
2		Hose burst at low pressure	A quick and fast leak with immediate loss of shifting function	8	Defective hose material. Hose damaged by debris on road.	3	Material burst test
3	Connectors-Attach hose to pressurized oil reservoir	Leaks around connector	A slow leak would not be noticed immediately. The driver would notice gear shifting delays over time and finally the vehicle will stop shifting gears.	8	Loose connector or omitted washer or uneven connector mating surface	3	100% leak test with air.
4		Connector disconnected from hose or reservoir	A quick and fast leak with immediate loss of shifting function	8	Connector not fully seated in the assembly	2	Pull out testing at final inspection. 100% visual checks during service
5	Conduit protects hose during operation from moving parts and stones sprayed by tires	Insufficient conduit wear strength	Possible damage to tube in service	6	Conduit material defective	1	Design verified during sign-off
6	Ferrule-Holds the connector on to the hose	Connector disconnected from hose or reservoir	A quick and fast leak with immediate loss of shifting function	8	Ferrule or hose not properly positioned in the crimping die or press failure	2	Pull out testing at final inspection. 100% visual checks during service
7	Barcode Label- Lot and part traceability	Missing label	Loss of traceability and identification	4	Label fell off, never affixed, not affixed properly	4	Visual inspection of label
8	Packing-Protect the part while shipping	Damaged part	Returned by the customer	4	Insufficient packing strength	1	Packaging standard used for all parts shipping

Detection Rating (D)	Risk Priority Number (RPN)	Recommended Action	Responsibility	Action Results				
				Action	S	O	D	RPN
1	24	No action required.						
7	168	Investigate if variations in the burst test results	T. James 07/20/16	Stronger exterior hose casing. Reduced hose extrusion temperature for consistency.	6	1	7	42
1	24	No action required.						
4	64	Investigate if a sensor could detect if the connector was fully seated at the hose end	T. Jack 07/20/16	Improved connector threads. Sensor not possible currently.	3	2	4	24
7	42	No action required.						
4	64	Investigate if parts presence sensor can be located on the die	T. James 07/20/16	Sensor installed on the die and press stops if sensor is not activated	8	1	2	16
4	32	No action required.						
4	16	No action required.						

Business Plan

A Business Plan is a proposal for creating or developing a new product. It is usually prepared internally within a company to obtain concurrence from the top management to approve the product program. The business plan is thus a document prepared to describe details of a proposed product, product program timing plan and corporate resource needs to develop the product. It is typically prepared jointly by the company's product planning, engineering, marketing and finance activities. The business plan provides basic information on the new vehicle program, and it serves as a guide in planning vehicle program activities to all departments including ergonomics.
The business plan should include the following:

1. Description of the Proposed Product:
   - Product configuration (e.g., for an automotive product: the body style of the vehicle such as a sedan, a coupe, a crossover, a sports utility vehicle (SUV), a pickup, or a multi-passenger vehicle (MPV) and its variations)
   - Size class (e.g., sub-compact, compact, intermediate, large)
   - Markets where the product will be sold and used (countries)
   - Market Segment (e.g., luxury, entry-luxury, or economy)
   - MSRP (Manufacturer's suggested retail price) and price range with different models and optional equipment
   - Production capacity and estimated sales volumes over the product life cycle
   - Makes, models and prices of leading competitors in the proposed market segment
2. Attribute rankings (i.e., how would the product be positioned in its market segment such as best-in-class, above the class average, average in the class or below the class average by considering each product attribute)
3. Pugh Diagram showing how the proposed product will compare with its current model (used as the datum) and other leading competitors' products by considering all important product attributes and changes in the proposed product.
4. Dimensions and Options:
   - Overall exterior dimensions (e.g., product package envelope with length, width, height, wheelbase, and cargo/storage volume)
   - Interior dimensions (e.g., people package with legroom, headroom, shoulder room, number of seating locations, luggage/cargo volume, and so forth)
   - Major changes in the product's systems as compared to the previous (outgoing) model, for example, drive options (front wheel drive [FWD], rear wheel drive [RWD], all-wheel drive [AWD]), powertrains (types, sizes and capacities of engines/motors and transmissions), descriptions of unique features (e.g., type of suspensions) and optional equipment
5. One paragraph description of the proposed product with several adjectives to describe its image, stance, and styling characteristics (e.g., futuristic, traditional, retro, fast, dynamic, aerodynamic, tough, or chunky – like a Tonka truck)

6. Program Schedule:
   - Program kick-off date, timings of major milestones
   - Job#1 date (i.e., date when first production unit will be out of the assembly plant) and model year
7. Projected Sales Volumes
   - Quarterly or yearly sales estimates of each model in each market segment
8. Financial Analysis:

   Curves of estimated cumulative costs, revenues and cash flow during product life cycle for different scenarios (e.g., best case, average, and worst case)

   Anticipated quarterly funding needed during product development and during revenue build-up

   Anticipated date of break-even point

   Return on investment
9. Product Life Cycle:
   - Estimated lifespan
   - Possible product refreshments, future models and variations
   - Recycling of the plant and products
10. Proposed plant location and plant investments
11. Potential sources of risks in undertaking the program
12. Justification (reasons) for approval of the proposed plan versus other alternatives

SYSTEMS ENGINEERING MANAGEMENT PLAN

The Systems Engineering Management Plan (SEMP) is a documentation created to plan the SE activities in terms of *what* needs to be done (e.g., types of analyses, methods, and procedures), *when* the activities need to be done (e.g., sequence of activities in the process), and *who* performs the activities (i.e., certain specified engineering departments or disciplines).

The SEMP is basically a planning document, and it should be prepared by the SE activity after the concept design phase. The SEMP is a higher-level plan (not very detailed) for managing the SE effort to produce a final operational product (or a system) from its initial requirements. Just as a project plan defines how the overall project will be executed, the SEMP defines how the engineering portion of the project (or program) will be executed and controlled. The SEMP describes how the efforts of system designers, test engineers, and other engineering and technical disciplines will be integrated, monitored, and controlled during the complete life cycle of the product. Figure 3.4 presents a flow chart illustrating the relationship of the SEMP to project work and project management.

For a small project, the SEMP might be included as part of the project plan document, but for any project or program of greater size or complexity, a separate document is recommended. The SEMP provides the communication bridge between the project management team and the technical teams. It also helps to coordinate and monitor work between and within different technical teams. It establishes the

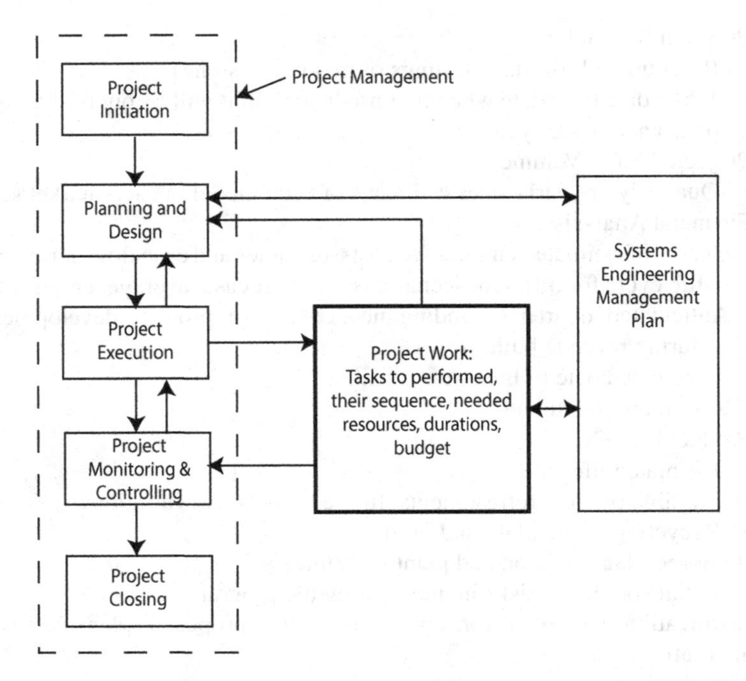

FIGURE 3.4 Relationship of the systems engineering management plan (SEMP) to project work and project management.

framework to realize the appropriate work (or tasks to be performed) that meet the entry and success criteria of the applicable project phases. The SEMP provides management with necessary information for making systems engineering decisions by focusing on requirements, design, development (detailed engineering), test, and evaluation; it also addresses traceability of stakeholder requirements and supportability across the project life cycle.

Contents of SEMP

The purpose of this section is to describe the activities and plans that will act as controls on the project's SE activities. For instance, this section identifies the outputs of each SE activity, such as documentation, meetings, and reviews. This list of required outputs will control the activities of the team and thus will ensure the satisfactory completion of the activities. Some of these plans may be completely defined in the SEMP (in the framework or the complete version). For other plans, the SEMP may only define the requirements for a particular plan. The plan itself is to be prepared as one of the subsequent SE activities, with a verification plan or a deployment plan as the case may be. Almost any of the plans described in the following discussion may fall into either category. It all depends on the complexity of the particular plan and the amount of up-front SE that can be done at the time the SEMP is prepared.

The first set of required activities relates primarily to the successful management of the project. These activities are likely to have already been included in the project/program plan but may need to be expanded in the SEMP (USDOT, 2024). Generally, they are incorporated into the SEMP but, on occasion, may be developed as separate documents. The items that can be included in the SEMP are listed as follows. The items and their descriptions provided in USDOT (2024) were modified to meet the needs of complex product development.

1. Work breakdown structure (WBS)consists of a list of all tasks to be performed on a project, usually broken down to the level of individually budgeted items.
2. *Task inputs*: It is a list of all inputs required for each task in the WBS, such as source requirements documents, interface descriptions, and standards.
3. *Task deliverables*: This is a list of the required deliverables (outputs) of each task in the WBS, including documents, and product configuration including software and hardware.
4. *Task decision gates*: These are critical activities that must be satisfactorily completed before a task is considered completed.
5. *Reviews and meetings*: This document contains a list of all meetings and reviews of each task in the WBS.
6. *Task resources*: This is the identification of resources needed for each task in the WBS, including personnel, facilities, and support equipment.
7. *Task procurement plan*: This is a list of the procurement activities associated with each task of the WBS, including hardware and software procurement and any contracted or supplier provided services such as SE services or development services.
8. *Critical technical objectives*: This is a summary of the plans for achieving any critical technical objectives that may require special SE activities. It may be that a new software algorithm needs to be developed and its performance verified before it can be used. Or a prototyping effort is needed to develop a user-friendly operator interface. Or a number of real-time operating systems need to be evaluated (verified) before a procurement selection or the level of assembly is made.
9. *Systems engineering schedule*: A schedule of the SE activities that shows the sequencing and duration of these activities. The schedule should show tasks (at least to the level of the WBS), deliverables, important meetings and reviews, and other details needed to control and direct the project. An important management tool is the schedule. It is used to measure the progress of the various teams working on the project and to highlight work areas that need management intervention.
10. *Configuration management plan*: This describes the development team's approach and methods to manage the configuration of the systems within the

products and processes. It also describes the change control procedures and management of the system's baselines as they evolve.

11. *Data management plan*: This describes how and which data will be controlled, the methods of documentation, and where the responsibilities for these processes reside.
12. *Verification plan*: This plan is always required. It is written along with the requirements specifications. However, the parts on tests to be conducted can be written earlier.
13. *Verification procedures*: These are developed by the core engineering experts, and they define the step-by-step procedures to conduct verification tests and must be traceable to the verification plan.
14. *Validation plan*: This is required. It helps to prove that the product being designed is the right product and would meet all the customer's needs.

The second set of plans is designed to address specific areas of the SE activities (USDOT, 2024). They may be included entirely in the SEMP, or the SEMP may give guidance for their preparation as separate documents. The plans included in the first set listed in the preceding list are generally universally applicable to any project. On the contrary, some of the plans included in this second set are only rarely required. The unique characteristics of a project will dictate their needs. The items that can be included in this second set are described as follows. The items and their descriptions provided in USDOT (2024) were modified to meet the needs of complex product development.

1. *Software development plan*: This describes the organization structure, facilities, tools, and processes to be used to produce the project's software. It also describes the plan to produce custom software and procure commercial software products.
2. *Hardware development plan*: This plan describes the organization structure, facilities, tools, and processes to be used to produce the project's hardware. It describes the plan to produce custom hardware (if any) and to procure commercial hardware products.
3. *Technology plan*: It describes (if needed) the technical and management process to apply new or untried technology. Generally, it addresses performance criteria, assessment of multiple technology solutions, and fallback options to existing technology.
4. *Interface control plan*: It identifies all important interfaces within and between systems (within the product and external to the product) and identifies the responsibilities of the organizations on both sides of the interfaces.
5. *Technical review plan*: Identifies the purpose, timing, place, presenters and attendees, topics, entrance criteria, and the exit criteria (resolution of all action items) for each technical review to be held for the project/ program.
6. *System integration plan*: Defines the sequence of activities that will integrate various product chunks involving components (software and hardware), subsystems, and systems of the product. This plan is especially important

if many subsystems and systems are designed and/or produced by different development teams from different organizations (e.g., suppliers).

7. *Installation plan or deployment plan*: It describes the sequence in which the parts of the product are installed (deployed). This plan is especially important if there are multiple different installations at multiple sites. A critical part of the deployment strategy is to create and maintain a viable operational capability at each site as the deployment progresses.
8. *Operations and maintenance plan*: This defines the actions to be taken to ensure that the product remains operational for its expected lifetime. It defines the maintenance organization and the role of each participant. This plan must cover both hardware and software maintenance.
9. *Training plan*: Describes the training to be provided for both maintenance and operation.
10. *Risk management plan*: Addresses the processes for identifying, assessing, mitigating, and monitoring the risks expected or encountered during a project's life cycle. It identifies the roles and responsibilities of all participating organizations for risk management.
11. *Other plans*: Other plans that might be included are, for example, a safety plan, an ergonomics plan, a security plan, and a resource management plan.

This second list is extensive and by no means exhaustive. These plans should be prepared when they are clearly needed. In general, the need for these plans becomes more important as the number of stakeholders involved in the project increases.

The SEMP must be written in close synchronization with the project plan. Unnecessary duplication between the project plan and the SEMP should be avoided. However, it is often necessary to put further expansion of the SE effort into the SEMP even if they are already described at a higher level in the project plan.

PROGRAM STATUS CHART

Product planners and program managers keep track of progress on product development programs by using a number of different techniques such as program timing charts and Gantt charts and cash flows. One chart that is very popular is the program status chart, which is typically used to track the status of problems encountered during a program (or a project).

The status charts are also called the "Red-Yellow-Green" charts as they indicate the problem status by use of colors as follows: a) Red indicates that the problem is not yet solved and it is a "job stopper" (i.e., it will stop the progress on the entire program until the problem is solved), b) Yellow indicates that the problem can introduce significant delays in the program unless it is solved quickly, and c) Green, indicates that the problem is no longer a timing threat to the program.

Table 3.12 presents an illustration of a program status chart. The status column in the chart uses letters R, Y and G to indicate the red, yellow and green colors when colored charts can be made on a normal "black ink only" printer. Such charts are typically used in the senior management level meetings to draw attention and get fast resolution on the job-stopper problems.

TABLE 3.12
Program Status Chart

Program Status Chart							
Program: XM25					**Program Manager: RJW**		**Date: 08/30/15**
Problem No.	**Problem Description**	**Status***	**Target Date**	**Expected Completion Date**	**System/ Subsystem/ Component**	**Product Attribute**	**Responsibility: Organization and Manager**
1	Center stack screen freezes unexpectedly during navigation operation	R	02/03/2016	02/04/2016	Navigation – Software	Driver communication	Driver Interface – JLM
2	Gloss level on the top resurface of the instrument panel should be reduced to meet veiling glare standard	Y	15/09/2015	30/10/2015	Instrument Panel – Crash pad	Safety	Body and Trim – JBM
3	Squealing noise during braking	Y	10/11/2015	30/12/2015	Braking System- Brake pad material	Safety	Chassis-Brakes- WLV

4	Premature wear in front bearings of turbo-boosters	R	15/11/2015	Unresolved	Powertrain- Turbo Boosters	Performance	Powertrain – EER
5	Noticeable jerk during transmission shifting	Y	25/11/2015	30/12/2015	Powertrain- X5 Transmission	Performance	Transmission- JJT
6	Wiper flutter at speeds over 70 mph	G	15/07/2015	26/08/2015	Body – Visibility	Safety	Body Electrical – RGK

Notes:

Key: * = R= Red: Critical issue (Job stopper – must be resolved before next program review meeting)Y= Yellow: Important customer need- must be resolved by next milestone)

G= Green: Issue resolved- no action needed.

DESIGN STANDARDS

Product design standards serve as very useful tools in saving time required to make design decisions. Properly developed design standards that incorporate rationale and assumptions used in their development can provide basic knowledge on whether the standard can be applied during the design process. When the standard meets the needs of the customers for the product, then use of the design requirements and design procedures provided in the standards can reduce the time required to get the necessary information and decide on how to design.

Some standards may only specify how the product should perform and thus they provide design flexibility (i.e., the designer can design by using any appropriate solution as long as it meets the required performance). Such performance standards can promote innovative product designs as they will not be restricting compliance to any given design configuration and its specifications.

CAD TOOLS

A number of Computer-Aided Design (CAD) tools are used to create three-dimensional solid models using software such as AutoCAD, CATIA, Pro/Engineer, SolidWorks, Rhino, and so forth. These tools not only perform the traditional engineering drawing and drafting work but they allow visualization of the product model from different eye points to evaluate issues such as: a) exterior and interior appearance (e.g., shapes, continuity/discontinuity between adjacent surfaces, tangents, reflections), b) spaces (clearances) between different entities within the product, c) postures of human occupants/operators (with digital human models) in the products (e.g., cars, airplanes, and boats) or workplaces, d) feeling of interior spaciousness and storage spaces, layouts of hardware placements (mechanical packaging), e) comparisons of alternate designs (by superimposition or side-by-side viewing of different product concepts and competitive products), f) assembly analyses to evaluate assembly feasibility (e.g., by detecting interferences between parts being assembled), g) alternate assembly methods and fit (e.g., gaps) between parts, and so on. The newer CAD models can also simulate movements of parts within the product, movements of the product in its work environment to aid in visualizing how the product will look and fit within other existing systems.

CAD models are also very useful for communications between different design studios, product engineering offices and supplier facilities. The CAD files for the products can be also used as inputs to a number of other sophisticated computer-aided engineering (CAE) analyses to evaluate structural/mechanical (e.g., strength, dynamic forces, deflections, vibrations during simulated operating environments), aerodynamic (using computer-aided fluid dynamics), thermal (temperature, heat build-up and heat transfer) aspects of the products, and so forth. The CAD files can also be used to facilitate manufacturing operations. For example, CAD files serve as inputs to computer-aided process planning (CAPP) as well as for creating machining programs for the computerized numerical control (CNC) machines.

CAD has become an especially important technology within the scope of computer-aided technologies, with benefits such as lower product development costs, a greatly shortened design cycle and reducing error in product dimensions shown in the drawings and CAD models. CAD enables the designers to create layouts and develop their work on a display screen, print it out and save it for future editing – thereby saving time in creating variations in designs and their drawings.

PROTOTYPING AND SIMULATION

Virtual and physical prototyped parts can be created for visual evaluations and physical mock-ups for use in design reviews. A number of computer simulation systems are also available for human factors testing of user interfaces. Three-dimensional parametric solid modeling requires the design engineer to input values of key parameters – what can be referred to as the "design intent". The objects and features created can be shown to the customers for their feedback and adjusted by creating many design iterations until an acceptable design is achieved. Further, any future modifications can be easily made by inputting parameter changes in a computer-controlled prototype. Many automotive manufacturers use computer-controlled adjustable vehicle models (or programmable vehicle bucks) during early concept phases to compare and evaluate a number of automotive designs by quick changes in many key vehicle package parameters (Richards and Bhise, 2004; Bhise and Pillai, 2006; Prefix, 2023).

A number of specialized computer software systems are increasingly used to simulate product testing and evaluation. For example, CAD is used to create accurate photo simulations that are often required in the preparation of environmental impact reports in which computer-aided designs of intended buildings, vehicles and other products are superimposed into photographs or videos of existing environments to represent what that locale will be like when the proposed facilities are allowed to be built. Potential blockage of view, corridors and shadow studies are also frequently made through the use of CAD. Vehicle designers often use such simulations to compare exterior designs of various vehicle concepts with their competitor products in various simulated road environments.

PHYSICAL MOCK-UPS

Physical mock-up of concepts of products such as cars, trucks, airplanes, and boats are useful during design reviews to get a better feel of the size, space and configuration of a product in its early phases. The mock-ups also can be shown to potential customers and users for their feedback during informal quick evaluations as well as structured market research clinics (see Volume 2, Chapter 19 for more information).

TECHNOLOGY ASSESSMENT TOOLS

Using new technologies to improve product designs has been a continuous process to gain improvements in performance, efficiencies, safety and costs. However, most new technologies cannot be immediately applied. It takes many years or even decades to solve problems in bringing some new technology applications to a state of readiness and implementation. Technical experts in various specialized areas generally follow advances in new technologies. Progress in the most promising technologies is closely followed, and research departments are asked to perform evaluations and undertake development projects to improve the technologies so that they can be quickly implemented in future products (e.g., advances in battery technologies used in electric vehicles).

Many methods to assess technologies have been developed. Forgie and Evans (2011) have provided an excellent review of available techniques for technology assessments.

INFORMATIONAL NEEDS IN DECISION MAKING

The key to making good decisions is to have sufficient information and a good understanding of issues related to the alternatives, outcomes, trade-offs, and payoffs associated with the decision situation. Therefore, it is important to select a decision maker carefully and make sure that the person is familiar with the product and its uses. In some situations, customers who have used similar products are asked to provide their ratings on each product (or alternative) used in the evaluation. On the other hand, experts who are very familiar with the product and have extensive knowledge about the product can be very discriminating (much more than even the most familiar customers) and can provide unbiased evaluations.

In addition, the experts can obtain additional information through other methods such as: (1) benchmarking other products; (2) literature surveys; (3) exercising available models (e.g., models to predict performance of products under different situations) and using the information obtained from the model results; and (4) conducting experiments.

Exercising available models under various "what if" scenarios (i.e., conducting sensitivity analyses) can provide more insights into the variability (or robustness) in the performance of the product and thus can prepare the decision maker to make more informed decisions. Design reviews with different groups, disciplines, and experts can also generate information on strengths and weaknesses of the product (or product concept) being reviewed.

KEY DECISIONS IN PRODUCT LIFE CYCLE

Some of the key decisions made in a product program typically involve the following:

1. *Program kick-off*: Top management of the organization decides that a new product (or revisions to an existing product) should be developed and a project should be kicked-off to plan the product development and budget planning activities.
2. *Program confirmation*: Based on the additional information obtained from the design team's presentation of the created product concepts, market research results, trends in the new technologies and design, and the competitors' capabilities, the top management confirms the decision to select a product

concept. Additional decisions are also made to allocate the budget and dates for the product introductions in the selected markets.

3. *Product concept freeze*: Management decides that the selected product concept is sufficiently developed (i.e., all design and engineering managers feel confident that the product could be produced [i.e., it is feasible] within the planned budget and schedule). Thus, the concept will be frozen (i.e., no major changes will be made) and succeeding program activities will be continued.
4. *Engineering sign-off*: All key managers of engineering activities sign a document stating that the product "as designed" will meet all applicable requirements with a high probability (e.g., 90%).
5. *Production release*: All product testing (verification and validation tests) is completed, and the product is determined to be ready for the market. Product is released for production; that is, the factories begin production of units for sale.
6. *Periodic reviews*: Periodic (monthly, quarterly, or annual) reviews of the product sales, customer satisfaction, and comparison of data with the competitors' products are conducted to determine whether any changes in the product volume or product characteristics are needed.
7. *Product discontinuation and replacement*: Based on the market data and the customer feedback, the management decides to terminate the production of the product on a certain date and requests the marketing department to plan for future product(s) or model(s) for its replacement.

Trade-offs during Design Stages

Teams involved in designing any product need to make a number of decisions involving trade-offs between a number of conflicting design considerations (e.g., product characteristics and attributes). Some examples of trade-off considerations in designing passenger cars are described below.

1. *Space for vehicle systems versus space for occupants*: The space within the vehicle is occupied by various vehicle systems and their components, and space is used to accommodate occupants in the passenger compartment and other items in the trunk (or cargo areas). In order to provide more space for the occupants (interior passenger space), the space occupied by vehicle systems (e.g., vehicle body structure sections, engine, chassis and suspension components, fuel tank) needs to be reduced. Thus, a vehicle designer can make trade-offs by designing more compact vehicle systems to allow more space for the occupants. This trade-off is commonly referred in the auto industry as "Machine Minimum and Man Maximum" (i.e., minimizing the space for mechanical components – hardware – and maximizing the space for the occupants).
2. *Fuel economy versus performance*: A vehicle with a high acceleration capability (commonly referred to as the time required for accelerating from 0 to

60 mph [called the 0-to-60 time in seconds]) requires higher engine power, which in-turn reduces fuel economy (measured in miles per gallon of gasoline consumed).

3. *Vehicle weight versus performance*: This trade-off is commonly referred to by considering the horsepower-to-weight ratio. An increase in vehicle weight reduces acceleration capability of the vehicle with the same engine power.
4. *Ride comfort versus handling*: Better (more comfortable) riding cars require softer suspensions that reduce handling (maneuvering) capability of the vehicle.
5. *Lightweight materials versus cost*: Lightweight materials (e.g., aluminum, magnesium, high-strength steels, and carbon-fiber materials) can reduce vehicle weight. However, these lightweight materials are several times more costly than the commonly used steel sheet (mild steel) material.
6. *High-raked windshield versus costs*: Windshields with higher rake angle (more sloping windshields; see Figure 3.5) can reduce aerodynamic drag, increase fuel economy, and provide sleeker more-aerodynamic appearance than conventionally styled vehicles (with more up-right windshield). The high-raked windshields are longer in length (see Figure 3.5 where $L_1 > L_0$) than more conventional low-raked windshields. The longer length (L_1) also requires thicker glass, longer wipers, more powerful wiper motor, higher capacity windshield defroster and higher capacity air-conditioning (due to

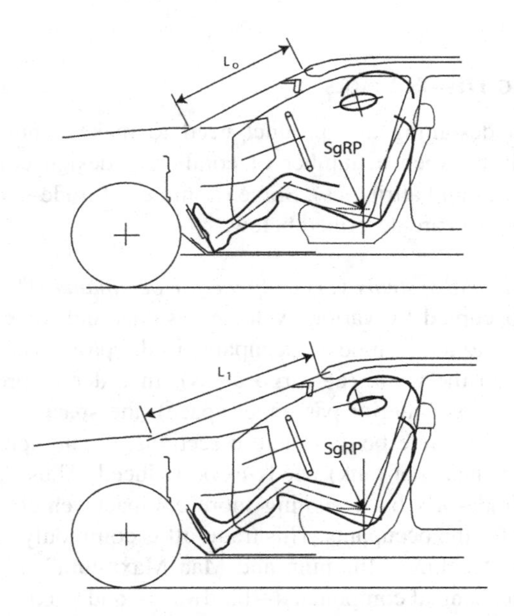

FIGURE 3.5 Comparison of conventional versus high-raked windshield.

Note: The upper picture shows a vehicle with a conventional windshield rake angle. The lower picture shows a vehicle with a higher rake angle windshield.

higher heat/sun load). The thicker glass also reduces light-transmission of the windshield, which in-turn reduces driver visibility. The thicker glass also increases vehicle weight, which in-turn can reduce fuel economy. The higher raked windshields thus can increase vehicle costs.

ROLE OF MARKETING RESEARCH AND SUPPORTING MARKET RESEARCH CLINICS

Market research is an important activity within Marketing and Sales corporate staff (see Table 2.5). It conducts research on customers and markets for new and existing products and provides estimates on vehicle sales potential for product development and senior management activities of the company. It helps vehicle development teams in (1) meeting and communicating with customers of products in various market segments; (2) designing and conducting market research clinics; and (3) presenting data and research on market research studies to senior management of the company.

The market research clinics are typically held during the following phases of a vehicle development program: (1) understanding customers and customer needs; (2) development of vehicle specifications; (3) selecting a vehicle concept; (4) evaluating alternate designs of the whole vehicle, vehicle systems or their sub-systems during detailed engineering; (5) vehicle validation; (6) customer inspection and feedback at different times after the vehicle purchase (e.g., 30 days, 90 days, 1 year and 3 years). The market researchers work closely with many vehicle development teams and team members to ensure that their needs are understood, and the clinics are designed to obtain bias-free results.

The market research clinics generally involve inviting customers or potential customers of vehicles in different market segments. The clinics are held in different cities that would have high concentration of customers of vehicles in the market segment being studied. Typically 75 to 300 customers are invited to participate in a market research study in a city. The customers are paid for their participation time. The customers are shown prototypes of the products along with other competitors' products and asked to evaluate the products in static and/or dynamic driving situations (see Chapter 19 for examples of vehicle evaluations).

Depending upon the type of clinic being designed a number of precautions must be taken to ensure that biases are not introduced in selecting the customers (e.g., the selected customers are not employees or related to the employees of the company and other similar companies in the auto industry; ensuring that the sponsor of the market research clinic and vehicle brand and model are not revealed to the customers), developing procedures for preparing and showing test properties, preparing and administering the questionnaire (e.g., instructions to the clinic participants), data collection, data processing (e.g., tabulating) and conducting statistical analyses to ensure that reliable information is provided to the vehicle program personnel for decision making. Chapters 19 and 22 provide descriptions of techniques used in collecting and analyzing the data collected in the market research clinics.

RISKS IN PRODUCT DEVELOPMENT AND PRODUCT USES

The product programs involve many risks. All important decisions in business and life involve some level of risk. A risk is considered to be present when an undesired event (which generally incurs substantial loss) is probable (i.e., likely to occur with some level of probability). Risks are possible any time during or after the product development process. If the decision maker takes too little risk by over-designing (or using too high a safety factor), the product will be more costly, and the extra cost most likely will be wasted. On the other hand, if the decision maker takes too much risk by under-designing (e.g., product will have insufficient strength, or use cheap low-quality materials and/or components), then the program will be too expensive due to high costs from product failures. Product failures can cause accidents that can incur additional costs due to occurrences of (1) injuries; (2) property damage; (3) loss of income; (4) interruptions or delays in work situations; (5) product liability cases (see Volume 2, Chapter 26), and so forth.

Definition of Risk and Types of Risks in Product Development

A risk is generally associated with an occurrence of an undesired event such as a financial loss and/or an injury resulting from a product-related failure. The risk can be measured in terms of the magnitude of the consequence due to the occurrence of an undesired event. The consequence due to a risk can be measured by costs associated with customer dissatisfaction, loss due to product defects or resulting accidents, loss due to interruption of work, loss of revenue, loss of reputation, and so forth.

The risk is generally assessed by consideration of the following variables: (1) probability of occurrence of the undesired event; (2) the consequence (or severity) of the undesired event (e.g., amount of loss or severity of injuries); (3) probability of detection of the undesired event before or when it occurs; and (4) preparedness of the risk fighting unit (e.g., fire department, emergency response units, and police) that can attempt to contain the severity of the loss or injury.
Risks during the product development process can be categorized as follows:

1. *Technical risk:* This type of risk occurs due to one or more technical problems with the design of the product. For example, a design flaw discovered during testing of an early production component. Such a problem may prevent the product from achieving the required technical capability or performance. To eliminate the technical problem or the flaw, additional analyses, engineering changes and/or technology changes, and testing may be needed. These additional tasks usually result in an increase in the costs and delays in the schedule. Adoption of new technologies before adequate developmental work often leads to serious delays (e.g., problems in developing carbon fiber components for airplanes [see Bhise (2023)], and manufacturing problems with lightweight materials to improve fuel consumptions in automotive products).
2. *Cost risk*: This risk is associated with the cost overruns due to technical problems and resulting delays in the schedule. The risk also may be due to

under-budgeting caused by assuming optimistic estimates or underestimation of required tasks, time, and costs (e.g., not providing sufficient allowance for rework).

3. *Schedule risk*: This risk is related to not being able to meet the schedule due to delays from a number of possible reasons [e.g., parts not delivered by suppliers on time, late changes made in the design due to failures uncovered in testing, or planned schedule may be too optimistic (see Bhise (2023)].
4. *Programmatic risk*: This risk is associated with the product development program (e.g., being over budget, delayed, modified, or even cancelled due to any number of reasons). Since most of the complex products have many components that are made and supplied by various suppliers, selection of suppliers with unproven or low technical capabilities often leads to program delays, lower quality, and cost overruns.

These above four categories of risks are generally interrelated; that is, a risk in any one of the categories also affects associated risks in other categories. The risks also cause backward cascading effects of the problems in the work completed in the early phases but discovered in the later phases. These problems affect the progress in the succeeding phases due to factors such as redesign, rework, retests, delays, and cost overruns.

Types of Risks during Product Uses

The risk after the product is introduced in the market and used by end-users can be categorized as follows:

1. *Loss of user confidence in the product*: The end-users may be afraid to use a product because of a defect in the product. The defect may be caused due to a design or manufacturing defect or some "hidden danger" that can cause an undesired event (e.g., a sudden loss of control, a fire or an explosion, an accident, or exposure to toxic substances).
2. *Loss in future sales*: The likelihood of an undesired event can cause loss in the reputation of the producer and thus can affect future sales.
3. *Excessive repair or recall costs*: The producer will need to fix the product problem by repairing it under warranty or by initiating a product recall.
4. *Product litigation costs*: The costs of defending the product in product liability cases and costs related to settlements before the court trials or payments of penalties, fines, and so forth.

RISK ANALYSIS

A risk analysis can be defined as a decision-making exercise conducted to determine the next course of action after a potential undesired event has been identified and the magnitude of the consequence of the undesired event has been estimated. The phase of identification of undesired event can be called the "risk identification" phase, and the phase of estimation of the magnitude of the consequence due to the undesired event can be called the "risk-assessment" phase.

Some commonly used methods for risk identification, risk assessment, and risk analysis are given here.

1. *Risk identification methods*: Brainstorming, interviewing experts, hazard analysis (see Bhise, 2023), failure modes and effects analysis (FMEA), and use of checklists and historic data (e.g., past records of product defects, warranty problems, and customer complaints).
2. *Risk-assessment methods*: Estimation of probability (or frequency) of occurrence of undesired events, magnitude of the consequence (or severity) of the undesired event and probability of detection of the undesired event by using brainstorming, interviewing experts, safety analysis (e.g., fault tree analysis [Bhise, 2023] and FMEA, and historic data (e.g., costs of past product failures).
3. *Risk analysis methods*: Risk matrix, risk priority number (RPN), nomographs, existing design and performance standards, and specialized risk models (Floyd et al., 2006).

RISK MATRIX

The risk matrix involves simply creating a matrix with combinations of relevant variables associated with the degree of risk due to the undesired outcomes. A risk matrix is a simple graphical tool. It provides a process for combining (1) the probability of an occurrence of an undesired event (usually an estimate) and (2) the consequence if the undesired event occurred (usually cost estimates in dollars).
The risk (in dollars) can be computed as follows:

$$\text{Risk}(\$) = [\text{Probability of occurrence}] \times [\text{Consequence of the undesired event}(\$)]$$

Figure 3.6 shows a plot of the above relationship of probability of occurrence and consequence (loss in dollars) to the risk (expected loss in dollars) due to an undesired event. The plot is made by using log scales (logarithm to the base of 10) for both the axes. Thus, the expected loss, which is the result of the multiplication of the values on its X and Y axes, is represented by slanting lines of risk on the logarithmic axes. The magnitude of the risk is the same on any given risk line. The risk lines for \$1 to \$1 million are shown in Figure 3.6.

A simplified form of the above relationship between the probability of occurrence and the magnitude of the consequence can be presented in a matrix format. Figure 3.7 presents an example of a risk matrix. The cells of the matrix represent different risk levels increasing from low risk to high risk from the lower left corner of the matrix to the top right corner of the matrix. The risk matrix thus allows for a quick assessment

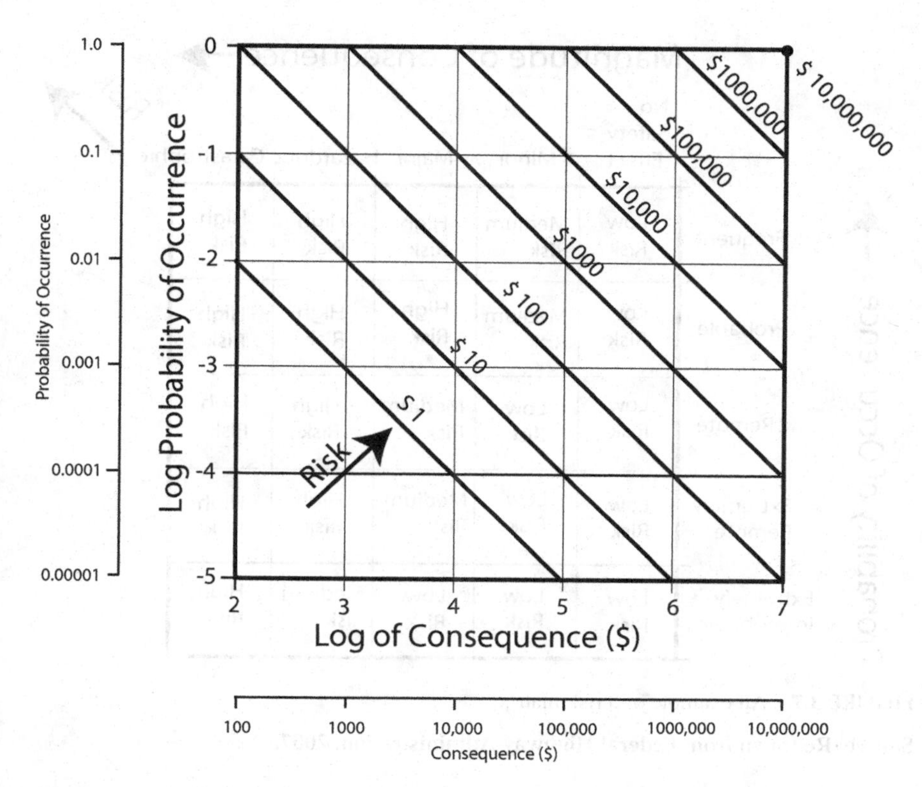

FIGURE 3.6 Relationship of risk to probability of occurrence and consequence of occurrence of an undesired event.

of the risk level after the occurrence probability, and the magnitude of the consequence due to an undesired event are estimated.

Risk Priority Number and Nomographs

RPN is another method used to assess the level of risk. It is based on multiplication of three ratings, namely: (1) severity; (2) occurrence; and (3) detection. This method is used in FMEA, which is presented in an earlier section of this chapter. Examples of rating scales used for severity, occurrence, and detection are presented in Tables 3.8, 3.9 and 3.10, respectively. Different definitions of the rating scales are generally used in different companies, industries, and government agencies.

Nomographs are also used as an alternate method for estimating the RPN. An example of a nomograph is shown in Bhise (2023). Other methods such as modeling and simulations are used to facilitate decision making. Exercising models under different assumptions (conducting sensitivity analysis) can provide a good understanding of the underlying variables and their effects on risks and subsequent decisions. Analyses under a range of possibilities with different levels of optimistic

Magnitude of Consequence →

Risk ↗

Probability of Occurrence ↑

	No Safety Effect	Minor	Major	Hazardous	Catastrophic
Frequent	Low Risk	Medium Risk	High Risk	High Risk	High Risk
Probable	Low Risk	Medium Risk	High Risk	High Risk	High Risk
Remote	Low Risk	Low Risk	Medium Risk	High Risk	High Risk
Extremely Remote	Low Risk	Low Risk	Medium Risk	High Risk	High Risk
Extremely Improbable	Low Risk	Low Risk	Low Risk	Medium Risk	High Risk

FIGURE 3.7 An example of a risk matrix.

Source: Redrawn from Federal Highway Administration, 2007.

and pessimistic assumptions are also useful to estimate the limits of risks. Historic data and judgments of experts can also play a major role in decision making.

Problems in Risk Measurements

Assessing the risks to users/customers involves identifying the hazards, assessing the potential consequences, and the occurrence probability of such consequences. Identification of hazards is particularly difficult when both the potential customers and product usages are difficult to predict. Products involving new technologies are also difficult to evaluate because very little failure data is generally available. It is especially hard to predict the risks during the early stages of product development when the product concept is also not fully developed.

The problems in risk measurements occur due to many reasons. Most problems occur due to (1) lack of data on different types of hazards and risks; (2) subjectivity involved in identification and quantification of the data; and (3) differences in assumptions made during the design phases about how the customers or users will use the product versus the actual uses of the product. The risk-assessment models used in this area are therefore not precise. But they can be used as guides along with the recommendations of multiple experts and discussions between the decision makers and the experts.

Subjective assessments of the three component areas (occurrence, severity, and detection) are also difficult and subject to a number of questions such as: Who would collect the data and conduct evaluations? Should the evaluations be conducted by experts, product safety advisory boards, teams, or individuals involved in the design process? Furthermore, the level of understanding and awareness of risks varies considerably between different evaluators. Costs are also another problem in collecting failure-related data as the product tests are generally costly and funds are usually limited to undertaking costly data collection studies.

There are trade-offs in the application of risk-assessment methods between the consistency in the data, the level of details related to the outcomes, and the time and resources (particularly human and financial) required for the analyses. Apparently, simple methodologies may contain implicit weightings that may not be appropriate for every product being assessed. Judgments may be intuitive, based on implicit assumptions, especially in relation to the boundaries between categories (or ratings). Taken together, these factors can result in a high degree of subjectivity in risk assessment, although the subjectivity can be reduced by the extent of guidance provided to the assessors in applying the various scales and ratings. In general, the potential for inconsistencies in the results will be directly related to the amount of subjectivity involved in the risk measurement process.

IMPORTANCE OF EARLY DECISIONS DURING PRODUCT DEVELOPMENT

"Designing right the first time" is very important, as reworking any product design in later phases is always very time consuming and costly. Early in product development, key decisions are generally made on what technologies to use and how the product should be configured. Any changes to these early assumptions made during the later stages of product development can increase the costs substantially. This is because such changes may require throwing away much of the early design work (and even some hardware development work) and redoing all the analyses again with a different set of assumptions and requirements.

Involvement of specialists from all key technical areas is very important aspect of the systems engineering process because it ensures that all possible technologies and design configurations are considered as possible alternatives before converging on one or a few alternatives. The subsequent decisions are dependent on the selected technologies and design configurations. For example, during the development of the Boeing 777 airliner, the management decided on a two-engine airplane as compared with the four-engine approach used in past long-distance commercial airplanes. The two-engine approach required substantially more work in designing bigger engines, improving reliability of the engines and performing additional flight tests to prove to the Federal Aviation Administration that the two-engine Boeing 777 aircraft was safe or safer than the four-engine airplanes in the long, trans-oceanic flights (PBS, 1995). Another example of a new material-related technology is as follows. Early during the program planning, Boeing decided to produce the Boeing 787 Dreamliner using carbon-fiber materials as compared to using aluminum for its structural components.

The development of large parts (e.g., airplane wings, tail, and fuselage) with the carbon fiber materials involved a number of developmental challenges related to understanding and implementing the carbon-fiber technology (see Case Study #7 in Bhise, 2023).

CONCLUDING REMARKS

This chapter covered many basic techniques and issues in decision making. Decision making in the real world involves consideration of many issues (both internal and external to the organization), many variables and their effects, likelihoods of outcomes, and associated costs that cannot be well quantified due to reasons such as missing facts, uncertainties in the readiness of new technologies, unknown future developments, and changes in global economy. Many models involving varied levels of complexity using many independent variables can be created to analyze the effects of many risk-related variables. The models can be exercised under different assumptions (conducting sensitivity analysis) to get a good understanding of underlying variables and their effect on the decisions. However, a good decision maker will also inject some subjectivity based on his/her intuition or judgment to make the final decisions. The decisions are never final and can be revisited after new and more reliable information is available.

REFERENCES

Akao, Y. 1991. *Development History of Quality Function Deployment. The Customer Driven Approach in Quality Planning and Deployment.* ISBN 92-833-1121-3. Minato, Tokyo, Japan: Asian Productivity Organization.

Bhise, V. D. 2023. *Designing Complex Products with Systems Engineering Processes and Techniques*. Second Edition. ISBN: 978-1-032-20369-0. Boca Raton, FL: CRC Press.

Bhise, V. D. and A. Pillai. 2006. A Parametric Model for Automotive Packaging and Ergonomics Design. *Proceedings of the International Conference on Computer Graphics and Virtual Reality*, Las Vegas, Nevada. June 28, 2006.

Blanchard, B. S. and W. J. Fabrycky. 2011. *Systems Engineering and Analysis*. Fifth Edition. Upper Saddle River, NJ: Prentice Hall PTR.

Federal Highway Administration, US Department of Transportation (USDOT). 2024. Systems Engineering management Plan Template. Website: www.fhwa.dot.gov/cadiv/segb/views/document/sections/section8/8_4_2.cfm (accessed March 9, 2024).

Floyd, P., T. A., Nwaogu, R. Salado, and C. George. 2006. Establishing a Comparative Inventory of Approaches and Methods Used by Enforcement Authorities for the Assessment of the Safety of Consumer Products Covered by Directive 2001/95/EC on General Product Safety and Identification of Best Practices. Final Report dated February 2006 prepared for DG SANCO, European Commission by Risk & Policy Analysts Limited, Farthing Green House, 1 Beccles Road, London, Norfolk, NR14 6LT, UK.

Forgie, C. C. and W. Evans. 2011. Assessing Technology Maturity as an Indicator of Systems Development Risk. *In Systems Engineering Tool and Methods* (Eds) Kamrani, A. K. and M. Azimi (pp.111–134) Boca Raton, FL: CRC Press.

Konz, S. 1990. *Work Design-Occupational Ergonomics*. Third Edition. Worthington, OH: Publishing Horizons.

Prefix, 2023. *Programmable Vehicle Modules*. Rochester Hills, MI: Prefix Corporation. www.prefix.com/design-engineering/ (accessed July 30, 2023).

Public Broadcasting Service (PBS). 1995. *21st Century Jet – The Building of the 777*. Producers: Karl Sabbagh, David Davis and Peggy Case. PBS Home Video (5 hours). Produced by Skyscraper Products for KCTS Seattle and Channel 4 London.

Richards, A. and V. Bhise.2004. Evaluation of the PVM Methodology to Evaluate Vehicle Interior Packages. SAE Paper no. 2004-01-0370. Also published SP-1877, SAE International, March 2004.

Satty, T. L.1980. *The Analytic Hierarchy Process*. New York, NY: McGraw-Hill.

4 Engineering Anthropometry and Biomechanics

INTRODUCTION

The process of vehicle design begins with a discussion of the size and type of the vehicle and the number of occupants the vehicle should accommodate. To ensure the required number of occupants can be accommodated, the designers must consider the dimensions of drivers and the passengers and their postures in the vehicle space. Therefore, in this chapter, we will review basic concepts, principles and data related to human anthropometric and biomechanical characteristics along with the considerations used in vehicle design with an emphasis on occupant packaging and seating design.

Anthropometry and biomechanics are related fields in the sense that both depend upon the dimensions of humans and the ability of humans to assume different postures while working or using vehicles. The two fields can be defined as follows.

Engineering anthropometry is the science of measurement of human body dimensions of different populations. It deals with skeletal dimensions (which are measured from certain reference points on the bones which are less flexible as compared to skin tissues), contours, shapes, areas, volumes, centers of gravity, weights, and so forth, of the entire human body and body segments. Engineering anthropometry involves applications of anthropometric measurement data to design and evaluate products to accommodate people.

Biomechanics deals primarily with dimensions, composition and mass properties of body segments, joints linking the body segments, muscles that produce body movements, mobility of joints, mechanical reactions of the body to force fields (e.g., static and dynamic force applications, vibrations, impacts), and voluntary body movements in applying forces (torques, energy/power) to external objects (e.g., controls, tools, handles, etc.). It is used to evaluate if the human body and body parts will be comfortable (i.e., when internal forces well below strength and tolerance limits) and safe (avoidance of injuries) while operating/using machines/equipment (or vehicles).

DOI: 10.1201/9781003485582-4

USE OF ANTHROPOMETRY IN DESIGNING VEHICLES

The very first step in designing a vehicle is to determine the user population(s) and their anthropometric and biomechanical characteristics. The anthropometric data of the user population will help in determining many basic dimensions of the vehicle. The biomechanical data will help design the vehicle such that the users will not be required to exert or be subjected to forces that are above their tolerance or comfort levels.

Figure 4.1 shows some basic dimensions of people in standing and seated postures. A number of populations from different countries and different types of vehicles are measured (available in many human factors books, e.g. Pheasant and Haslegrave, 2006). Such data are also obtained by automotive companies by measuring dimensions of participants invited to attend market-research clinics to evaluate new vehicle concepts or early prototypes.

Table 4.1 provides anthropometric data on US adults for dimensions that are useful in accommodating occupants and in evaluating spaces and clearances within

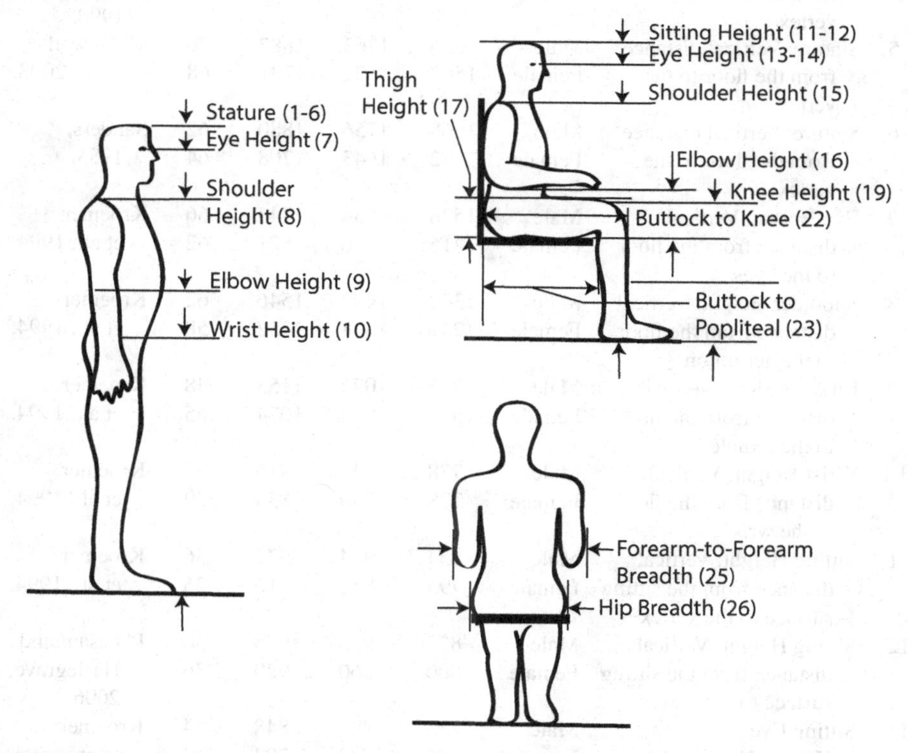

FIGURE 4.1 Static anthropometric measurements in standing and seated postures.

Note: The numbers in the parentheses refer to the dimension presented in Table 4.1.

TABLE 4.1
Static Body Dimensions of United States Adults (Values Are in mm)

Sr. No.	Measurement	Gender	5th	50th	95th	Std. Dev.	Reference
1	Stature: Vertical distance from the floor to the vertex	Male Female	1634 1498	1756 1619	1881 1735	75 72	National Center for Health Statistics, 2016.
2	Stature: Vertical distance from the floor to the vertex	Male Female	1647 1415	1756 1516	1855 1621	67 63	Kroemer et al.,1994.
3	Stature: Vertical distance from the floor to the vertex	Male Female	1640 1520	1755 1625	1870 1730	71 64	Pheasant and Haslegrave, 2006.
4	Stature: Vertical distance from the floor to the vertex	Male Female	1670 1540	1790 1650	1900 1760	70 67	Jurgens, Aune & Pieper, 1990.
5	Stature: Vertical distance from the floor to the vertex	Male Female	1636 1507	1763 1622	1887 1731	76 68	McDowell et al., 2008.
6	Stature: Vertical distance from the floor to the vertex	Male Female	1665 1572	1756 1643	1880 1708	62 64	Sanders, 1983.
7	Eye Height: Vertical distance from the floor to the eyes	Male Female	1528 1415	1634 1516	1743 1621	66 62	Kroemer et al., 1994.
8	Shoulder Height: Vertical distance from the floor to the acromion	Male Female	1342 1241	1442 1334	1546 1432	62 58	Kroemer et al., 1994.
9	Elbow Height: Vertical distance from the floor to the radiale	Male Female	995 926	1072 998	1153 1074	48 45	Kroemer et al., 1994.
10	Wrist Height: Vertical distance from the floor to the wrist	Male Female	778 728	846 790	915 855	41 39	Kroemer et al., 1994.
11	Sitting Height: Vertical distance from the sitting surface to the vertex	Male Female	854 795	914 852	972 910	36 25	Kroemer et al., 1994.
12	Sitting Height: Vertical distance from the sitting surface to the vertex	Male Female	855 800	915 860	975 920	36 36	Pheasant and Haslegrave, 2006.
13	Sitting Eye Height: Vertical distance from the sitting surface to the eyes	Male Female	735 685	792 739	848 794	34 33	Kroemer et al., 1994.

TABLE 4.1 (Continued)
Static Body Dimensions of United States Adults (Values Are in mm)

Sr. No.	Measurement	Gender	5th	50th	95th	Std. Dev.	Reference
14	Sitting Eye Height: Vertical distance from the sitting surface to the eyes	Male	749	811	863	38	Sanders, 1983.
		Female	736	761	849	41	
15	Sitting Shoulder Height: Vertical distance from the sitting surface to the acromion	Male	548	598	646	30	Kroemer et al., 1994.
		Female	509	555	604	29	
16	Sitting Elbow Height: Vertical distance from the sitting surface to the underside of the elbow	Male	184	232	274	27	Kroemer et al., 1994.
		Female	176	220	264	27	
17	Thigh Height: Vertical distance from the sitting surface to highest top of thigh surface	Male	149	168	190	13	Kroemer et al., 1994.
		Female	140	159	180	12	
18	Seated Stomach Depth: Horizontal depth of trunk at the level of abdominal extension	Male	229	299	374	45	Sanders, 1983.
		Female	195	247	309	48	
19	Knee Height (sitting): Vertical distance from the floor to the upper surface of the knee	Male	514	559	606	28	Kroemer et al., 1994.
		Female	474	515	560	26	
20	Popliteal Height (sitting): Vertical distance from the floor to the underside of the knee	Male	395	434	476	25	Kroemer et al., 1994.
		Female	351	389	429	24	
21	Forward Thumb tip Reach: Horizontal distance from the back of the shoulder blade to thumb tip with arm raised at shoulder level	Male	739	801	867	39	Kroemer et al., 1994.
		Female	677	735	797	36	
22	Buttock to Knee Distance (sitting): Horizontal distance from the back of uncompressed buttock to the front of the kneecap	Male	596	616	667	30	Kroemer et al., 1994.
		Female	542	589	640	30	

(continued)

TABLE 4.1 (Continued)
Static Body Dimensions of United States Adults (Values Are in mm)

Sr. No.	Measurement	Gender	5th	50th	95th	Std. Dev.	Reference
23	Buttock to Popliteal Distance (sitting): Horizontal distance from the back of uncompressed buttock to the back of the knee	Male	458	500	545	27	Kroemer et al., 1994.
		Female	440	482	528	27	
24	Elbow to Fingertip Distance: Horizontal distance from the back of the elbow to middle-finger tip with lower arm horizontal in sitting position	Male	448	484	524	23	Kroemer et al., 1994.
		Female	406	443	482	23	
25	Forearm to Forearm Breadth: Horizontal distance between the outermost points on the forearms in sitting position	Male	474	546	620	44	Kroemer et al., 1994.
		Female	415	468	528	35	
26	Hip Breadth (sitting): Maximum horizontal distance across the hips in the sitting position	Male	329	367	412	25	Kroemer et al., 1994.
		Female	342	384	432	27	
27	Foot Length: Distance from the back of the heel to the tip on the longest toe measured in longitudinal (forward) axis	Male	249	270	292	13	Kroemer et al., 1994.
		Female	224	244	264	12	
28	Shoe Length: Distance from back of heel to front edge of sole	Male	277	299	319	13	Sanders, 1983.
		Female	241	264	286	15	
29	Foot Breadth: Maximum horizontal breadth across the foot perpendicular to the longitudinal axis	Male	92	101	109	53	Kroemer et al., 1994.
		Female	82	90	98	49	
30	Shoe Breadth: Maximum breadth of shoe at outside edges of sole	Male	98	107	116	11	Sanders, 1983.
		Female	89	96	107	10	

TABLE 4.1 (Continued)
Static Body Dimensions of United States Adults (Values Are in mm)

Sr. No.	Measurement	Gender	5th	50th	95th	Std. Dev.	Reference
31	Hand Length: Distance from the crease of the wrist to the tip of the middle finger with hand held straight and stiff	Male Female	179 165	194 180	211 197	98 97	Kroemer et al., 1994.
32	Hand Length: Distance from the crease of the wrist to the tip of the middle finger with hand held straight and stiff	Male Female	183 165	197 179	212 193	9 9	Garrett, 1971.
33	Hand Breadth: Maximum breadth across the palm of the hand at distal ends of the metacarpal bones	Male Female	84 73	90 79	98 86	4 4	Kroemer et al., 1994.
34	Hand Breadth: Maximum breadth across the palm of the hand at distal ends of the metacarpal bones	Male Female	83 71	90 77	97 83	4 4	Garrett, 1971.
35	Hand Depth: Measured at Thenar Pad	Male Female	55 45	62 52	70 58	5 4	Garrett, 1971.
36	Hand Thickness: Measured at Metacarpale III	Male Female	30 25	33 28	36 30	2 2	Garrett, 1971.
37	Thumb Breadth: Measured at interphalangeal joint	Male Female	22 19	24 21	26 23	4 1	Kroemer et al., 1994.
38	Thumb Breadth: Measured at interphalangeal joint	Male Female	21 17	23 19	25 21	4 1	Garrett, 1971.
39	Digit 2 Breadth: Measured at interphalangeal joint	Male Female	17 14	18 15	20 17	1 1	Garrett, 1971.
40	Digit 3 Breadth: Measured at interphalangeal joint	Male Female	17 14	18 15	20 17	1 1	Garrett, 1971.
41	Digit 3 Depth: Measured at interphalangeal joint	Male Female	14 12	16 13	18 15	1 1	Garrett, 1971.
42	Digit 1 (Thumb) Length: Fingertip to crotch level	Male Female	51 47	59 54	66 61	5 4	Garrett, 1971.
43	Digit 2 Length: Fingertip to crotch level	Male Female	68 61	75 69	82 78	5 5	Garrett, 1971.
44	Digit 3 Length: Fingertip to crotch level	Male Female	78 70	86 78	95 87	5 5	Garrett, 1971.

(continued)

TABLE 4.1 (Continued)
Static Body Dimensions of United States Adults (Values Are in mm)

Sr. No.	Measurement	Gender	5th	50th	95th	Std. Dev.	Reference
45	Weight (kg)	Male	62	86	125	19	National Center for Health Statistics. 2016.
		Female	50	72	116	20	
46	Weight (kg)	Male	58	79	99	13	Kroemer et al., 1994.
		Female	39	62	85	14	
47	Weight (kg) (with shoes)	Male	69	92	120	17	Sanders, 1983.
		Female	53	72	90	19	

automotive interiors. The table presents 5th, 50th and 95th percentile values and standard deviations of various anthropometric dimensions of females and males compiled from different sources. It should be noted that these values do not consider the effects of attire (i.e., clothes, shoes, caps, etc.). The data from the U.S. National Center for Health Statistics [NCHS (2016), Kroemer et al. (1994), Pheasant and Haslegrave (2006), Jurgens et al. (1990) and McDowell et al. (2008) are for U.S. civilian adult population. The hand finger data from Garrett (1971) are from the U.S. Air Force flight personnel, and the data by Sanders (1983) are from measurement surveys of U.S. heavy truck drivers. The parameters of the normal distributions provided by different sources will also differ somewhat due to differences in the samples (due to differences in ages, races, ethnicities, year when the samples were measured, and so forth) used to measure the dimensions and the differences in the measurement process used by different sources. Further, since the anthropometric data needed for designing vehicles for different market segments differ from the U.S. civilian population, most automotive manufacturers maintain anthropometric databases on customers in different market segments (e.g., economy passenger cars, luxury passenger cars, pickup trucks). Males-to-females ratios are also different for vehicles in different market segments (see Volume 2, Chapter 24. Additional anthropometric data on populations from different countries are provided in Appendix 2 and Volume 2, Chapter 24).

The majority of human anthropometric characteristics have been found to be normally distributed. Therefore, the normal distribution is used to compute the percentile values of populations that can be accommodated by a given vehicle dimension.

The normal distribution of a random variable x is defined as follows:

$$f(x) = \frac{1}{\sigma\sqrt{2\pi}} e^{-(x-\mu)^2/2\sigma^2}$$

Where

x = variable defining an anthropometric dimension (e.g., standing height)
$f(x)$ = probability density function of x
μ = mean of the normal distribution of x
σ = standard deviation of the distribution of x

Thus, the mean (μ) and standard deviation (σ) are the two parameters of the normal distribution (i.e., they define the location and spread of the distribution, respectively).

The cumulative distribution of $f(x)$ is denoted as F(x) and is defined as follows:

$$\mathrm{F}(x) = \int_{-\infty}^{x} f(x)\,dx = \frac{1}{\sigma\sqrt{2\pi}} \int_{-\infty}^{x} e^{-(x-\mu)^2/2\sigma^2}\,dx$$

Since the normal distribution is symmetrical about mean, F(x) = 0.5 defines the 50th percentile value. Thus, the 50th percentile value equals the mean (μ).

Percentile values are used to evaluate accommodation with respect to a given variable (x) (i.e., what percentage of the population can fit within a given value of an anthropometric variable x). For example, 95th percentile value of x will be defined at F(x) = 0.95. Thus, if x is the stature of individuals, then the 95th percentile value of x will mean that only 5% of the individuals in that population would be taller than that value. From Table 4.1 Row #1, the 95th percentile value of stature of males is 1881 mm. This means that only 5% of US adult males are taller than 1881 mm. Thus, if a door (e.g., for an adult movie theatre) needs to be designed so that 95% of the males can walk through the door opening without ducking their heads, then the door opening height must be at least 1881 mm. Generally an additional 50–100 mm clearance will be provided to account for the increase in accommodation height due to shoes and caps.

COMPUTATION OF PERCENTILE VALUES

Let us assume that you want to determine the percentile value of an adult male who is 1778 mm (5 ft, 10 inches) tall. From Table 4.1 Row #3, the mean and standard deviation values of stature of males are 1756 mm and 67 mm, respectively. Next, we need to determine the value of the standardized normal variable (z) as follows:

$$z = (x - \mu)/\sigma = (1778-1756)/67 = 0.3284$$

The standardized normal variable (z) has a mean value equal to 0.0 and standard deviation equal to 1.0. The values of F(z) for different values of z (generally ranging from about -3.0 to +3.0 are available in the normal distribution tables provided in textbooks of statistics. Referring to the table of cumulative normal distribution, the F(z) value for z = 0.330 is 0.6293. (Note: For z = 0.3284, F(z) = 0.6287).

If you are familiar with the Microsoft Excel application, you can obtain its value by using NORMDIST function (Go to Insert, Function, and Select the statistical function called NORMDIST) [e.g., for our problem, NORMDIST (1778,1756,67,TRUE) = 0.628679].

Thus, the percentile value of a male with a stature of 1778 mm is 62.87. Which means that 62.87% of the males will be shorter than 1778 mm; or conversely, 37.13 = (100-62.87) percent of males will be taller than 1778 mm.

The anthropometric data such as those provided in Table 4.1 can be used to come up with approximate values of various vehicle dimensions. The values obtained will be "approximate" because human "functional" dimensions in actual postures (that differ from the static postures of sitting and standing, as shown in Figure 4.1, used for anthropometric measurements) used in interacting with the vehicle (e.g., entering into and exiting from the occupant compartment, sitting in a vehicle, loading or unloading items in the trunk, accommodating three adults in the rear seat of a vehicle, and so forth) cannot be easily predicted from the static anthropometry based data. This is because of many reasons such as: human posture angles between different body segments vary among individuals in performing different tasks, the human joints are not like simple pin joints, human body tissues deflect (e.g., compression of tissues under the buttocks and back while sitting in a chair), change within a person over time (e.g., slumping or leaning in the seat), and so forth.

Some examples of the use of static anthropometric dimensions for vehicle design are provided below.

1) Maximum seat cushion width can be estimated by considering the 95th percentile hip width of females. (The value is 432 mm from Row #26 of Table 4.1).
2) Minimum seat cushion length can be estimated from the 5th percentile value of buttock-to-popliteal length of females. (The value is 440 mm from Row #23 of Table 4.1).
3) Space above the driver's head can be estimated by considering the 99th percentile value of sitting height of males, torso angle and top of deflected seat.
4) Interior shoulder width (W3) can be evaluated by comparing (W3/2-W20) with half shoulder width of the 95th percentile male [or 95th percentile male forearm-to-forearm breadth of 620 mm, Row #25 of Table 4.1]. (Note: The SAE dimensions W3 and W20 are defined in Chapter 5. Dimension W3 in SAE J1100 is defined as the cross-car distance between door trim panels at shoulder height; see Figure 5.13.W20 is defined as the lateral distance between the driver centerline and the vehicle centerline. See Figure 5.20).
5) The length of interior grab handles, and exterior door handles can be estimated by considering the 95th percentile value of palm width without thumb. (The value is 98 mm from Row #33 of Table 4.1).

To improve the accuracy of predicting key dimensions used to develop occupant package, the SAE Occupant Packaging Committee has developed a number of SAE standards (available in the SAE Handbook. SAE, 2009) (e.g., SAE J1516, J1517, J941, J1052, J287, J4004) based on "Functional Anthropometric" measurements of a large number of drivers seated in actual vehicles or vehicle bucks. In functional anthropometry, the measurements of relevant dimensions of people are measured under the actual postures of people while using the vehicle. This avoids the problem of taking the static anthropometric data (measured under the standard standing or sitting erect postures) and assuming posture angles of different body segments. The functional anthropometric tools used in the occupant packaging are covered in Chapter 5.

APPLICATIONS OF BIOMECHANICS IN VEHICLE DESIGN

Biomechanics is applied here to study and evaluate vehicle design issues in the following four problem areas:

1) Seating comfort (designing seats)
2) Comfort and convenience during entry and egress (see Chapter 10)
3) Evaluating non-seated postures during loading and unloading of items in trunks or cargo areas, changing tires, servicing vehicles, refueling, etc.(see Chapter 11)
4) Protecting occupants in impacts with interior hardware during accidents (e.g., use of crash-test dummies)

Most biomechanical research studies reported in the technical literature have been conducted to understand how humans get injured due to (a) cumulative trauma (i.e., repetitive movements and stresses in body tissues while performing industrial tasks) and (b) accidents that subject the human body to high levels of deceleration and forces during impacts. Since impact protection (also called "crashworthiness") is generally considered as a specialized field (within safety attribute) in the automobile industry, it is not covered in this book. Further, during the normal usage of vehicles, the drivers and passengers do not perform tasks that require application of forces at higher magnitudes (i.e., near maximum voluntary strength levels) and/or at higher rates as compared to many industrial tasks. Thus, the problems of cumulative trauma injuries in vehicle usages are not at all as common or severe as in industrial tasks. But since discomfort can be associated with stresses at sub-maximal levels in the human body and in awkward postures, many of the biomechanical considerations and principles are useful in improving comfort and convenience problems in vehicle usages.

BASIC BIOMECHANICAL CONSIDERATIONS

Human strength depends upon many factors. They include gender, age, duration of the exertion, static versus dynamic nature of exertion, anthropometry (lengths of body segments), posture (angles of various body segments), training, motivation, and so forth. A few basic biomechanical considerations can be summarized as follows:

1. *Male versus Female Strength*: Women typically have 65–70% of the strength of men (See Figure 4.2).

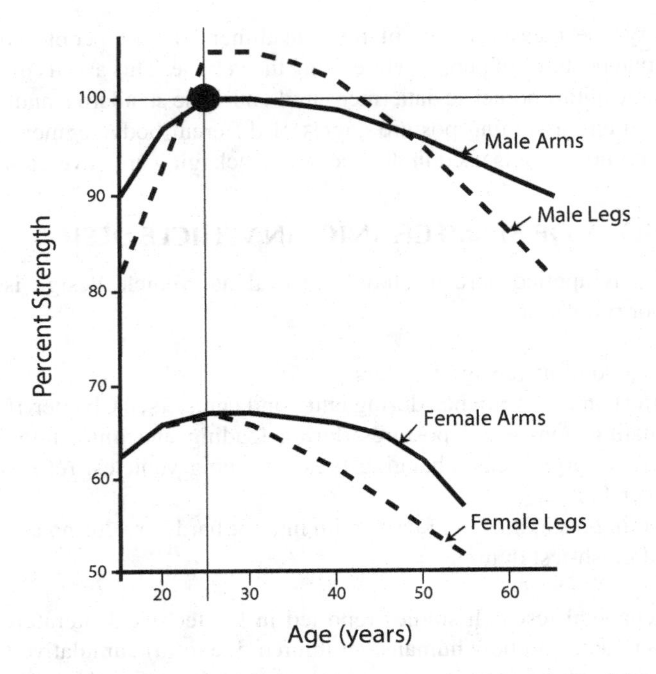

FIGURE 4.2 Strength of arms and legs of men and women as functions of age.

Note that male strength at age 25 is set as 100%. The strength values are for isometric conditions where the muscles remain in static postures.

Source: Redrawn from data in Asmussen and Heebol-Nielsen, 1962, and Konz and Johnson, 2004.

2. *Effect of Age*: The maximum force-producing capabilities (i.e., muscular strength) of adults decrease with age (about 5–10% decrease on average every decade after about age 25). (See Figure 4.2).
3. *Muscular Contraction*: A muscle generates its strength during contraction. The maximum strength is reached at about 4 s after muscular contraction begins.
4. *Endurance Time and Strength Trade-Off*: The time over which a human can continuously exert force (called the endurance time) increases with a decrease in the level of exerted force (strength). At about 15–20% of the maximum voluntary contraction (MVC) strength, a human can maintain exertion for a long period of time (See Figure 4.3). The shape of the endurance time curves varies depending upon factors such as individual differences, particular muscles tested, work conditions, exertion rate, rest period between exertions, and training (Chaffin et al., 1999).
5. *Third Class of Lever*: Most body segments related to large limb motions involve a "third class of lever system" in which the fulcrum is at one end of the

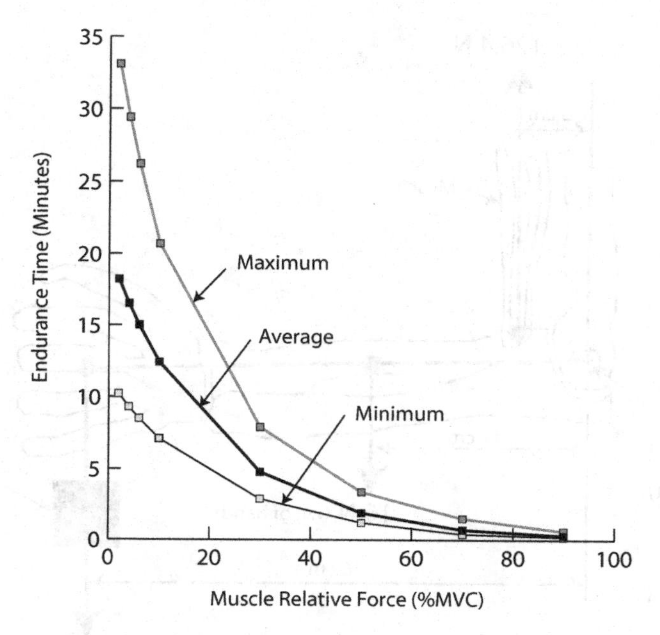

FIGURE 4.3 Tradeoff between muscle relative force and endurance time.Source: Plotted from data presented in Chaffin, D.B., G.B.J. Andersson and B.J. Martin, 1999.

lever and the external load is at the other end with the activating force (muscle) in between, usually close to the fulcrum. This arrangement of lever requires a larger force to be exerted by the muscle in comparison to the external load. The following example will illustrate this.

Figure 4.4 shows the lower arm held in the horizontal position by the muscle (i.e., Biceps Brachii) in the upper arm with the elbow point as the fulcrum. The amount of force required in the muscle to hold a weight grasped by the hand can be calculated by computing moments around the elbow point. Let's assume that 10N of load is held in the hand and at a distance of 36 cm (measured from the load in the hand to the elbow point). The center of gravity of the lower arm with the hand is about 17 cm from the elbow point and the weight of the arm with the hand is 16 N. The moment rotating the hand down or clockwise around the elbow point would be (10x36 + 16x17) = 632 Ncm. Assuming the muscle holding the lower arm is attached 5 cm from the elbow, then the force in the muscle would be (632/5) = 126.4 N. Thus, in this case, to hold a load of 10N in the hand, the reactive force in the muscle to hold the hand in equilibrium would be 12.6 times that of the load.

6. *Design Loads*: For tasks involving large internal loads (e.g., during lifting), the job should be designed around the 5th percentile load exertion capability. Thus, 95% of the population can perform the task. For highly repetitive jobs, multiply the 5th percentile load capability by 0.15 to 0.2 (see consideration

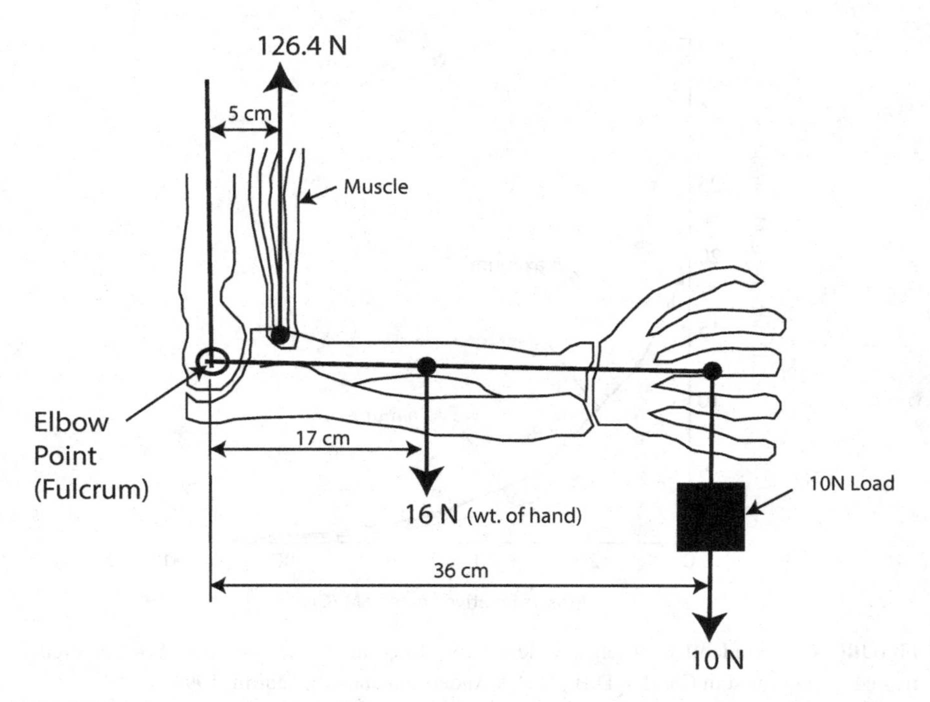

FIGURE 4.4 Upper hand lever and force configuration while holding a weight of 10 N in hand.

(4) above) to obtain a comfortable exertion level for accommodating the largest proportion of the population over longer exertion durations.

BIOMECHANICAL CONSIDERATIONS IN SEAT DESIGN

1. *Load in the L5/S1 Region*: One important consideration in seat design is to reduce the load on the spinal column of the seated person. Biomechanical research (Chaffin, Andersson and Martin, 1999) has shown that during seating, the L5/S1 (the joint between the 5th lumbar vertebra and the first sacral vertebra) experiences the highest concentration of stress due to compressive force. The stresses in the L5/S1 can be reduced by providing a lumbar support that maintains the natural shape of the spinal column in the lumbar region (the natural shape of the spinal column is observed when a person is standing erect). The natural shape of the spinal column in the lumbar region is convex toward the front of the body (i.e., protruding forward; called lordosis). If the seatback provides the right amount of protrusion in the lumbar region at the correct height, and if the user can recline and support his torso on the seatback, the natural shape of the lumbar can be maintained.

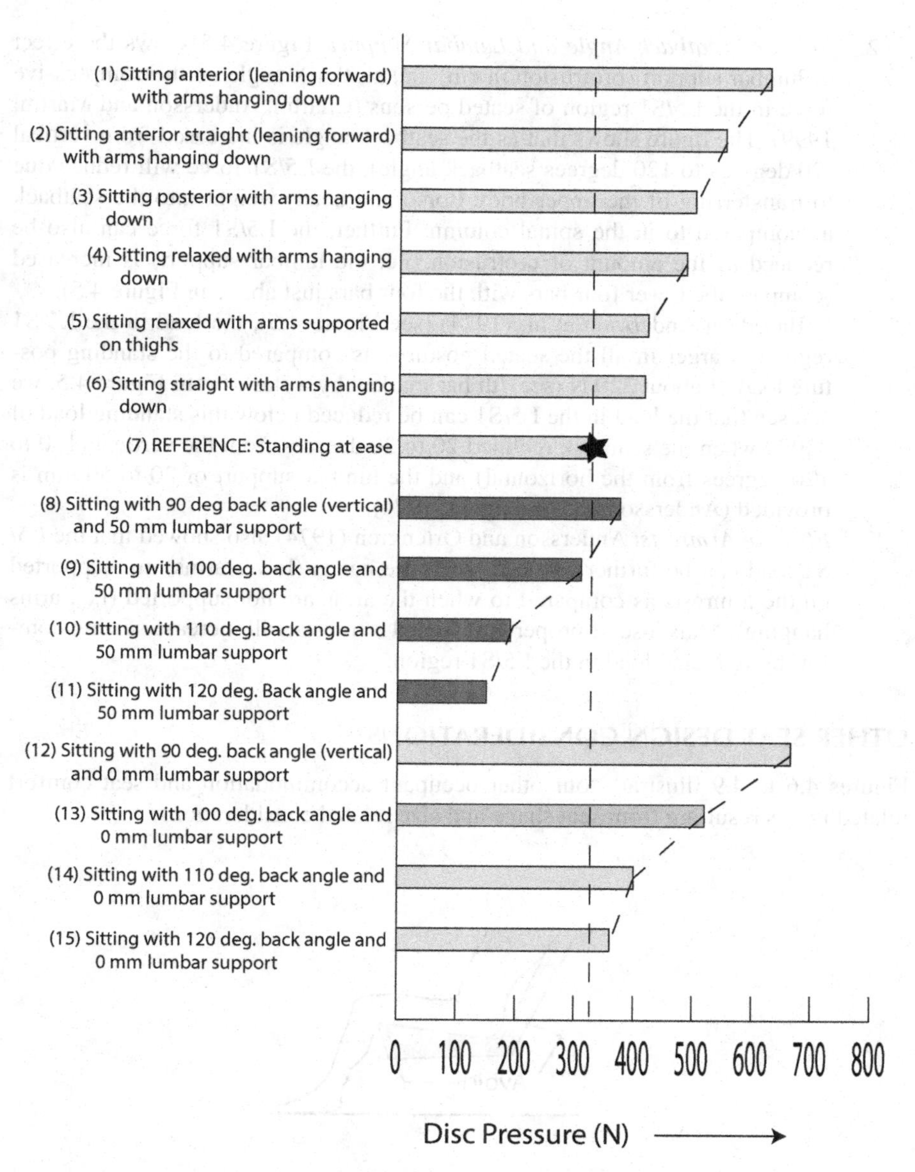

FIGURE 4.5 Effect of different sitting postures, seatback angle and lumbar support on disc pressure at L5/S1.

Source: Plotted from Andersson et al., 1974 and Andersson and Ortengren, 1974 data presented in Chaffin, D.B., G.B.J. Andersson, and B.J. Martin, 1999.

2. *Effect of Seatback Angle and Lumbar Support*: Figure 4.5 shows the effect of lumbar support (protrusion in cm) and seatback angle on the compressive force in the L5/S1 region of seated persons (Chaffin, Andersson and Martin, 1999). The figure shows that as the seatback angle is increased from vertical (90 degrees to 120 degrees seatback angle), the L5/S1 force will reduce due to transferring of the upper body (torso and head) weight into the seatback as compared to in the spinal column. Further, the L5/S1 force can also be reduced as the amount of protrusion (i.e. the lumbar support) is increased (compare the lower four bars with the four bars just above in Figure 4.5).

   Based on Andersson et al. (1974) (see Figure 4.5), the load in the L5/S1 region is larger in all the seated postures as compared to the standing posture load of about 320 N (see 7th bar marked by a star). From Figure 4.5, we can see that the load in the L5/S1 can be reduced below this standing load of 320 N when the seatback reclined 20 to 30 degrees from the vertical (110 to 120 degrees from the horizontal) and the lumbar support of 20 to 50 mm is provided (Andersson and Ortengren, 1974).
3. *Effect of Armrests*: Andersson and Ortengren (1974) also showed that the L5/S1 load can be further reduced when the hands (lower arms) are supported on the armrests as compared to when the arms are not supported (i.e., arms hanging). Thus, use of properly designed armrests will increase seating comfort by reducing load in the L5/S1 region.

OTHER SEAT DESIGN CONSIDERATIONS

Figures 4.6 to 4.9 illustrate four other occupant accommodation and seat comfort related issues resulting from seat shape and size related considerations.

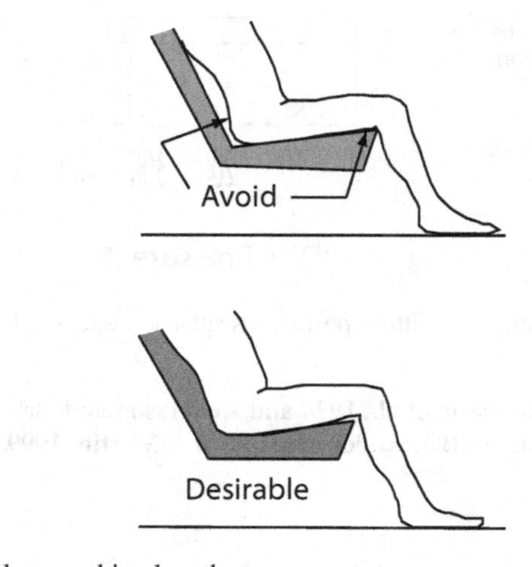

FIGURE 4.6 Avoid long cushion length.

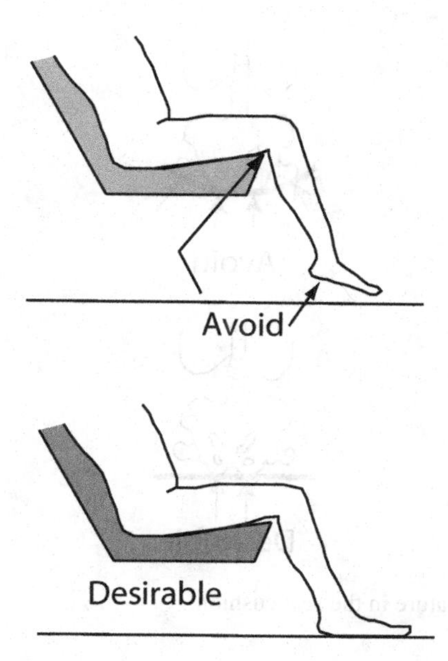

FIGURE 4.7 Avoid dangling feet.

4. *Avoid Long Cushion Length*: Figure 4.6 (upper figure) shows that with a very long seat cushion, a seated person will leave a gap behind the user's buttocks and the lumbar region of the seatback. The seat cushion length should be shorter than the buttock-to-popliteal length of the person so that the user can support some of his/her upper body weight on the seatback, and thus, reduce L5/S1 load (see Figure 4.6 lower figure). Thus, if the seat cushion length is not adjustable, then it is best to reduce the seat cushion length for the shorter (5th percentile) female buttock-to-popliteal length.
5. *Avoid Dangling Feet*: Figure 4.7 (upper figure) shows that if the seat cushion is too high then the user's feet will dangle and as a result the pressure under the thighs will increase, creating discomfort (due to pinched veins and nerves in the back side of the knees during extended periods of driving). Thus, dangling feet should be avoided by either providing a footrest or reducing the seat height so that user's feet can be supported on the floor (see Figure 4.7 lower figure).
6. *Avoid Curvature in the Seat Cushion*: The curvatures in the seat cushions (see Figure 4.8), in general, should be avoided as they will put higher pressure on the body tissues surrounding the ischial tuberosities and restrict body movements in the seat. A flatter seat cushion will allow the seat occupants to make small movements and postural changes that can increase overall seat comfort, especially during long trips (see Figure 4.8 lower figure). Provision of side bolsters in the seat cushion can increase effective curvature of the seat cushion. Therefore, tall (or heavily padded) bolsters should be avoided to improve long term seat comfort. Increased seat curvatures and thicker bolsters

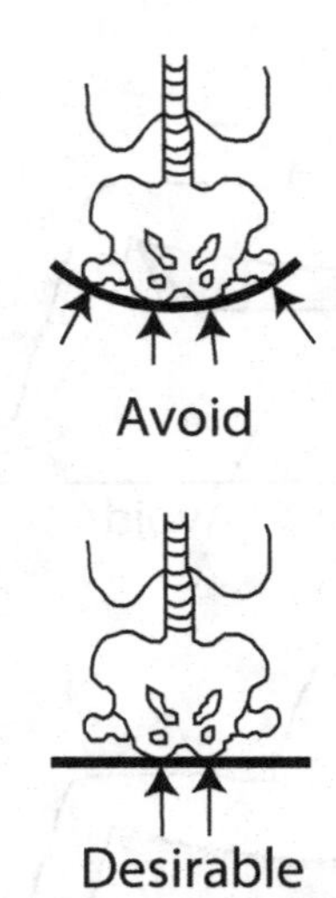

FIGURE 4.8 Avoid curvature in the seat cushion.

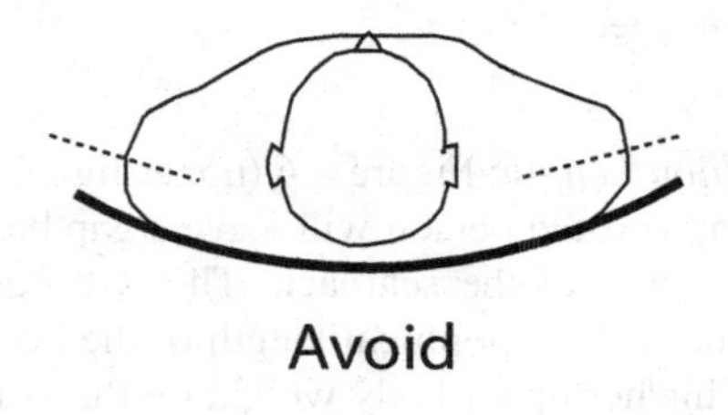

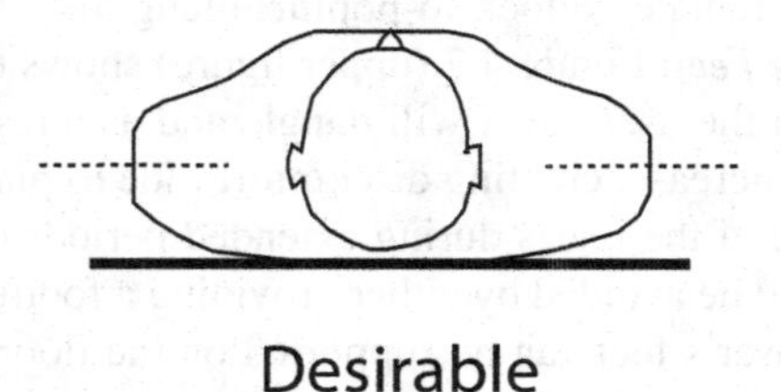

FIGURE 4.9 Avoid curvature in the seatback.

also make the tasks of entry into the vehicle and exiting from the vehicle more difficult (see Chapter 10).

7. *Avoid Curvature in the Seatback*: Provision of curved seatbacks (see Figure 4.9) or taller (protruding) bolsters on the seat backs can force the driver's shoulders forward and also restrict small body movements, which, especially during long trips, can reduce comfort and increase driver fatigue.

Thus, flatter seatbacks should be provided (see Figure 4.9 lower figure) to allow small body movements and increase overall seating comfort especially during longer trips. Provision of side bolsters in the seatbacks can also increase effective curvature of the seatback. Therefore, heavily padded bolsters should be avoided to improve long term seat comfort. Increased seatback curvatures and thicker bolsters also make the tasks of entry into the vehicle and exiting from the vehicle more difficult (see Chapter 10).

SEAT DESIGN CONSIDERATIONS RELATED TO DRIVER ACCOMMODATION

The following design considerations will improve the driver's accommodation and comfort (Kolich, M., 2006, Chaffin et al. 1999; Konz, S. and S. Johnson, 2004).

1) *Seat height*: The seat height in vehicle package is measured using dimension H30 which is defined as the vertical height of the seating reference point (SgRP) from the accelerator heel point (AHP) (see Chapter 5 for more details). The H30 dimension defines the driver's seated posture (defined by pedal plane angle, knee angle, torso angle, seatback angle). Vehicles with overall low height (e.g., sports cars) typically have very low H30 (about 150–250 mm); whereas commercial trucks have large H30 (over 405 mm). If the seat is too high, the short driver's feet will dangle and if the driver is unable to rest his/her heels on the vehicle floor/carpet (or on a footrest), the driver will find the seating posture to be very uncomfortable. Therefore, based on the comfort of the 5th percentile female seated popliteal height of 351 mm, the top of the seat from the vehicle floor should not be more than about 320 mm. Power seats generally allow adjustment of the seat height so that drivers with different leg lengths can be accommodated.

   The horizontal distance between the AHP (accelerator heel point) and SgRP (defined as L53) increases as the H30 value is decreased (see Chapter 5 for more details). Thus, to minimize the horizontal space required to accommodate the driver in commercial vehicles (truck products), the seat height is increased (as compared to the passenger cars) and the driver sits more erect (seatback angle typically is more vertical, around 12–18 degrees from the vertical). In sports cars, the seatback angle will be more reclined, to about 22 to 28 degrees from the vertical.

2) *Adjustable seat*: To accommodate the largest percentage of drivers at their preferred driving posture, it is very important to allow them to adjust: 1) seat height, 2) seat cushion angle, 3) seatback angle (reclining seatback), 4) height and protruding fore-aft length of the lumbar support, 5) headrest height and fore-aft location, 6) seat cushion length, 7) armrest height and length and its lateral location from the driver centerline, 8) seat cushion and seatback bolster heights and/or angles. Power seats (which allow easy adjustments of many of the above mentioned parameters with rocker or multi-function switches)

are generally more comfortable than non-powered seats, especially during long trips.

3) *Seat cushion length*: The seat cushion length should not be longer than the driver's buttock-to-popliteal (back of knee) distance. Thus, if this length is restricted to the 5th percentile female's buttock to popliteal distance (about 440 mm), then most drivers can use the seat and still use the back rest. Drivers with longer, upper legs would prefer longer seat cushions, but shorter females will not be able to use the seatback without a pillow on the seatback. Further, in case of longer seat cushion lengths, shorter females will find operation of the pedals difficult, as they will be compressing the seat cushions with their thighs while depressing the pedals. Thus, an adjustable cushion length will reduce such problems and accommodate a larger percentage of the drivers.
4) *Seat cushion angle*: The seat cushion should slope backward by about 5 to 15 degrees. This will allow the user to slide back and allow the transferring of torso weight on to the seatback. The provision of an adjustable seat cushion angle will allow the user to find his/her preferred seat cushion angle.
5) *Seat width*: Since females have larger hip widths (breadths), the seat cushion width should be greater than 95th percentile female sitting hip width (about 432 mm; see #26 in Table 4.1). In addition, clearance should be provided for clothing, thus, a width of 500 to 525 mm at the hips can be recommended.
6) *Seatback angle*: The seatback angle (called A40 in SAE J1100; see Chapter 5 for more details) in automotive seating is defined by the angle of the torso line (back line) of the SAE H-point machine or the two dimensional (manikin) template (refer to SAE standards J826 and J4002 (SAE, 2009)) with respect to the vertical. The seatback angle (seat recline angle) should allow drivers to assume their preferred back angles. For passenger cars, drivers generally prefer to set the seatback angle between about 20–26 degrees. In trucks, due to the higher seat height (H30), drivers prefer to sit more erect with seatback angles between about 12–18 degrees.
7) *Seatback height*: From an anthropometric accommodation viewpoint, the maximum seatback height can be selected as the 5th percentile female acromial height, which is about 509 mm above the seat surface. However, considering the Federal Motor Vehicle Safety requirements on head restraints, the seatback height is dictated by the headrest design.
8) *Lumbar area*: The seat contour in the lumbar area affects the shape of the seated person's spinal column. The most important characteristic of the seat contour in the lumbar region is that it should maintain the natural curvature (bulging forward, i.e., convex, called lordosis) of the spinal column in the lower back region of the seated person. An adjustable lumbar support that allows setting its height (i.e., up and down adjustment) and protrusion location (i.e., fore-aft adjustment) would allow accommodation of different individuals while maintaining their natural lordosis.
9) *Lateral location of the seat*: The dimension W20-1 defines the lateral distance between the vehicle centerline and the driver's SgRP. It should be designed so that the driver will have sufficient elbow clearance from the driver's door trim

panel between the shoulder and elbow heights. This lateral distance from the driver centerline to the door trim panel should be larger than half of 95th percentile elbow-to-elbow width of males plus elbow clearance to avoid elbows rubbing against the door trim panel while holding the steering wheel.

10) *Armrest height*: A properly designed and adjusted armrest can reduce the load on the driver's spinal column and thus increase the perception of comfort and reduce driver fatigue. The preferred height of the armrest will depend upon the lateral location of the armrest from the driver centerline. Since it is difficult to provide an armrest at optimal locations for most drivers, the armrest height and lateral distance from the driver centerline should be adjustable. If the armrests are provided on both sides (i.e., on the door trim panel as well as on the seat or on the center console), both the armrests should be at the same height to reduce discomfort (due to leaning on one side).
11) *Bolster height*: The bolsters on the sides of the seat cushion and seat back can provide the driver feeling of sitting "snug or cuddled" (like in a contoured seat) in the seat and provide a sense of stability and security while negotiating curves and driving on winding roads. The bolsters restrict the seated person's movements in the seat and, therefore, especially on long trips, such seats will be perceived to be less comfortable. (Smaller postural movements can increase the comfort of seated persons especially during longer trips.) The taller bolsters on the seatback may also move the minimum reach distance to controls and door handles more forward (due to forward shifting of driver's elbows when touching the bolsters. See Chapter 5 for the minimum reach envelopes). Further, taller bolsters will increase the difficulty in "sliding" on the seat during entry and egress.
12) *Padding*: Cushioning/padding is desirable because it reduces pressure by increasing support area (Konz and Johnson, 2004). Seats should be covered with padded material to allow a deflection of about 25 mm and distribute the pressure under the buttocks and thighs. In general, the seat should be designed to allow higher pressure under the ischial tuberosities (i.e., the sitting bones – the lower protruding parts of the pelvic bones) and gradually decrease in outward directions. For long term comfort, the pressure on the body tissues should not be constant. Changes in the pressures (due to deliberate massaging actions or postural movements) will reduce discomfort and fatigue. The padding also helps in reducing discomfort caused by vehicle body vibrations under dynamic driving conditions.
13) *Seat track length*: The locations of hip points of different drivers as they adjust the seat fore and aft define the length of the seat track. The foremost and rearmost hip points on the seat track define the seat track length. It should be long enough and placed at a horizontal distance from the ball of foot (BOF) on the accelerator pedal of 2.5 percentile to 97.5 percentile hip point locations (defined as $X_{2.5}$ and $X_{97.5}$ in SAE J1517 and J4004). Based on the SAE J 4004, a seat track length of about 240 mm would be needed to accommodate 95% of the drivers in passenger cars (see Chapter 5 for more details).

RECENT ADVANCES IN DIGITAL MANIKINS

A number of two- and three-dimensional (3D) anthropometric and biomechanical models are presented in the literature, and many are commercially available for design and evaluation purposes. These models can be configured to represent individual males and females in different percentile dimensions for different populations. Many of the models have built-in human motion, posture simulations, and biomechanical strength, as well as percentile force exertion prediction capabilities. Crash-test dummies resembling human biomechanical characteristics are also used to evaluate the crashworthiness of vehicles in accident situations (Seiffert and Wech, 2003).

Integrated digital workplaces and digital manikins and visualization tools are available in several software applications. Computer-aided design (CAD) tools with manikin models (digital human models), such as Jack/Jill, SAFEWORK, RAMSIS, SAMMIE, and UM 3DSSP, are being used by different designers to assist in the product development process (Chaffin, 2001, 2007; Reed et al., 1999, 2003; Badler et al., 2005; Human Solutions, 2010). Many of these tools are being updated to incorporate additional capabilities.

Before using any of the models in the design process, the ergonomics engineer should conduct validation studies to determine if the population of the particular users of the product being designed can be accurately represented in terms of their dimensions, postures, motions, strength, and comfort. The postures assumed by the selected digital human model and their outputs should match closely with the postures and dimensions of real users under different actual usage situations.

CONCLUDING COMMENTS

Anthropometric and biometric characteristics of drivers and passengers are most important in physically accommodating people in the vehicle space in terms of seating, entering in and exiting from the vehicle and during loading and unloading items. The population of the customers of the proposed vehicle must be identified at the earliest phases of the vehicle development and their anthropometric and biomechanical characteristics must be determined and used during the vehicle package development process. In addition, provision of adjustable features in locating the driver, finding preferred seating posture with respect to the primary vehicle controls and providing required fields of view are important in developing safer and comfortable vehicles. Next chapters in this book provide more details on various vehicle design considerations.

REFERENCES

Andersson, G. B. J., and R. Ortengren. 1974. Lumbar Disc Pressure and Myoelectric Back Muscle Activity during Sitting: II. Studies on an Experimental Chair. *Scandinavian Journal of Rehabilitation Medicine*, 3: 122–127.

Andersson, G. B. J., R. Ortengren, A. Nachemson, and G. Elfstrom. 1974. Lumbar Disc Pressure and Myoelectric Back Muscle Activity during Sitting: I. Studies on an Experimental Chair. *Scandinavian Journal of Rehabilitation Medicine*, 3: 104–114.

Asmussen, E. and K. Heebol-Nielsen.1962. Isometric Muscle Strength in Relation to Age in Men and Women. *Ergonomics*, 5: 167–169.

Badler, N., J. Allbeck, S.-J. Lee, R. Rabbitz, T. Broderick, and K. Mulkern. 2005. New Behavioral Paradigms for Virtual Human Models. *SAE Transactions Journal of Passenger Cars – Electronic and Electrical Systems*. Paper 2005-01-2689. Presented at the 2005 SAE Digital Human Modeling Conference, Iowa City, Iowa.

Chaffin, D. B. 2001. *Digital Human Modeling for Vehicle and Workplace Design*. SAE International. ISBN: 978-0-7680-0687-2, New York City, NY Association of Computer Machinery.

Chaffin, D. B. 2007. Human Motion Simulation for Vehicle and Workplace Design: Research Articles. *Human Factors in Ergonomics & Manufacturing*, 17(5): 475–484.

Chaffin, D. B., G. B. J. Andersson, and B. J. Martin. 1999. *Occupational Biomechanics*. New York City, NY: John Wiley & Sons, Inc.

Garrett, J. W. 1971. The Adult Human Hand: Some Anthropometric and Biomechanical Considerations. *Human Factors*, 13(2): 117–131.

Human Solutions. 2010. RAMSIS model applications. www.human-solutions.com/automotive/index_en.php (accessed May 3, 2013).

Jurgens, H., I. Aune and U. Pieper. 1990. *International Data on Anthropometry*. Geneva, Switzerland: ILO.

Kroemer, K. H. E., H. B. Kroemer and K. E. Kroemer-Elbert. 1994. *Ergonomics – How to Design for Ease and Efficiency*. Englewood Cliffs, NJ: Prentice Hall.

Kolich, M. 2006. Applying axiomatic design principles to automobile seat comfort evaluation. *Ergonomia, IJE & HF*, 28(2): 125–136.

Kolich, M. 2009. Repeatability, Reproducibility, and Validity of a New Method for Characterizing Lumbar Support in Automotive Seating. *Human Factors: The Journal of the Human Factors and Ergonomics Society*, 51(2): 193–207.

Konz, S. and S. Johnson. 2004. *Work Design-Industrial Ergonomics*, Sixth Edition. Scottsdale, Arizona: Holcomb Hathaway.

McDowell, M. A., C. D. Fryor, C. L. Ogden and K. M. Flegal. 2008. Anthropometric Reference Data for Children and Adults: United States 2003–2006. National Health Statistics Reports, Vol. 10 . www.cdc.gov/nchs/data/nhsr/nhsr010.pdf

National Center for Health Statistics (NCHS, U.S.). 2016. Anthropometric reference data for children and adults: United States, 2011–2014. Department of Health and Human Services (DHHS) publication; no. (PHS) 2016-1423.

Pheasant, S. and C. M. Haslegrave. 2006. *BODYSPACE: Anthropometry, Ergonomics and the Design of Work*. Third Edition. London: CRC press, Taylor and Francis Group.

Reed, M. P., R. W. Roe and L. W. Schneider. 1999. Design and Development of the ASPECT Manikin. Technical Paper 990963. *SAE Transactions: Journal of Passenger Cars*, 108: 1800–1817.

Reed, M. P., M. B. Parkinson and D. B. Chaffin. 2003. A New Approach to Modeling Driver Reach. Technical Paper 2003-01-0587. *SAE Transactions: Journal of Passenger Cars – Mechanical Systems*, 112: 709–718.

Sanders, M. S. 1983. U.S. Truck Driver Anthropometric and Truck Work Space Data Survey. Report no. CRG/TR-83/002. Canyon Research Group, Inc., West Lake Village, CA.

Seiffert, U. and L. Wech. 2003. *Automotive Safety Handbook*. Warrendale, PA: SAE International.

Society of Automotive Engineers, Inc. 2009. *SAE Handbook*. Warrendale, PA: Society of Automotive Engineers, Inc.

5 Occupant Packaging

WHAT IS VEHICLE PACKAGING?

"Packaging" is a term used in the automobile industry to describe the activities involved in locating various vehicle systems and components and occupants in the vehicle space. Thus, it is about space allocation for various vehicle systems (i.e., hardware), accommodating "People" (i.e., the driver and the passengers), and providing storage spaces for various items (e.g., suitcases, boxes, golf bags, cups, and sunglasses) that people store in their vehicles.

The term "packaging" was used because the task of the package engineering is essentially "bringing in systems and components" produced by others (e.g., different suppliers) and fitting them into the vehicle space so that they will function properly to satisfy the customers and users of the vehicle. The vehicle package engineering department is generally responsible for creating the overall vehicle package consisting of mechanical packaging of all hardware of vehicle systems and occupant packaging. In addition, individual specialty departments and suppliers providing assembled systems such as instrument panels, seats, and engines also have their own package engineering teams developing detailed CAD models with every component and fastener (e.g., a bolt) included in their supplied entities. These teams closely coordinate their work with the vehicle package engineering department. The overall goal is to maximize the space provided for the occupants and their storage areas and minimize the space occupied by the hardware.

OCCUPANT PACKAGE OR SEATING PACKAGE LAYOUT

The occupant package includes drawings and three dimensional graphic representations of the occupant compartment with the position of the occupants (key reference points, e.g., accelerator heel point [AHP], ball of the foot [BOF] point on the accelerator pedal, seating reference point [SgRP]), manikins (e.g., specified in SAE standards J826 and J4002. Note: The SAE standards are available in the *SAE Handbook* [SAE, 2009] or the *SAE International*), the outputs of the occupant packaging tools (e.g., tools and procedures provided in the SAE standards such as J1517, J4004, J941, J1052, J287, J1050), the primary vehicle controls (e.g., steering wheel,

DOI: 10.1201/9781003485582-5

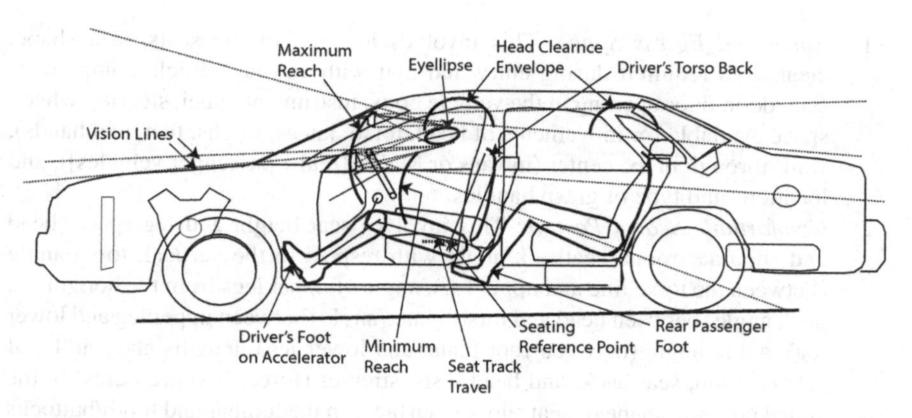

FIGURE 5.1 Illustration of a vehicle occupant package layout.

pedals, and gear shifter) and vehicle body and trim components (e.g., seats, instrument panels, center console, door trim panels, mirrors, and so forth).

The occupant (or seating) package layout is, thus, a drawing or a three-dimensional model shown in a CAD application (e.g., CATIA, IDEAS, ALIAS) showing locations and positioning of the driver and all other occupants (mostly in the form of mannequins), *eyellipses* (drivers' eye locations specified in SAE J941), various reach, clearance, and visibility zones (e.g., hand reach envelopes, head clearance contours, fields of view and obstructions caused by the vehicle components such as the steering wheel, pillars and headrests), and other relevant vehicle details (e.g., steering wheel, floor, instrument panel, pedals, seats, arm rests, gear shifter, parking brake, mirrors, hard points, fiducial marks/points, eye points, sightlines, hood, roof, and doors) and vehicle dimensions. Figure 5.1 illustrates a side view of an occupant package drawing.

It should be noted that most of the occupant packaging tools and practices used in the automotive industry were developed by ergonomics engineers by working through various subcommittees of the SAE Human Factors Committee such as the Human Accommodation and Design Devices Subcommittee. Since the SAE practices are followed in the automotive industry during the vehicle design process, the following part of the chapter will describe key dimensions, reference points and procedures specified in a number of relevant SAE standards.

DEVELOPING THE OCCUPANT PACKAGE: DESIGN CONSIDERATIONS

In developing a vehicle package, a number of design considerations related to the functioning of various vehicle systems and interfaces (e.g., joints and fasteners and links such as wires, tubes, and hoses) between the systems, vehicle spaces and occupant comfort, convenience and safety issues are considered. The occupant packaging considerations can be grouped into the following areas:

1. *Entry and Egress Space*: This involves location of the seats, seat shape, clearances required during entry and exit with various vehicle components (i.e., door, door opening in the vehicle body, instrument panel, steering wheel, space available for movements of head, torso, knees, thighs, feet, and hands), walk-through in the center (in vans or MVPs [multi-passenger vehicles]), and locations and type of grasp handles.
2. *Comfortable Seated Posture*: This involves seat height and leg space, head and shoulder room, seatback angle with respect to the vertical, torso angle (between the torso line and upper leg), angle of upper legs from the horizontal, neck angle (between head and torso), knee angle (between upper leg and lower leg), ankle angle (between foot plane and lower leg), lengths and widths of seat cushion, seat back, and head rests, stresses (forces and pressures) in the spinal column, shape of seat support surfaces in the lumbar and thigh/buttocks regions with respect to the steering wheel and pedal locations.
3. *Operating Controls and Reading Displays*: Locations of hand and foot controls and displays involve driver's body movements and postures (hands, feet, head, and torso) during viewing, reaching, grasping and operating controls, head, eye and ear positions (for acquiring information/reading displays), natural versus awkward postures, and using other in-vehicle items (e.g., cup holder, map pockets, entertainment and information systems).
4. *Visibility of Interior and Exterior Areas*: This involves drawing sightlines from eye points (eye locations) and by considering movements of eyes, head and torso during gathering of visual information from the road and inside the vehicle (e.g., visibility of displays) and available fields of view (by considering obstructions caused by vehicle structures [e.g., pillars] and components and in-direct fields from mirrors, sensors and cameras).
5. *Storage Spaces*: These involve providing convenient and safe storage spaces to accommodate items brought into the vehicle during trips, for example, sunglasses, papers, cups, water bottles, tools, luggage, grocery bags, and boxes.
6. *Vehicle Service*: This involves providing convenient access and space for performing vehicle service and maintenance tasks (e.g., cleaning glass areas, refueling gas, checking engine oil, filling windshield washer fluid, replacing bulbs, and changing flat tires)

The challenge to the occupant package engineer is to ensure that the largest percentage of the user population is accommodated in performing all the tasks involved in the above areas during vehicle uses --while driving and not driving. The following portion of this chapter will cover issues associated with accommodating the driver (primarily item 2 above). Other areas are covered in later chapters of this book.

SEQUENCE IN DEVELOPMENT OF VEHICLE PACKAGE

Advanced Vehicle Design Stage

In many automotive companies, the advanced design departments are given the responsibility to develop new vehicle concepts. The concept development generally begins with brainstorming activities of a multidisciplinary team involving market researchers, designers (industrial designers in the styling/design studios), engineers and researchers who attempt to predict trends in future designs (e.g., fashions, shapes, features in luxury products, and benchmarking competitors' products), technologies (e.g., materials, electronics, telematics, and manufacturing processes), economy (e.g., energy costs and availability), markets (e.g., consumer desires and expectations in different markets and countries), government regulations (e.g., safety, fuel economy and emissions requirements), manufacturing capabilities (availability of manufacturing and assembly plants and equipment) and customer feedback (from past vehicle models and competitors). The team defines the vehicle type, market segments, major exterior and interior dimensions, and desired product characteristics and features. The description of the proposed vehicle is written in a document (sometimes called the "Product Assumptions") and is continuously updated as new information is gathered by the team. The designers usually take the lead in illustrating sketches of the future product concepts. The package engineering members of the team begin creation of the vehicle layout in a CAD system which is continuously shared with the design team members. The team members discuss many different design ideas, features, engineering issues, trade-offs between different issues, engineering feasibility, costs and timing issues and arrive at several alternate vehicle concepts.

The vehicle concepts are illustrated by creating sketches, three-dimensional computer models with different levels of details, from wireframes to fully rendered vehicles (showing color, texture, reflections, and shadow effects) that can be shown in realistic images (static or dynamic) on backgrounds of roadways and in showroom environments, and/or creating physical bucks or models (e.g., full-size vehicle models with representation of interior and exterior surfaces). The vehicle concepts are generally shown to samples of prospective customers in market research clinics at one or more sites at different geographic locations in the selected markets. Competitive products are also included in the market research clinics to understand how the proposed concepts will be perceived by the customers in relation to selected leading competitive products. The customer reactions and responses to the product concepts, their characteristics and features are documented.

The product planners also prepare a business plan for the proposed vehicle. A comprehensive business plan generally includes sections involving vehicle description, market and time schedule of producing the vehicle, proposed manufacturing facilities (e.g., where to build – in an existing plant or a new plant, in what state or country), corporate product plans (i.e., how this vehicle fits in the overall corporation plan to produce other vehicles and vehicle lines), supplier capabilities (e.g., who would supply major systems, subsystems and key components), and estimates of costs, revenues and cash flows based on a number of assumptions (e.g., sales volumes, assembly plant location, costs of building tools and manufacturing facilities, cost of capital, competitors and their projected vehicle development and introduction plans) (see Chapter 3 for more details on the business plan).

The above outputs (i.e., the product concepts, market research findings and the business plan of the proposed vehicle) are presented to the higher management of the company to decide if the proposed product concept should be accepted.

Development of the "Accepted" Vehicle Concept

Figure 5.2 presents a flow diagram showing different tasks involved in occupant packaging and ergonomics evaluations. The process begins with Task #1, which involves defining the vehicle to be designed. As described in the previous section, this task involves obtaining inputs from a number of disciplines to prepare assumptions for the vehicle program. It is extremely important to first define the intended customer population, that is, who would buy and use the proposed vehicle. The characteristics, capabilities, desires and needs of the customers/users must be understood. The market researchers along with the ergonomics engineers and the designers must make every effort to gather information about the intended population. A representative sample of

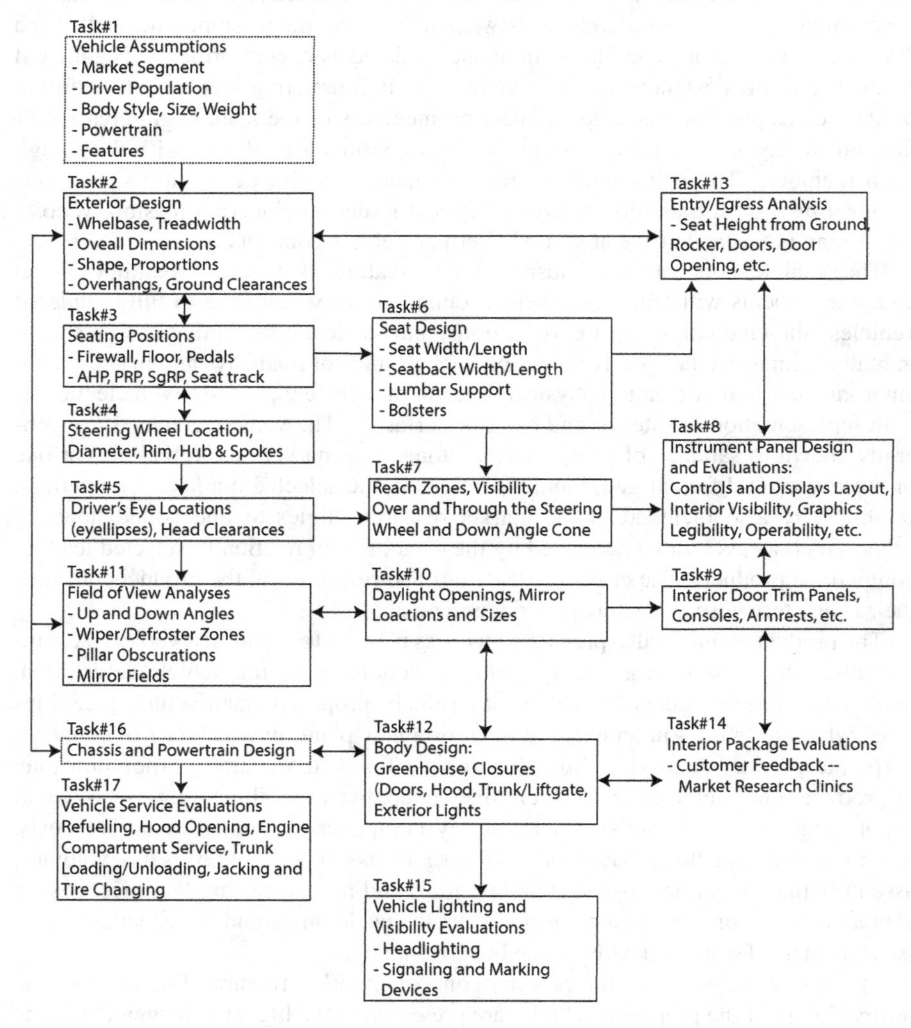

FIGURE 5.2 Flow diagram showing tasks involved in occupant packaging and ergonomic evaluations.

owners and users of the type of vehicle and from the intended market segment (e.g., luxury small 4-door car, economy 2-door hatchback, and mid-size luxury SUV) can be invited and shown early product concepts. They can be extensively interviewed and asked to respond to a number of questions related to how well they like or dislike the product concepts and details on their features and preferences. Their relevant anthropometric dimensions can also be measured to create a database for evaluation of various vehicle dimensions. The quality function deployment (called the QFD in the quality management field) is an excellent tool and it can be used at this early stage to translate the customer needs of the vehicle being designed into functional (engineering) specifications of the vehicle (Besterfield et al., 2003; Bhise et al., 2010, Bhise, 2017). The QFD is described in Chapter 3.

The exterior design as shown in Task #2 usually leads the design process. The vehicle package engineers work concurrently with the design and other engineering teams by positioning the driver (by assuming locations of the vehicle floor and the firewall) and the occupants in the vehicle space (Task #3) and conduct other analyses such as determining locations of primary controls (Tasks #3 and #4), locating driver's eyes (Task #5), designing seats (Task #6), determining maximum and minimum reach zones and visible areas (Task #7) to develop instrument panels, door trim panels, and consoles (Tasks #8 and #9). Tasks # 10 and #11 are conducted to ensure that the driver can obtain the fields of view needed to safely drive the vehicle. The mechanical body design (Task #12) and packaging of chassis and powertrain components (Task #16) are accomplished simultaneously by other engineering departments.

At this early design phase, many analyses are performed to ensure that the key vehicle parameters that define the vehicle exterior (e.g., wheelbase, tread width, overall length, width and height, overhangs, cowl point, deck point, and tumblehome) and vehicle interior (e.g., seat height, seating reference point, and locations of seat track, steering wheel and pedals) are evaluated simultaneously by involving experts from different disciplines. The key areas that link the exterior of the vehicle to the interior such as entry/egress (Task #13), fields of view and window openings (Tasks #10 and #11) are resolved in the very early stages as the exterior and interior surfaces of the vehicle are created in the CAD models. The goal, of course, is to ensure that the largest percentage of occupants can be accommodated, and functional aspects of the vehicle are not compromised. Further, the vehicle lighting design (Task #15) and illumination of lighted graphics and components (Task #8) are studied to ensure that the vehicle can be used safely during nighttime.

A number of special evaluations are also conducted to ensure that the drivers and the passengers can enter the vehicle and exit from the vehicle comfortably (Task #13) and extensive customer feedback on the interior package parameters and vehicle features are obtained (Task #14) by conducting market research clinics. Various evaluation methods used in the entire vehicle development process are summarized in Chapter 19.

The next sections of this chapter will cover details related to dimensions and positioning procedures related to Tasks #3, #4, #5 and #7. Design considerations related to seat dimensions and designs were covered in Chapter 4. And details related to controls and displays in Tasks # 8 and #9 will be covered in Chapters 6 and

7. Chapters 8 and 9 will cover considerations related to Tasks #10, #11 and #15. The entry/exit issues will be covered in Chapter 10, and other exterior issues in Tasks #12 and #17 will be covered in Chapter 11.

DEFINITION OF KEY VEHICLE DIMENSIONS AND REFERENCE POINTS

Units, Dimensions, and Axes

All vehicle and occupant linear dimensions are measured in millimeters (mm). The prefixes "L", "H", and "W" denote length (horizontal), height (vertical), and width (lateral) related dimensions, respectively. All angles are designated by prefix "A" and are measured in degrees (deg). (See SAE J1100 standard in the *SAE Handbook* [SAE, 2009] for more details on the nomenclature and dimensions).

The three-dimensional Cartesian coordinate system used to define locations of points in the vehicle space is generally defined as follows: (a) the positive direction of the longitudinal X-axis is pointing from the front to the rear of the vehicle; (b) the positive direction of the vertical Z-axis is pointing from the ground to upward; (c) the positive direction of the lateral Y-axis is pointing from the left side of the vehicle to the right side; and (d) the origin of the coordinate system is located forward of the front bumper (to make all x-coordinate values positive), below the ground level (to make all z-coordinate values positive) and at the mid-point between the vehicle width. (Refer to SAE standard J 182 (SAE, 2009); Figure 5.18 shows the XYZ coordinate system with its origin, called the "Body Zero").

INTERIOR PACKAGE DIMENSIONS, REFERENCE POINTS AND SEAT TRACK RELATED DIMENSIONS

Figure 5.3 presents a side view drawing showing important interior reference points and dimensions.

The reference points used for the location of the driver and their relevant dimensions are described below.

1) *The Accelerator Heel Point (AHP)*: is the heel point of the driver's right shoe that is on the depressed floor covering (carpet) on the vehicle floor when the driver's foot is in contact with the undepressed accelerator (gas) pedal (see Figure 5.3). The SAE standard J1100 defines it as "A point on the shoe located at the intersection of the heel of shoe and depressed floor covering when the shoe tool (specified in SAE J826 or J4002) is properly positioned. (Essentially, with the ball of the foot contacting the lateral centerline of the undepressed accelerator pedal, while the bottom of the shoe is maintained on the pedal plane)".
2) *The Pedal Plane Angle (A47)*: is defined as the angle of the accelerator pedal plane in the side view measured in degrees from the horizontal (see Figure 5.3). The pedal plane is not the plane of the accelerator pedal, but it is the plane representing the bottom of the manikin's shoe defined in SAE J826

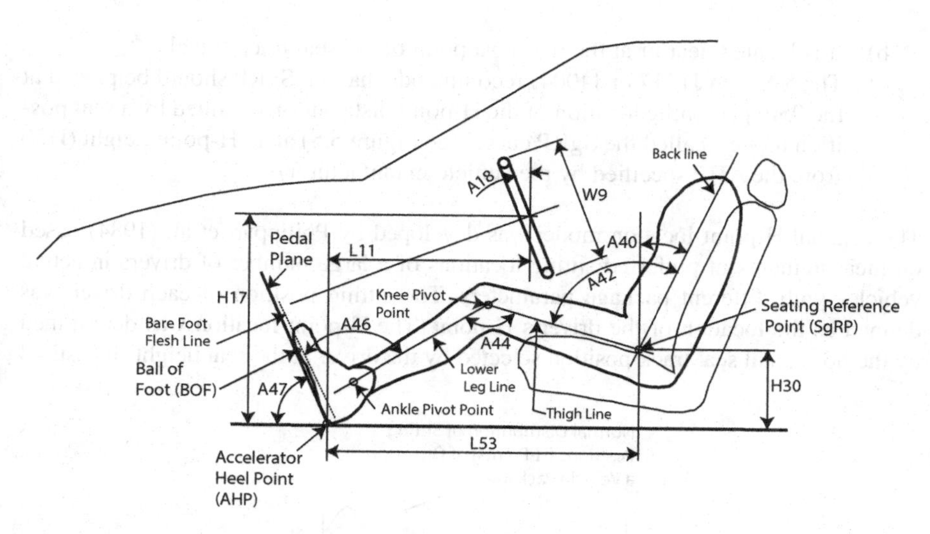

FIGURE 5.3 Interior package reference points and dimensions.

or J4002. (As described later in this chapter, A47 can be computed by using equations provided in SAE J1516 or J4004. Or it can be measured by using the manikin tools described in SAE J 826 or J4002).

3) *The Ball of Foot (BOF)*: on the accelerator pedal is the point on the top portion of the driver's foot that is normally in contact with the accelerator pedal. The BOF is located 200 mm from the AHP measured along the pedal plane (Ref. SAE J4004).
4) *The Pedal Reference Point (PRP)*: is on the accelerator pedal lateral centerline where the ball of foot (BOF) contacts the pedal when the shoe is properly positioned (heel of shoe at AHP, bottom of shoe on pedal plane). The SAE J4004 provides a procedure for locating PRP for curved and flat accelerator pedals using SAE J4002 shoe tool. If the pedal plane is based on SAE J826 and J1516 is used, the BOF point should be taken as the PRP.
5) *The Seating Reference Point (SgRP)*: is the location of a special hip point (H-point) designated by the vehicle manufacturer as a key reference point to define the seating location for each designated seating position. Thus, there is a unique SgRP for each designated seating position (e.g., the driver's seating position, front passenger's seating position, and left rear passenger's seating position). An H-point simulates the hip joint (in the side view as a hinge point) between the torso and the thighs, and thereby it provides a reference for locating a seating position. In the plan view, the H-point is located on the centerline of the occupant.

The SgRP for the driver's position is specified as follows:

a) It is designated by the vehicle manufacturer.

b) It is located near or at the rearmost point of the seat track travel.
c) The SAE (in J1517 or J4004) recommends that the SgRP should be placed at the 95th percentile location of the H-point distribution obtained by a seat position model (called the SgRP curve, see Figure 5.5) at an H-point height (H30 from the AHP specified by the vehicle manufacturer).

The original H-point location model was developed by Philippart et al. (1984) based on measurements of preferred sitting locations of a large number of drivers in actual vehicles with different package parameters. The sitting position of each driver was defined as the location of the driver's H-point. The H-point location was determined by the horizontal seat track position selected by the driver at the seat height (measured

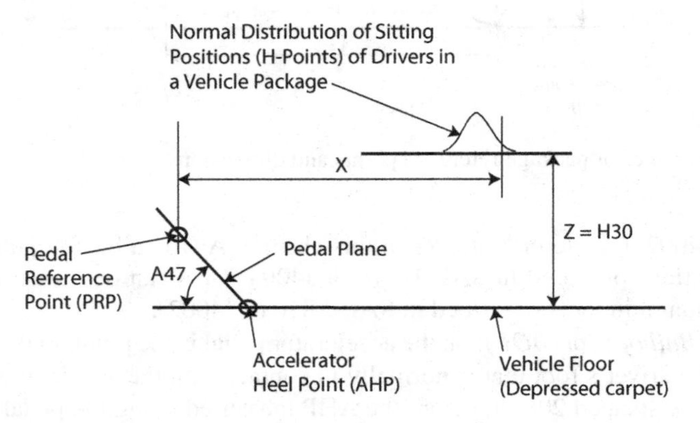

FIGURE 5.4 Distribution of horizontal location of H-points.

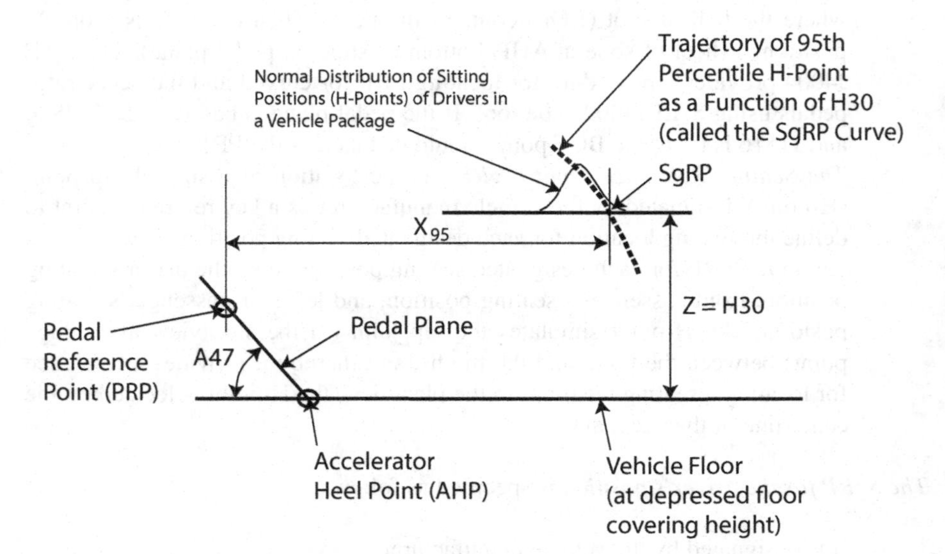

FIGURE 5.5 95th percentile H-point location curve for class A vehicles.

by H30, see Figure 5.3) in the vehicle. For any given vehicle, the H-point locations of a population of drivers can be represented by their distribution of horizontal locations. Figure 5.4 shows the distribution of the horizontal location (X) of the H-points. The 95th percentile value of H-point location distribution is generally selected as the location of the SgRP as shown in Figure 5.5. The SgRP is defined as the point located at X_{95} horizontal distance from the BOF point and H30 vertical distance from the AHP. The trajectory of X_{95} locations as a function H30 is called the "SgRP curve" (see Figure 5.5). The equation of the SgRP curve is provided in SAE J1516 and J4004 is as follows.

$$X_{95} = 913.7 + 0.672316\,(H30) - 0.0019533\,(H30)^2$$

Note: The values of H30 and X_{95} are in mm in the above equation.

The driver's SgRP is the most important and basic reference point in defining the driver package. The SgRP must be established early in the vehicle program and should not be changed later in the vehicle development process because

a) It determines the driver locations in the vehicle package;
b) All driver-related design and evaluation analyses are conducted with respect to this point, e.g., location of eyes, interior and exterior visibility, specifications of spaces (e.g., headroom, legroom, and shoulder room), reach zones, locations of controls and displays, and door openings (for entry/exit).
c) The SgRP can be located in a physical property (i.e., an actual vehicle or a package buck) by placing the SAE H-point machine (HPM) specified in J826 or an H-point device (HPD) specified in J4002. The HPM and HPD are three-dimensional fixtures, and they can be placed in a seat at any designated seating location to measure or verify the location of the SgRP at the seating location. Both HPM and HPD require a number of specified weights (at hip [pelvic], thighs, torso and legs) that should be placed on these fixtures to simulate the compression of the seat with weight of the occupant. Figure 5.6 shows

FIGURE 5.6 SAE H-point Machine placed in the driver's seat.

a picture of the SAE H-point machine in a vehicle. The HPM is referred to in the auto industry and by some seat manufacturers as "OSCAR". Since the seat is compressible and flexible, the HPM is placed on the seat and used as a development and verification tool by seat manufacturers and vehicle manufacturers to determine if the SgRP of the seat that is built and installed in an actual vehicle falls within the manufacturing tolerances from the design SgRP location (shown in the vehicle drawings or CAD model). The description and procedure for location of the H-point machine is provided in SAE J826. The HPD is designed with a three-segmental back pan to account for the effect of shape of the seatbacks (especially in the lumbar region).The SAE 4002 provides drawings, detailed specifications, and procedures for the use of the H-point device (HPD).

The SAE H-Point Machine and the H-point Device (HPM in SAE J826; HPD in SAE 4002) are designed such that when they are placed on a seat they deflect the seat somewhat in the way a real person would deflect the seat. Each device weighs 76 kg (167 lbs, which is 50th percentile of U.S. male weight) and has the torso contour of 50th percentile U.S. male. The devices use 95th percentile legs (10th and 50th percentile legs lengths are also available).

6) *The Seat Track Length*: is defined as the horizontal distance between the foremost and rearmost location of the H-point of the seated drivers. To accommodate 95 percent of the driver population with 50:50 males-to-females ratio, the foremost point and the rearmost points can be defined by determining 2.5 and 97.5 percentile H-point locations from the ball of foot (BOF). The computation procedures for determining different percentile values are specified in SAE J1517 and J4004. SAE J1517 has now been replaced by SAE J4004 standard, and the SAE recommends that J4004 should be used to determine the seat track length and the accommodation levels for the U.S. driving population. It should be noted that since the introduction of SAE J4002, 4003 and 4004, the package engineering community within various automotive companies are slowly transitioning from the old (J826, J1516 and J1517) procedures to the revised (J4002, 4003 and J4004) procedures. Therefore, relevant information from both procedures is provided below.

Figure 5.7 shows the original seat position location model developed by Philippart et al. (1984) and included in SAE J1517. The SAE standard J1517 was developed by measuring actual seated positions of a large number of drivers in vehicles with different H30 values (after they had driven the vehicles and adjusted the seat location at their preferred position) (Philippart et al., 1984). The H-point location model is thus based on functional anthropometric data (i.e., real drivers seated in actual vehicles at their preferred driving posture). The SAE standard J1517 entitled "Driver Selected Seat Position" provides statistical prediction equations for seven percentile values ranging from 2.5 to 97.5 of H-point locations in the vehicle space. The 2.5 and 97.5 percentile H-point location prediction equations are generally used to establish

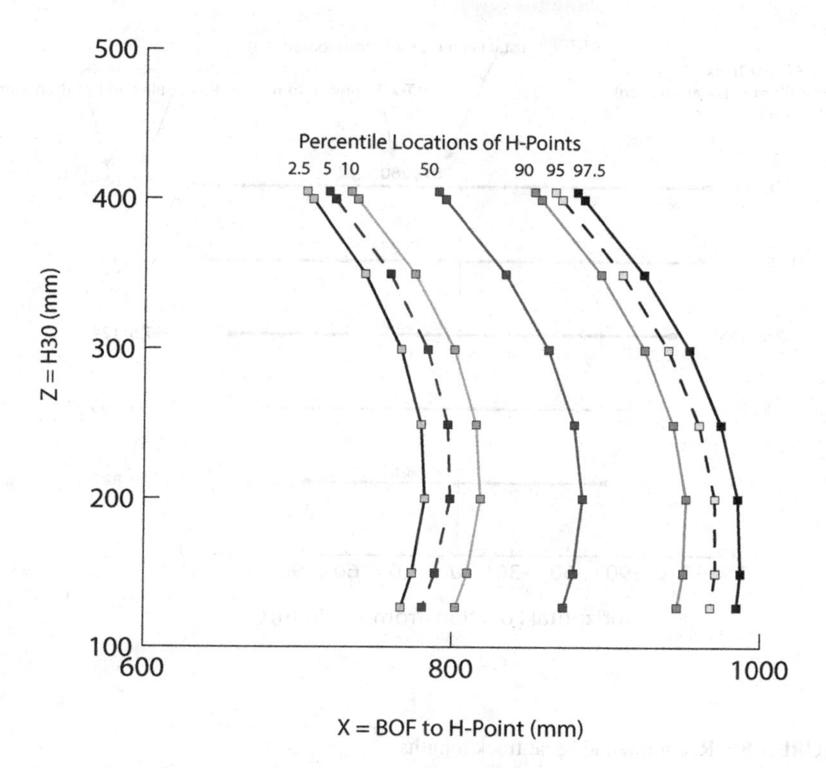

FIGURE 5.7 H-point location curves for 2.5 to 97.5 percentile H-points as functions of H30 for Class A vehicles.

Source: Drawn from equations provided in SAE J1517.

seat track travel to accommodate 95 percent of drivers. The equations are quadratic functions of H30 for Class A vehicles (passenger cars and light trucks) and linear functions of H30 for Class B vehicles (medium and heavy commercial trucks). Class A vehicle equations are based on a 50:50 male-to-female ratio. Figure 5.7 presents seven percentile curves of H-point locations from equations presented in SAE J1517 for 50:50 male-to-female ratio for class A vehicles. The 95th percentile curve shown in Figure 5.7 is called the SgRP curve. The Class B vehicle driver selected seat position lines are specified in SAE J1517 for 50:50, 75:25, and 90:10 to 95:5 male-to-female ratios.

The SAE J4004 presents the H-point location procedure based on more recent work by Flanngan et al. (1996 and 1998) for Class A vehicles. The recommended seat track lengths to accommodate different percentages of drivers are presented in Figure 5.8. The horizontal (X) locations of the front and rear locations of the seat track are specified with respect to a reference location called, X_{ref}. This reference distance is measured aft of the PRP. The X_{ref} is a linear function of H30, steering wheel location (L6) and type of transmission (with or without a clutch pedal). The SAE J4004 standard suggests that until the year 2017 the ball-of-foot and accelerator heel

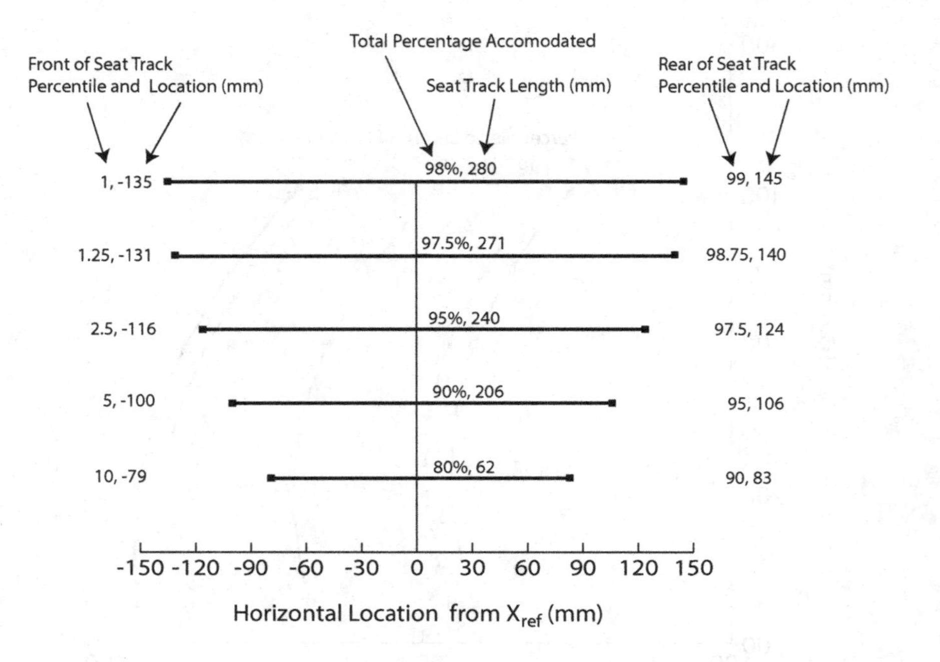

FIGURE 5.8 Recommended seat track lengths.

Source: Drawn from data provided in SAE J 4004, SAE, 2009.

point determined according to SAE J1517 may be used in lieu of the pedal reference point cited in J4004 document. However, SAE J4004 should be used to determine the seat track length and the accommodation levels for the U.S. driving population. The equations illustrating the above described two procedures are provided in a later section of this chapter.

INTERIOR DIMENSIONS

A number of interior package dimensions shown in Figure 5.3 are described in this section. The dimensions are defined using the nomenclature specified in SAE J1100.

1. *AHP to SgRP Location*: The horizontal and the vertical distances between the accelerator heel point (AHP) and the seating reference point (SgRP) are defined as L53 and H30, respectively (see Figure 5.3).
2. *Posture Angles*: The driver's posture is defined by the angles of the H-point machine (HPM) or the H-point device (HPD). The angles shown in Figure 5.3 are defined as follows:

a) Torso angle (A40). The angle between the torso line (also called the back-line) and the vertical. It is also called the seatback angle or back angle.
b) Hip angle (A42). The angle between the thigh line and the torso line.
c) Knee angle (A44). The angle between the thigh line and the lower leg line. It is measured on the right leg (on the accelerator pedal).
d) Ankle angle (A46). The angle between the (lower) leg line and the bare foot flesh line, measured for the right leg.
e) Pedal plane angle (A47). The angle between the accelerator pedal plane and the horizontal.

3. *Steering Wheel*: The center of the steering is specified by locating its center by dimensions L11 and H17 in the side view. The steering wheel center is located on the top plane of the steering wheel rim (see Figure 5.3). The lateral distance between the center of the steering wheel and the vehicle centerline is defined as W7. The diameter of the steering wheel is defined as W9. The angle of the steering wheel plane with respect to the vertical is defined as A18 (see Figure 5.3).
4. *Entrance Height (H11)*: The the vertical distance from the SgRP to the upper trimmed body opening (see Figure 5.9). The trimmed body opening is defined as the vehicle body opening with all plastic trim (covering) components installed. This dimension is used to evaluate head clearance as the driver enters the vehicle and slides over the seat during entry and egress.
5. *Belt Height (H25)*: It is vertical distance between the SgRP and the bottom of the side window DLO (daylight opening) at the SgRP X-plane (plane perpendicular to the longitudinal X-axis and passing through the SgRP) (see Figure 5.10). The belt height is important to determine the driver's visibility to the sides. It is especially important in tall vehicles such as heavy trucks and buses to evaluate if the driver can see vehicles in the adjacent lane on the right

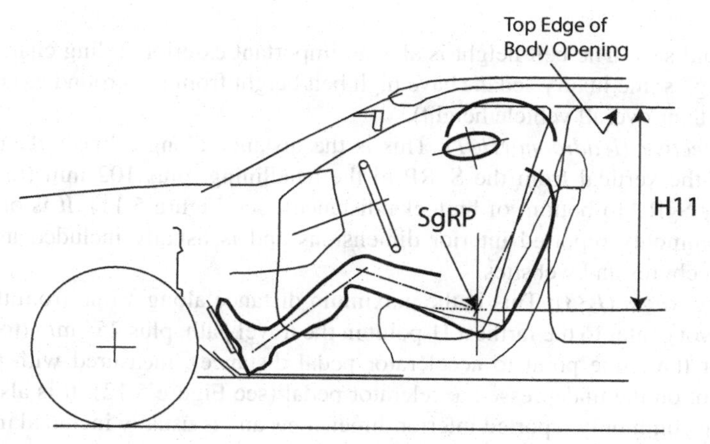

FIGURE 5.9 Entrance height (H11).

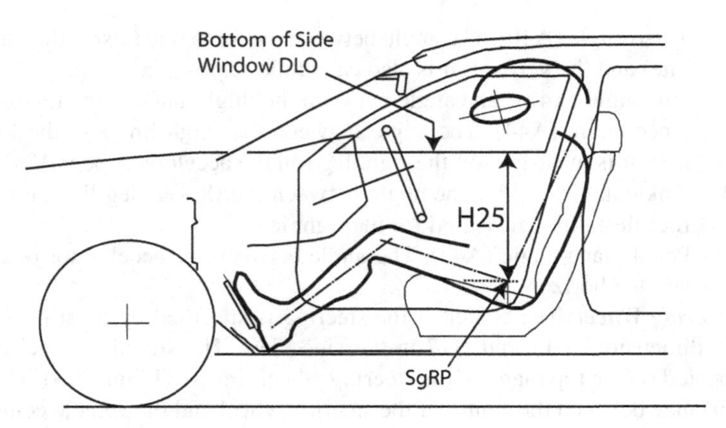

FIGURE 5.10 Belt height (H25).

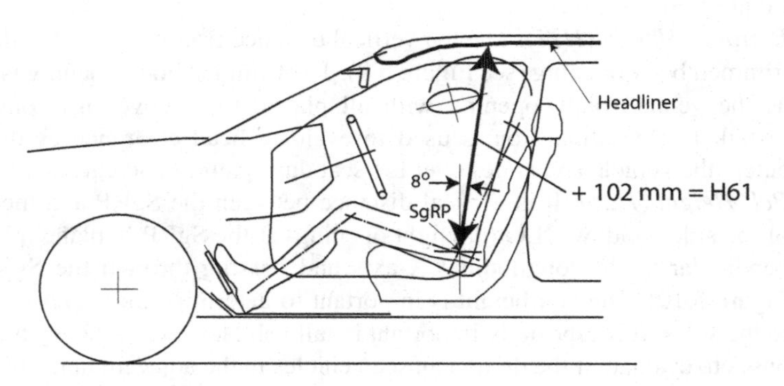

FIGURE 5.11 Effective headroom (H61).

hand side. The belt height is also an important exterior styling characteristic (e.g., some luxury sedans have high belt height from the ground as compared to their overall vehicle height).

6. *Effective Headroom (H61)*: This is the distance along a line 8 degrees rear of the vertical from the SgRP to the headlining, plus 102 mm (to account for SgRP to bottom of buttocks distance) (see Figure 5.11). It is one of the commonly reported interior dimensions and is usually included in vehicle brochures and websites.
7. *Leg Room (L33)*: This is the maximum distance along a line from the ankle pivot center to the farthest H-point in the travel path, plus 254 mm (to account for the ankle point to accelerator pedal distance), measured with the right foot on the undepressed accelerator pedal (see Figure 5.12). It is also one of the commonly reported interior dimensions and is usually included in vehicle brochures and websites.
8. *Shoulder Room (W3)* (Minimum cross-car width at beltline zone): It is the minimum cross-car distance between the trimmed doors within the measurement

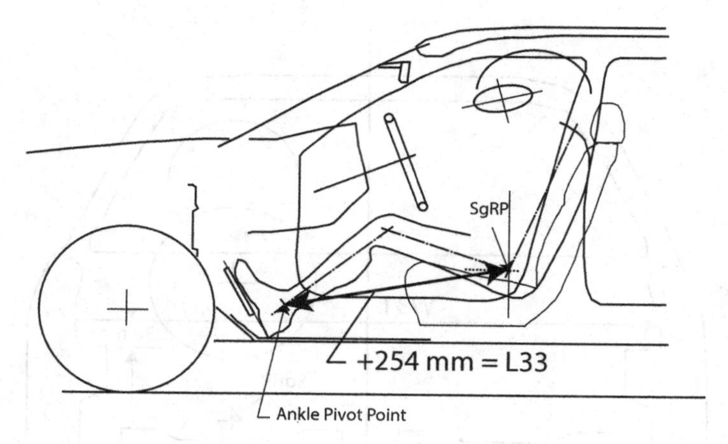

FIGURE 5.12 Leg room (L33).

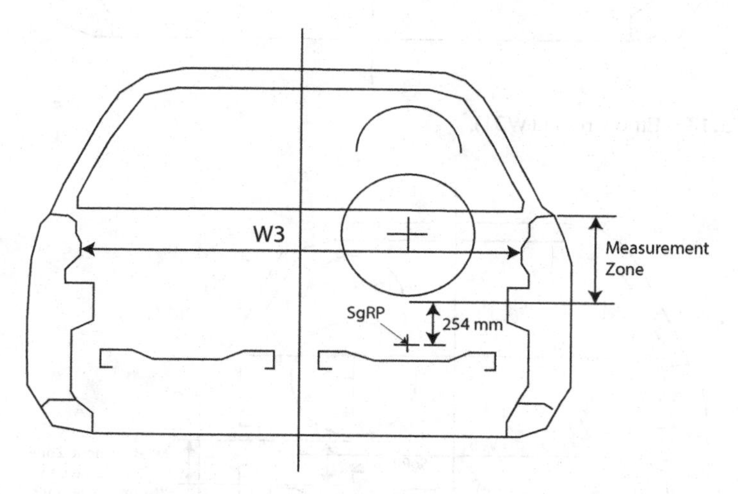

FIGURE 5.13 Shoulder room (W3).

zone. The measurement zone lies between the beltline and 254 mm above SgRP, in the X-plane through SgRP (see Figure 5.13). It is also one of the commonly reported interior dimensions and is usually included in vehicle brochures and websites.

9. *Elbow Room (W31)* (Cross-car width at armrest): It is the cross-car distance between the trimmed doors, measured in the X-plane through the SgRP, at a height 30 mm above the highest point on the flat surface of the armrest. If no armrest is provided, it is measured at 180 mm above the SgRP (see Figure 5.14).
10. *Hip Room (W5)* (Minimum cross-car width at SgRP zone): It is the minimum cross car distance between the trimmed doors within the measurement zone.

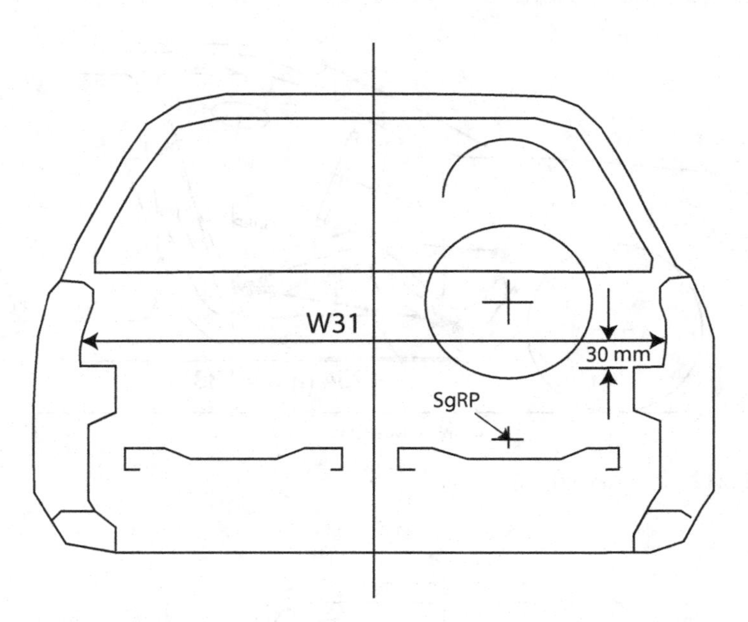

FIGURE 5.14 Elbow room (W31).

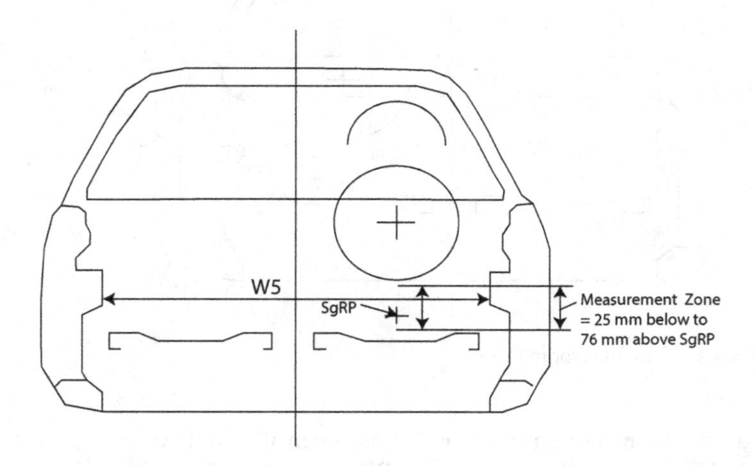

FIGURE 5.15 Hip room (W5).

The measurement zone extends 25 mm below and 76 mm above SgRP, and 76 mm fore and aft of the SgRP (see Figure 5.15).

11. *Knee Clearance* (L62) (Minimum knee clearance: front): It is the minimum distance between the right leg K-point (knee pivot point) and the nearest interference, minus 51 mm (to account for the knee point to front of the knee distance) measured in the side view, on the same Y-plane as the K-point, with the heel of shoe at FRP (floor reference point) (see Figure 5.16).
12. *Thigh Room (H13)* (Steering wheel to thigh line): It is the minimum distance from the bottom of the steering wheel rim to the thigh line (see Figure 5.17).

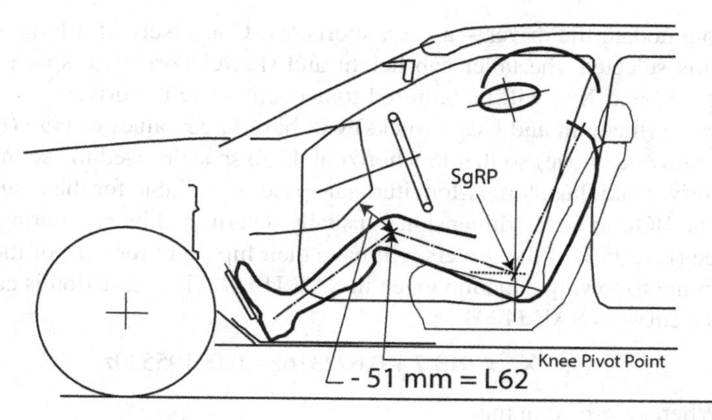

FIGURE 5.16 Knee clearance (L62).

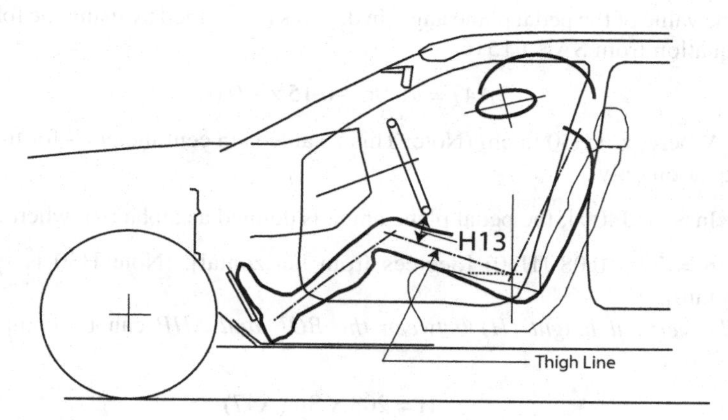

FIGURE 5.17 Thigh room (H13).

DRIVER PACKAGE DEVELOPMENT PROCEDURES

In this section, we cover basic steps involved in positioning the driver, determining the seat track length, positioning eyellipse and head clearance envelopes, determining maximum and minimum reach envelopes and positioning the steering wheel.

1. *Determine H30* = height of the seating reference point (SgRP) from the accelerator heel point (AHP).

   The H30 value is usually selected by the package engineer based on the type of vehicle to be designed. The H30 dimension is one of the dimensions used in the SAE standards to define Class A vehicles (passenger cars and light trucks) and Class B vehicles medium and heavy trucks). The values of H30 for Class A vehicles range between 127 to 405 mm. It should be noted that smaller values of H30 will allow lower roof height (measured from the vehicle floor) and will require longer horizontal space (dimension L53 and X_{95}) to accommodate the driver – as in a sports car. Conversely, if a large value of H30 is selected, the taller cab height and shorter horizontal space (dimension L53 and X_{95}) will be required to accommodate the driver. The Class B vehicles (medium and heavy trucks) will have large values of H30 (typically 350 mm and above) so that less horizontal cab space is used to accommodate the driver and thus, longer longitudinal space is available for the cargo area.

The BOF to SgRP dimension is usually determined by computing the X_{95} value (i.e., 95% of the drivers will have their hip point forward of the SgRP) from the following equation given in SAE J1517. (This equation is called the SgRP curve in SAE J4004.)

$$X_{95} = 913.7 + 0.672316z - 0.00195530z^2$$

Where, z = H30 in mm

2. *Determine pedal plane angle (A47).*

The value of the pedal plane angle in degrees is obtained by using the following equation from SAE J 1516.

$$A47 = 78.96 - 0.15\ z - 0.0173\ z^2$$

Where, z = H30 in cm (Note: This z value is in centimeters – for the above equation only).

In SAE J4004, the pedal plane angle is defined as alpha (α), where:

α = 77 - 0.08 (H30) [degrees from horizontal]. (Note H30 is specified in mm).

3. *The vertical height (H) between the BOF and AHP* can be computed as follows:

$$H = 203 \text{ x } \mathrm{Sin}(A47)$$

It should be noted that distance between AHP to BOF is specified as 203 mm in SAE J1517 and 200 mm in SAE J4004.

4. *The horizontal length (L) between the BOF and AHP* can be computed as follows:

$$L = 203 \text{ x } \mathrm{Cos}(A47)$$

5. *The horizontal distance between the AHP and SgRP (L53)* can be computed as follows:

$$L53 = X_{95} - L$$

6. *The seat track length* is defined by the total horizontal distance of the fore and aft movement of the H-point (for a seat that does not have vertical movement of the H-point). The foremost H-point and rearmost H-point on the seat track are defined by the vehicle manufacturer. To accommodate 95 percent of the drivers (with 50% males and 50% females), the foremost point is defined as at $X_{2.5}$ horizontal distance from the rearward of the BOF and the rearmost point is defined as at $X_{97.5}$ horizontal distance from the rearward of the BOF. The SAE J1517 defines $X_{2.5}$ and $X_{97.5}$ distances as follows:

$$X_{2.5} = 687.1 + 0.895336z - 0.00210494z^2$$

$$X_{97.5} = 936.6 + 0.613879z - 0.00186247z^2$$

Where, z = H30 in mm

$TL23 = X_{95} - X_{2.5}$

= Horizontal distance between the SgRP and the foremost H-point

$TL2 = X_{97.5} - X_{95}$

= Horizontal distance between the SgRP and the rearmost H-point

Total Seat track Length to accommodate 95 percent of the drivers = TL1

Where, $TL1 = TL23 + TL2 = X_{97.5} - X_{2.5}$

If SAE J4004 is used to locate the seat track, then the x distance of the H-point reference point aft the pedal reference point (PRP) is computed as follows:

$$X_{ref} = 718 - 0.24(H30) + 0.41 (L6) - 18.2 t$$

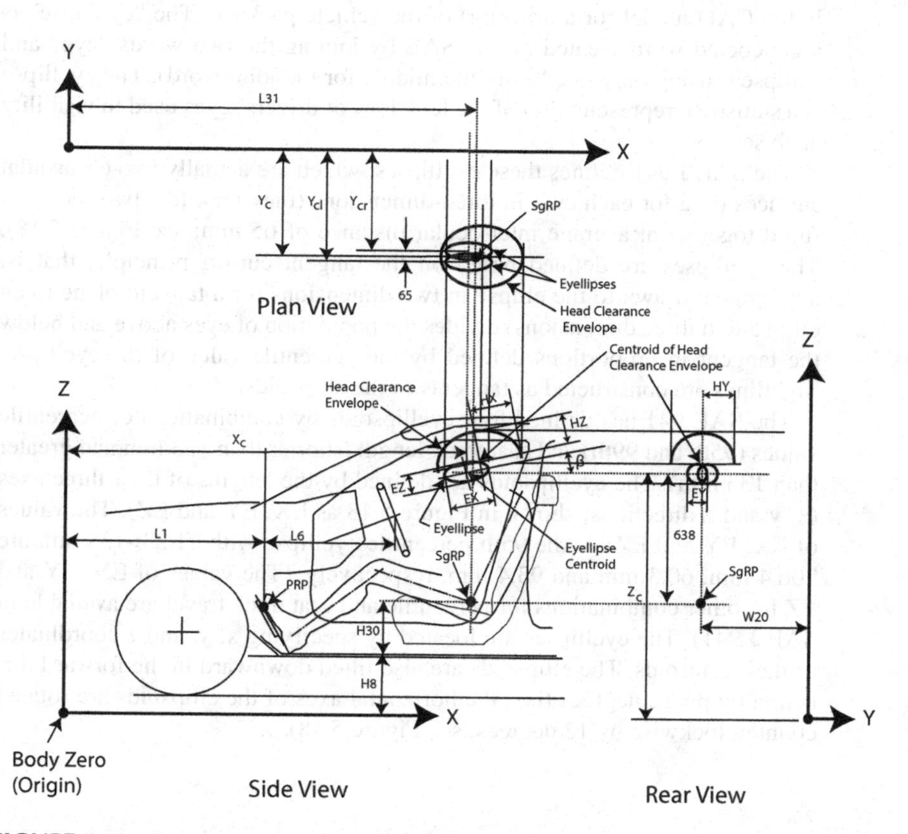

FIGURE 5.18 Location of eyellipses and head clearance envelope.

where, L6 is the horizontal distance from the PRP to the steering wheel center (see Figure 5.18) and "t" is the transmission type (t=1 if a clutch pedal is present, and t=0 if no clutch pedal).

The foremost and rearmost points on the seat track are obtained from data presented in Figure 5.8. It should be noted that the x-axis of Figure 5.8 presents distances of the foremost and rearmost points with respect to X_{ref}. From Figure 5.8, for 95 percent accommodation the TL1 would be 240 mm.

7. *The seatback angle* (or what is also called the torso angle) is defined by dimension A40 (measured in degrees with respect to the vertical). With the reclinable seatback feature, a driver can adjust the angle to his/her preferred seatback angle. The seatback angle in the 1960s and 1970s was defined as 24 or 25 degrees by many manufacturers (due to bench seats which were not reclinable). But with the reclinable seatback features, most drivers prefer to sit more upright with angles of about 18 to 22 degrees in most passenger cars, and about 15 to 18 degrees in pickups and SUVs. The seatback angles selected by the Class B (medium and large commercial trucks) drivers are generally more upright – about 10–15 degrees.
8. *The driver's eyes* are located in the vehicle space by positioning "eyellipses" in the CAD model (or a drawing) of the vehicle package. The "eyellipse" is a concocted word created by the SAE by joining the two words "eye" and "ellipse" (using only one "e" in the middle for the joint word). The eyellipse is a statistical representation of the locations of drivers' eyes used in visibility analyses.

   The SAE J 941 defines these eyellipses, which are actually two ellipsoidal surfaces (one for each eye) in three-dimensions (they look like two footballs fused together at average inter-ocular distance of 65 mm; see Figure 5.18). The eyellipses are defined based on the tangent cut-off principle, that is, any tangent drawn to the ellipse in two dimensions (or a tangent plane to an ellipsoid in three dimensions) divides the population of eyes above and below the tangent in proportions defined by the percentile value of the eyellipse. Sightlines are constructed as tangents to the ellipsoids.

   The SAE 941 has defined four eyellipsoids by combinations of percentile values (95th and 99th) and seat track length (shorter than 133 mm and greater than 133 mm). The eyellipsoids are defined by the lengths of their three axes (x, y and z directions; shown in Figure 5.18 as EX, EY and EZ). The values of EX, EY and EZ for the 95th percentile eyellipse with TL23>133 mm are 206.4 mm, 60.3 mm and 93.4 mm, respectively. (The values of EX, EY and EZ for other combinations for percentile and seat track travel are available in SAE J 941). The eyellipses are located by specifying x, y, and z coordinates of their centroids. The ellipsoids are also tilted downward in the forward direction by β =12 degrees (i.e., the horizontal axes of the ellipsoids are rotated counterclockwise by 12 degrees; see Figure 5.18).

The coordinates of the left and right eyellipse centroids [(X_c, Y_{cl}, Z_c) and (X_c, Y_{cr}, Z_c), respectively] with respect to the body zero are defined in SAE J941 as follows (see Figure 5.18):

$$X_c = L1 + 664 + 0.587\,(L6) - 0.178(H30) - 12.5t$$

$$Y_{cl} = W20 - 32.5$$

$$Y_{cr} = W20 + 32.5$$

$$Z_c = 638 + H30 + H8$$

Where, (L1, W1, H1) = coordinates of the PRP, L6 = Horizontal distance between the BOF (or PRP) and the steering wheel center, and t = 0 for vehicle equipped with automatic transmission and t = 1 for vehicle with clutch pedal (manual transmission). [Note: The SgRP coordinates with respect to the body zero are L31, W20, H8+H30. See Figure 5.18. And L1 = L31- $X_{95.}$ W20 is the lateral distance between the driver centerline and the vehicle centerline. The value of coordinate of centroid is negative as the driver in a left hand drive vehicle sits at W20 distance to the left of the vehicle centerline.]

9. *Locations of tall and short driver eyes*: The tall and short driver's eyes on the 95th percentile eyellipse are located at approximately 46.7 mm (half of EZ = 93.4 mm) above and below the eyellipse centroid. By considering that the eyellipses are tilted 12 degrees forward, the height of the 46.7 mm can be adjusted to 46.7/Cos 12 (i.e., 47.74) mm.
10. *The head clearance envelopes* are defined in SAE J1052 (see Figure 5.18). They were developed to provide clearance for the driver's hair on the top, front and side of the head. They are defined as ellipsoidal surfaces (above the centroid only) in three-dimensions with specified dimensions of three axes from the centroid. The dimensions are shown in Figure 5.18 as HX, HY and HZ. The values of HX, HY and HZ for the 99th percentile head clearance ellipsoid are 246.04, 166.79 and 151 mm, respectively for seat track lengths over 133 mm.

The head clearance envelopes are also defined as tangent cut-off ellipsoids and clearances from vehicle surfaces such as the roof, header or roof rails can be measured by determining magnitude of movements (in the three directions defined by the vehicle coordinate system) of the head clearance envelope needed to touch different interior surfaces. The centroid of the head clearance contour is (x_h, y_h, z_h) distance from the cyclopean centroid (mid-point of the left and right centroids) of the eyellipse. For seat track travel (TL23) greater than 133 mm, the values of (x_h, y_h, z_h) coordinates in mm are (90.6, 0.0, 52.6).

The SAE standard J1052 provides four head clearance ellipsoids for the combinations of two percentile values (95th and 99th) and two seat track lengths (below 133 mm, and above 133 mm). In addition, to accommodate horizontal head shift of occupants seated in the outboard (toward the side glass) locations, J1052 requires an additional lateral shift of 23 mm of the

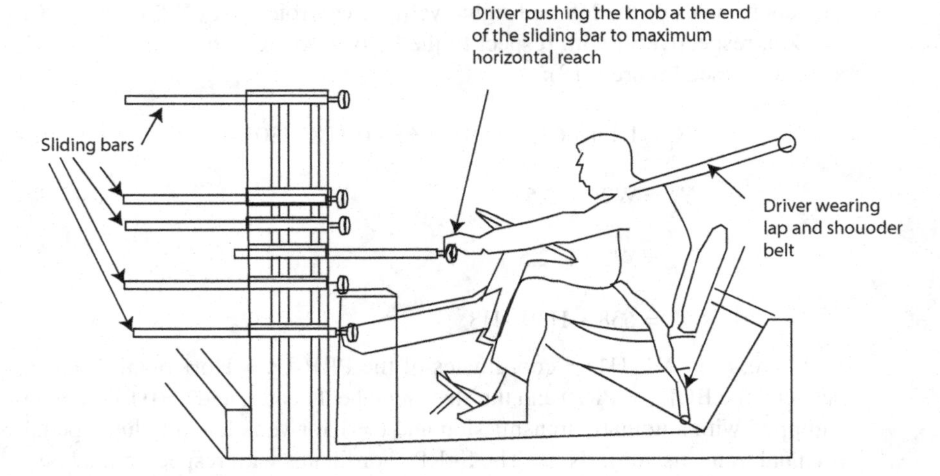

FIGURE 5.19 Maximum hand reach study buck (the buck shown in the above picture was configured to represent a heavy truck package).

ellipsoid on the outboard side. The ellipsoids are also tilted downward in the forward direction by 12 degrees.

11. *The maximum reach* data are provided in SAE J287. The reach distances are based on the controls reach studies conducted by the SAE (Hammond and Roe, 1972; Hammond, Mauer and Razgunas, 1975). In these studies, each subject was asked to sit in an automotive buck at his preferred seat track position with respect to the steering wheel and the pedals. The subject was then asked to grasp each knob (like the old push-pull head lamp switch knob) with three fingers and slide the knob (mounted at the end of horizontally sliding bar) as far forward as he could comfortably reach at each of the vertical and lateral bar locations (see Figure 5.19). The experimenters were looking for the maximum rather than the preferred reach distances. The SAE standard J287 provides tables that present horizontal distances forward from a "HR" (hand reach) reference plane at combinations of different lateral and vertical locations.

   The longitudinal location of the HR plane from the AHP is established by computing the value of 786-99G, where G=General Package Factor. If the above computed value of HR is greater than L53, then the HR plane is located at the SgRP. The G value is computed by using the following formula in SAE J287 FEB2007:

$$G = 0.00327\ (H30) + 0.00285\ (H17) - 3.21$$

   Where, H17 = height of the center of the steering wheel (on the plane placed on the driver's side the steering wheel rim) from the AHP (see Figure 5.20).

   The values of G vary from -1.3 (for a sports car package) to +1.3 (for a heavy truck package).

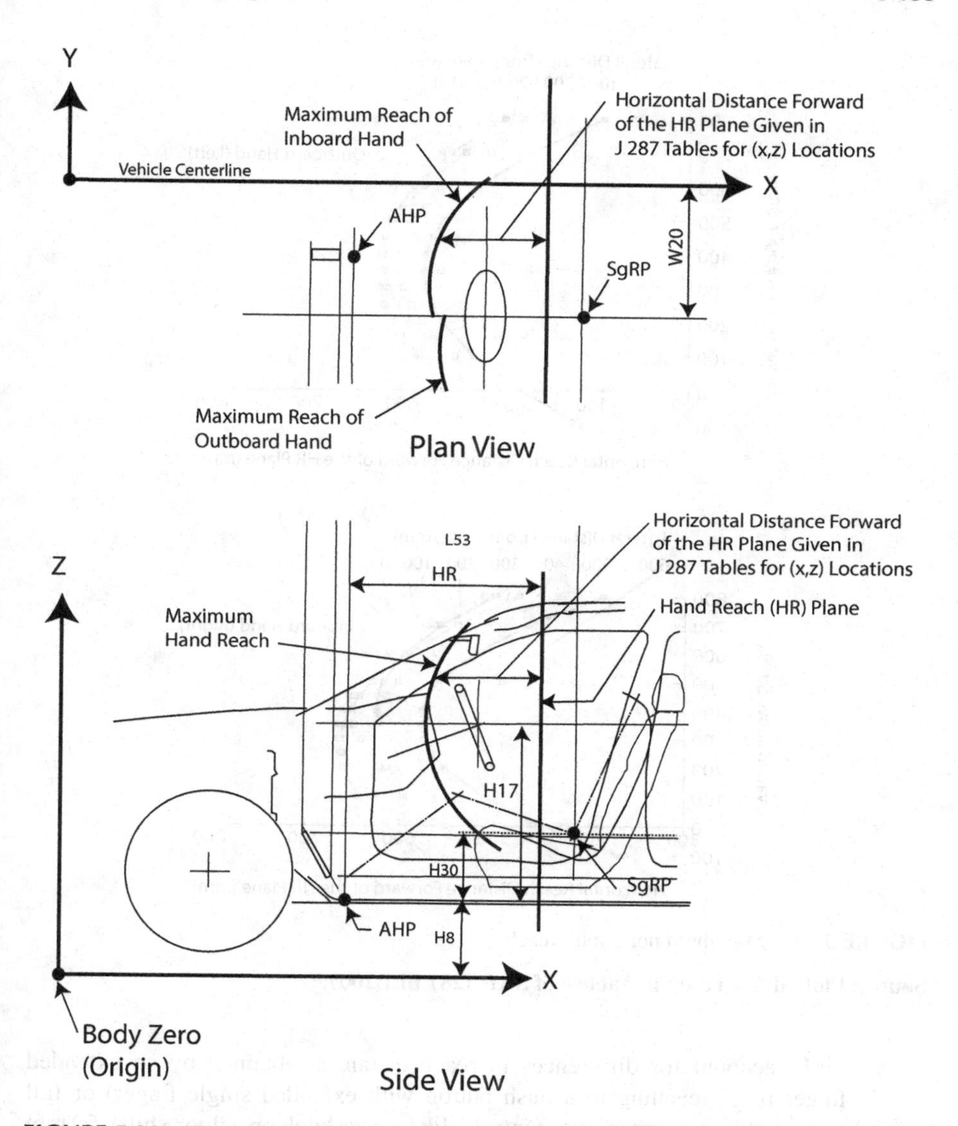

FIGURE 5.20 Plan and side views showing the HR plane and horizontal distances forward of the HR plane given in tables of SAE J287.

The reach tables are provided for combinations of the three variables: a) type of restraints used by the driver (unrestrained = lap belt only; restrained = lap and shoulder belt), b) G value, and c) male-to-female population mix. Figure 5.20 presents a side and plan view showing the reach contours.

The reach contours actually generate two complex surfaces, one for each hand, in the three dimensions. Figure 5.21 presents cross sections of the reach surfaces at different lateral locations for the left hand (top figure) and right hand (bottom figure) from a reach table (Table 4 from the SAE J287 standard).

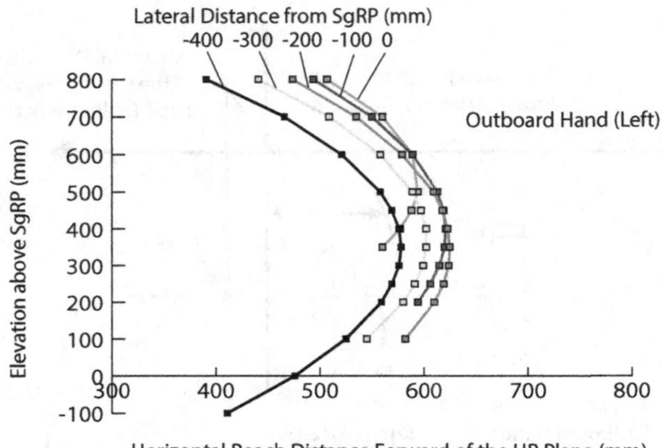

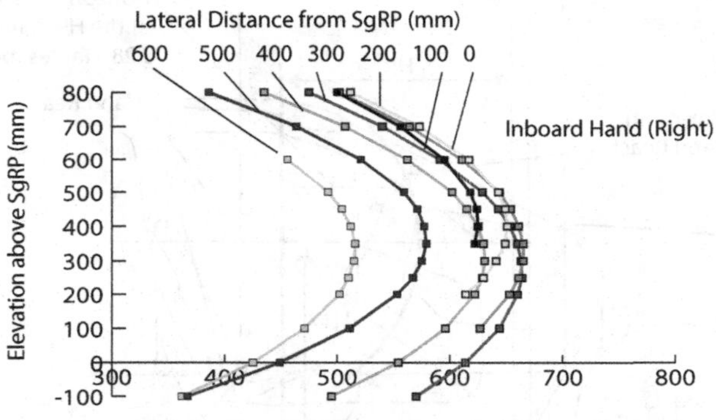

FIGURE 5.21 Maximum horizontal reach.

Source: Plotted from data in Table 4 of SAE J287 FEB2007.

To account for differences in reach distances obtained by an extended finger (e.g., reaching to a push button with extended single finger) or full grasp (all fingers grasping a control – like a gear knob on a floor-shift), 50 mm is added or subtracted, respectively, from the value obtained from the tables.

12. *The minimum reach* is the shortest distance (i.e., closest to the driver) that a short driver seated at the foremost point on the seat track (i.e., her H-point located at the forward-most point of the seat track) will be comfortable in reaching for a control. The procedure for the minimum comfortable reach is covered in Chapter 7.
13. *Steering Wheel Location.* The steering wheel location is constrained by the maximum and minimum reach envelopes, visibility of the roadway and thigh clearance. The steering wheel should be placed rearward of the maximum reach (J287) and forward of the minimum reach envelopes. The sight line (or

the visibility) over the top of the steering wheel rim from the short driver's (5th percentile) eye-point should allow the driver to view the road surface. The ground intercept distance of about 20 to 70 feet in front of the front bumper is generally considered acceptable. The thigh clearance between bottom of the steering wheel and top of seat should allow accommodation of at least a 95th percentile thigh thickness during entry and egress.

In additional to meeting the above requirements illustrated in Figure 5.22, the nominal location of the steering wheel is also determined by benchmarking steering wheel locations of other vehicles (e.g., superimposing steering wheel locations of other vehicles using common SgRP and/or BOF) and by using subjective assessment techniques in vehicle bucks (see Volume 2, Chapter 19). Further, use of a tilt and telescopic steering column would allow most drivers to adjust the steering wheels to their preferred positions.

14. *Steering Wheel and Pedals Placement*: It is important to realize that the relative locations of the accelerator pedal (AHP or BOF), center of the steering wheel, and the driver's SgRP are critical in accommodating the highest percentage of drivers. After getting into the driver's seat, the most drivers will adjust their seat position (fore/aft and up/down with a power seat option) to their preferred driving position. After the driver selects his/her seating position, the steering wheel position is usually adjusted at its preferred location through tilt and telescopic adjustments. This process of adjustments is also affected by the following: a) closeness (or reach distance) of the instrument panel, b) height of the instrument panel, c) height of the beltline, d) closeness of the header, and e) upper edge of the outside mirrors and lower edge of the inside mirror. It is therefore important to conduct preferred driving position studies in a buck or a drivable vehicle to ensure that most drivers can find their preferred driving position such that they do not feel that "the pedals are too far forward, and the steering wheel is too close to their torso" or "the pedals are

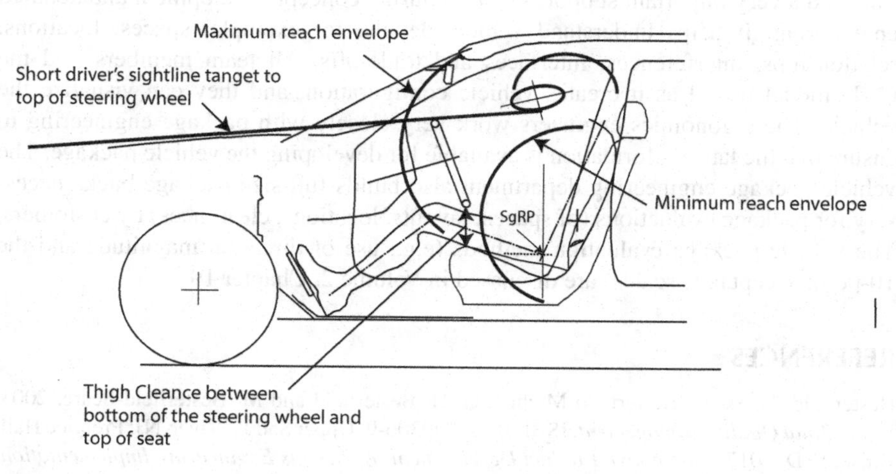

FIGURE 5.22 Considerations related to location of the steering wheel.

too far rearward, and the steering wheel is too far forward". The driver who cannot adjust the seat height to his/her preferred height will complain, saying that "seat is too low and I feel like I'm sitting in a well"; or "the seat is too high and I feel naked due to too much openness of the glass area around me", or "the steering wheel rim blocks the view of the speedometer or the roadway". Incorporating either an adjustable steering column or adjustable pedals along with a power seat will reduce these driver accommodation problems. Package engineers generally perform such driving position validation studies to ensure that the drivers are most comfortable in finding their preferred driving position. Otherwise, the drivers will find that the vehicle does not fit them well and it may result in driver discomfort, complaints or lost vehicle sales.

15. *SgRP Couple Distance (L50).* It is the longitudinal distance between the SgRPs of adjacent rows.

    L50-1= SgRP couple distance between the front to second rows
    L50-2= SgRP couple distance between the second to third rows

OTHER ISSUES, DIMENSIONS, AND CONCLUDING COMMENTS

Other package and ergonomics design-related issues such as locating controls and displays, entry/exit, field of view, opening the hood and servicing the engine, opening the trunk (or liftgate) and loading and unloading, are covered in subsequent chapters. An Excel-based spreadsheet program is provided in the publisher's website for the readers to better understand the various inputs and calculate the resulting package dimensions. The program can also be used to set up different driver packages, analyze an existing package or conduct sensitivity analyses by changing combinations of different input parameters and studying the resulting driver packages.

The vehicle package engineering department generally maintains the latest CAD models and dimensions of the vehicle. The package engineering department is "central" and a very important support function during concept development and detailed engineering. It helps understand vehicle details, for example, spaces, locations, relationships, interferences, interfaces and trade-offs. All team members find the CAD model useful as it creates vehicle configuration, and they can visualize the vehicle. The ergonomics engineers work very closely with package engineering to ensure that the latest information is available for developing the vehicle package. The vehicle package engineering department also builds full-size package bucks necessary for package evaluations (of spaces, layouts, locations, clearances) by customers. The vehicle package evaluation methods (e.g., use of direction magnitude and the 10-point acceptance scales) are discussed in Volume 2, Chapter 19.

REFERENCES

Besterfield, D. H., C. Besterfield-Michna, G. H. Besterfield and M. Besterfield-Scare. 2003. *Total Quality Management*. ISBN 0-13-099306-9. Upper Saddle River, NJ: Prentice Hall.

Bhise, V. D. 2017. *Automotive Product Development: A Systems Engineering Implementation*. ISBN 978-1-4987-0681-0. Boca Raton, FL: CRC Press.

Bhise, V. D., H. Dandekar, A. Gupta and U. Sharma. 2010. Development of a Driver Interface Concept for Efficient Electric Vehicle Usage. A paper presented at the 2010 SAE World Congress held in Detroit, Michigan.

Flannagen, C. C., L. W. Schneider and M. A. Manary. 1996. Development of a Seating Accommodation Model. SAE Technical Paper# 960479. Warrendale, PA: Society of Automotive Engineers, Inc.

Flannagen, C. C., L. W. Schneider and M. A. Manary. 1998. An Improved Seating Accommodation Model with Application to Different User Populations. SAE Technical Paper# 980651. Warrendale, PA: Society of Automotive Engineers, Inc.

Hammond, D. C. and R. W. Roe.1972. SAE Controls Reach Study. SAE Paper no. 720199. Warrendale, PA: Society of Automotive Engineers, Inc.

Hammond, D. C., Mauer, D. E. and L. Razgunas. 1975. Controls Reach – The Hand Reach of Drivers. SAE Paper no. 750357. Warrendale, PA: Society of Automotive Engineers, Inc.

Philippart, N. L., Roe, R. W., Arnold, A. J., and T. J. Kuechenmeister. 1984. Driver Selected Seat Position Model. SAE Technical Paper# 840508. Warrendale, PA: Society of Automotive Engineers, Inc.

Society of Automotive Engineers, Inc. 2009. *The SAE Handbook*. Warrendale, PA: Society of Automotive Engineers, Inc.

6 Driver Information Acquisition and Processing

INTRODUCTION

Driving a vehicle is an information processing activity. During driving, the driver continuously acquires information from various senses (vision, hearing, tactile, vestibular, kinesthetic, and olfactory), processes the acquired information, makes decisions and takes appropriate control actions to maintain vehicle motion on the roadway, and navigates to an intended destination. Vision is essential for driving. It is estimated that during driving, a driver receives over 90% of the input from his eyes. Therefore, in this chapter we will begin with understanding the structure of the human eye and the capabilities of the human visual system (visual capabilities are also called human visual functions). The acquired visual information is sent to the brain and the brain processes the information along with the information stored in the memory to make numerous decisions.

It is important to understand that most driver failures occur due to failure to obtain the necessary information in the right amounts at the right time and the right place. When a driver is asked to describe how he/she got involved in an accident, the examples of common responses are: "I did not see the target" (a pedestrian, a car, a curve, a sign, etc.), or "I did not realize that the other vehicle was approaching so fast, or I misunderstood the situation". Thus, the vehicle designer should constantly think about designing the vehicle to reduce the chances of driver information processing failures and errors.

On many occasions, the driver may find that he/she has too many tasks to do within a very short time interval due to traffic, roadway situations, state of his/her vehicle, and/or other non-driving tasks or distractions (e.g., answering a cell phone). To understand the various demands placed on the driver and how the driver should prioritize, and time-sharing between different tasks are also areas of great importance to the vehicle designers.

IMPORTANCE OF TIME

Understanding the amount of time that the driver needs to perform different tasks is probably the most important concept in designing the driver-vehicle interface. Most drivers take about 0.5 to 1.2 s to read speed from an analog speedometer with a

DOI: 10.1201/9781003485582-6

moving pointer on a fixed scale (Rockwell et al., 1973). To view the objects in a driver's side view mirror, most drivers will make about 0.8 to 2.0 s glances. In operating more complex devices such as radios and climate controls, the drivers typically make 2 to 4 glances; and each glance takes about 1.0 s in performing tasks such as selecting a radio station, changing temperature or fan speed (Jackson et al., 2002; Bhise, 2002).

A vehicle traveling on the highway at 100 km/h (62 mph) is equivalent to traveling 28 m (90 feet) per second. Thus, when a driver takes time away from the forward scene to make a 1 s glance, the vehicle travels 28 m on the roadway. If the driver takes a glance for more than 2.5 s time away from the roadway to perform other tasks, the driver will have difficulty in maintaining his/her vehicle within the lane. And if the driver takes more than 4.0 s away from the road, he/she is almost guaranteed to drift outside the driving lane (Senders et al., 1966).

Thus, it is important to design equipment inside the vehicle that drivers can use with glances no longer than about 1.5 s, and the total number of glances away from the road should be as few as possible. The time that the driver takes to perform a task depends upon the complexity of the task (e.g., number of items to search and read from a display, number of sequential actions or steps to perform, number of decisions to make, number of hand and/or finger motions to make to operate a control, etc.) and the capabilities of the driver to obtain the necessary information, make decisions, and to execute the necessary responses.

The simplest model of human information processing is based on a series of four steps that involve: (1) acquiring information available from the sensors (e.g., primarily visual receptors inside the eyes); (2) processing the sensed information to understand the situation; (3) selecting what to do (i.e., selecting a response); and (4) executing the response. To perform each of the steps, the driver will need time, and the total time taken to complete all the above four steps will thus depend upon the complexity associated with each of the steps. The four-step model and other information processing issues are covered later in this chapter.

UNDERSTANDING DRIVER VISION CONSIDERATIONS

Since drivers obtain most of their information visually, we will begin with the structure of the human eye. The structure of the eye will provide a basic understanding of the visual information that can be available for processing.

STRUCTURE OF THE HUMAN EYE

The human eye is somewhat like a camera. It has a lens, an adjustable iris (diaphragm with an adjustable diameter aperture in the middle to allow the light inside), and a surface where the image is formed. The sensor surface is called the retina in the eye (similar to the film/image surface in a camera. See Figure 6.1).

The retina contains photo-sensitive receptors. The receptors are of two types: cones and rods. The rod and cone receptors have different sensitivity to light. The cone receptors are sensitive under daytime lighting conditions (called photopic vision),

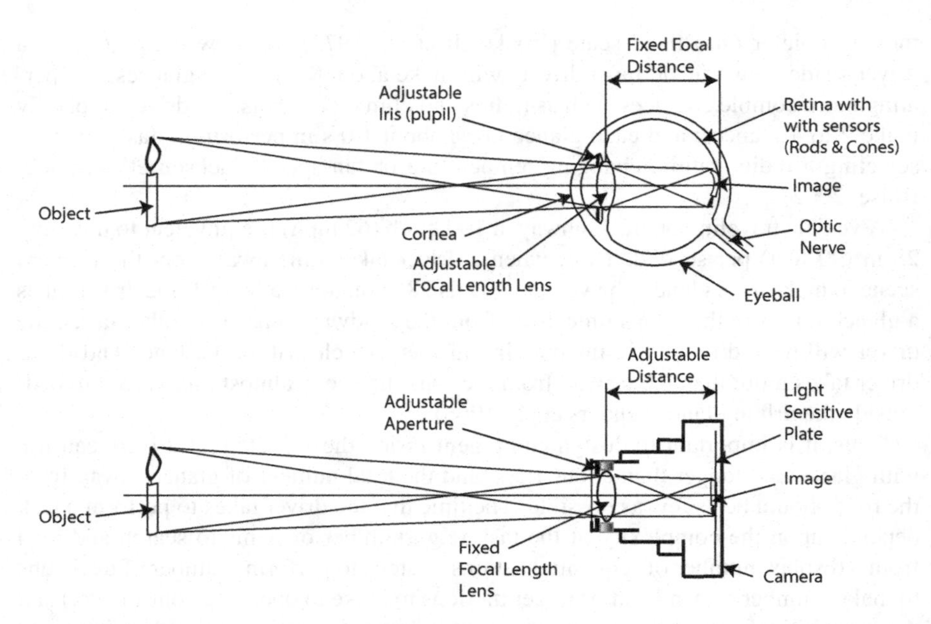

FIGURE 6.1 Comparison of the human eye with a camera.

and they also provide color vision. The rod receptors provide vision under dark visual conditions called the scotopic vision. The mesopic vision is when both the rods and cones are active, that is, when the day and night visions overlap under dusk or dawn, and under most night driving conditions with low beam headlamps. Under very dark scotopic conditions, vision is only possible with the rod receptors. Under scotopic vision with the rod receptors, a human cannot perceive any color and the vision consists of different shades of grey from white to black. Thus, the photopic, mesopic and scotopic vision are defined in terms of level of luminance (physical brightness measured in cd/m^2) as follows (Konz and Johnson, 2004):

Photopic Vision: Cones only – daytime vision: above 3 cd/m^2
Mesopic Vision: Both cones and rods function: 10^{-3} to 3 cd/m^2
Scotopic Vision: Rods only – nighttime vision: 10^{-6} to 10^{-3} cd/m^2

There are about 7 million cones in the fovea in the center of the visual field. The center of the visual field is defined by the visual axis of the eye, and the fovea is a region of about 1.0 to 1.8 degrees centered at the visual axis. The fovea is packed with all cone receptors, and it covers a field approximately the size of a penny held at an arm's length. There are about 125 million rods scattered over the entire area of the retina except for the fovea (which does not contain any rods). The distribution of the rods and cones are presented in Figure 6.2.

It is important to note that the density of the cones in the center (at the fovea) is the highest, about 140,000 cones per sq.mm of the retinal surface. The cone density drops

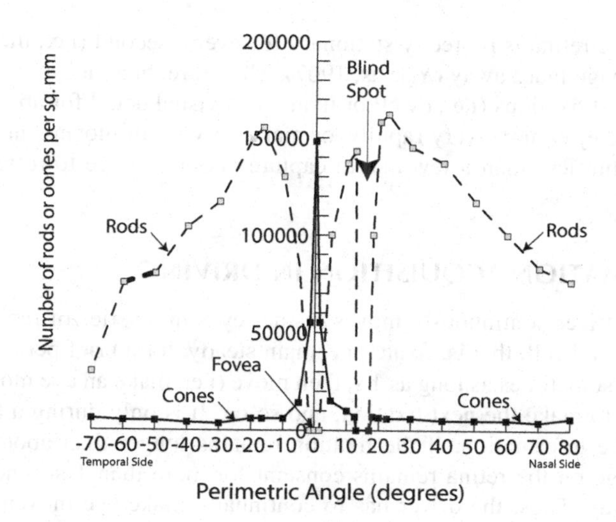

FIGURE 6.2 Distribution of ro and cone receptors in the human eye.
Source: Redrawn from: Boff and Lincoln, 1988.

rapidly as the angular distance from the visual axis (called the perimetric angle or the eccentricity angle) increases (see Figure 6.2). At 10 degrees and 30 degree perimetric angles, the cone density is about 10,000 and 5,000 cones per sq.mm, respectively. The visual ability to distinguish small details (called the visual acuity) is directly proportional to the cone density. Thus, to see a small detail, the driver would need to aim his/her eyes (i.e., to point the visual axis of each eye) at the detail so that the image of the detail falls on the foveas. Figure 6.2 also shows that in the outer regions of the visual field (called the peripheral visual field), there are relatively too few cones (about 5000 cones/sq.mm) in the peripheral field. Therefore, in the peripheral visual field, the drivers can only detect large objects and have overall awareness of the objects but cannot see finer details.

There are different types of cone receptors that respond to different wavelengths of light. The cones are specialized in detecting red, green and blue colors. Thus, there are (a) red-sensitive cones; (b) green-sensitive cones; and (c) blue sensitive cones. The color vision characteristics and color perception are thus related to relative sensitivities of the cones and also to the presence or absence (in color deficient persons) of the above three types of cones.

The very high concentration of cones (about 140,000 per sq.mm along the visual axis) in the foveal region provides high visual resolution capability (i.e., provides clearest/sharpest vision) and also higher speed of information transfer to the brain during photopic and mesopic ambient lighting conditions. Thus, in order to see small visual details (e.g., reading signs or graphic elements in displays) the driver will move his/her eyes and fixate (i.e., aim axes of both eyes) on the visual detail and dwell at that location for a short period, typically about 100–300 ms.

If an image on the retina is perfectly stationary for over a second (i.e., the image is stabilized), the image fades away (Yarbus, 1967). Therefore, human eyes normally make a series of rapid fixations (i.e., dwell or fixate on a visual detail for about 200–300 ms and then the eyes move very rapidly, or make very small movements called micro-saccades within less than a few ms, to capture another image to refresh and continue vision).

VISUAL INFORMATION ACQUISITION IN DRIVING

While driving, the driver continuously moves his/her eyes in a series of rapid jerky movements. The eyes dwell, that is, fixate or remain steady, for a brief period typically 0.2 to 0.3 s and sometimes as long as 1 s, then move (i.e., make an eye movement called the saccade) to make the next fixation, and so on. It is only during a fixation that the eyes receive information. If the fixation remains longer than about 1–3 s (i.e., when the image on the retina remains constant for more than 1 second), then the image fades away. Thus, the driver has to continually make eye movements to refresh or to maintain vision and also to gather information from different parts of the visual fields. Many of the fixations made to maintain vision are involuntary and unconscious. The eye movements made on the driving scene can be measured by an eye movement measurement system (generally called an eye-marker system).

The speed of eye travel between fixations is very fast – from about 200 degrees/s for a 5-degree eye movement to about 450 degrees/s for about 20-degree eye movement (Yarbus, 1967). The eye movement time T (in seconds) for an eye movement of magnitude α (degrees) is estimated to be $T = 0.021\ \alpha^{2/5}$. Thus, 5 degree and 20 degree eye movements take about 40 ms and 70 ms, respectively (Yarbus, 1967).

The fovea, because of greater temporal and spatial resolution, can also acquire information faster than in the peripheral parts of the visual field. Thus, during driving the driver moves his/her line of sight or the visual axes toward details in the road scene that he/she needs to recognize or use as a reference while obtaining information from extra-foveal (outside the fovea) regions.

ACCOMMODATION

The lens in the eye can change its power (i.e., its convexity) to focus a sharp image on the retina and to clearly see objects located at different distances. The process of changing focusing power is called the "accommodation". A person with a normal range of accommodation can focus on objects located at a long distance away (far distance, i.e., infinity) to objects as close as about 90 mm (near distance). The human accommodation ability decreases with age due to hardening of the lens. After about 45 years of age, the closest (near) distance at which a person can focus (or clearly see small details) increases beyond his hand reach distance (beyond about 800 mm). Thus, for an older person to see objects at closer distances, reading glasses are required. Reading glasses are generally made to allow the reader to see objects (primarily reading material) at near distances of about 350–400 mm. Bi-focal lenses are generally designed so that the person can read at closer distances from the lower part of the bi-focal lenses and the upper part of the lenses (with lower power) are used for far distance viewing.

Therefore, it is important to realize that if a display is provided at a viewing distance greater than the near reading distance (about 400 mm), older people will not be able to clearly see the display with the lower part of their bi-focal lenses. The traditional instrument clusters (i.e., speedometer and other gauges) are generally located at about 800–900 mm from the driver's eyes; and they cannot be clearly viewed from the lower or the upper parts of the bi-focals unless the driver uses tri-focal or continuously changing focal distance lenses. The header-mounted displays can especially cause reading difficulties to older drivers, because to read the displays located higher and at close distances through the lower parts of their bi-focal lenses, the drivers will need to tilt their heads up by a large angle. The convex mirrors located within hand reach distances are also difficult for older drivers to use because the images of far away objects in the convex mirror are located (i.e., focused) very close to the mirror. The head-up displays (HUD) are designed to focus display images at larger distances (beyond the front bumper of the vehicle) and are, therefore, useful for older drivers.

INFORMATION PROCESSING

The visual images available to the driver from his/her eyes are processed by the brain. During the information processing, relevant details from the visual images are extracted (i.e., recognized and placed in the short term memory) and used along with the information retrieved from the long term memory (stored information from learning, practice and experience) to interpret the present situation and a decision is made on whether a response action is needed at that time. If a response is needed, then the information is further processed to decide on what to do (i.e., to select a response). Based on the selected response, signals are sent to appropriate muscles to generate movements to execute the response action (e.g., a hand movement to operate a control).

SOME INFORMATION PROCESSING ISSUES AND CONSIDERATIONS

In the cognitive sciences, a number of studies have been conducted to understand how humans process information and make decisions (Kantowitz and Sorkin, 1983; Sanders and McCormick, 1993; Wickens et al., 1997). Some of the challenging questions and issues studied are presented below:

a) Is the human a single or a parallel-channel information processor? Can a human process information coming from different sources or sensors simultaneously or is the information queued up for a single processor to handle, one chunk at a time? In the early stages of learning or when a driver is under stress, most of the conscious information processing will be performed under the single channel mode. As the driver learns and acquires skills, many simpler tasks can be performed simultaneously.

b) What is a human's channel capacity? At what rate can information be processed? Hick's Law and Hick's-Hyman Law described in the next section allow us to estimate human information processing rates.
c) Why is it important to measure reaction times? The duration of the reaction time allows us to determine the amount of information processed by the brain to decide on an output action.
d) What is the role of attention in information processing? If the driver is not attentive or not paying attention, the information available at the sensors (or the sensory level) may not be processed by the brain. Thus, an inattentive or distracted driver will fail to understand what is going on at a given moment.
e) How does a human operator time-share between tasks? Normally, the driver performs many tasks while driving and shares resources (e.g., sensors located in different parts of visual fields or different sense modalities) between different tasks (e.g., driving and listening to the radio or talking to other passengers). Due to limitations in capabilities of the information processing resources, the driver generally attends to the important tasks and disregards other lesser priority tasks.
f) How does the driver select what he should attend to? The driver must be processing information available from different sources to decide which resource to concentrate on and what to disregard.
g) How did I get to a destination while I was preoccupied with something else? While the brain is actively processing some other information, the driver's pre-attentive (or unconscious) information processing activities (if they are highly learned) can take over and perform the driving tasks without the driver being fully aware of the situation.

To understand human information processing capabilities, we will review a few simple but important human information processing models and concepts in the next sections.

HUMAN INFORMATION PROCESSING MODELS

a) Basic Four-stage Model of Human Information Processing

A simple four stage serial information processing model is presented in Figure 6.3. The model shows that the information from an input stimulus is processed through four stages, namely: (1) stimulus encoding and decoding; (2) central processing; (3) response selection; and (4) response execution. The model shows that an input stimulus (or an event) in the first stage is received at the sensor (e.g., eye). For example, the stimulus may be a "traffic light" which turns "red". Some relevant features or cues (the traffic signal device and its signal color) from the image are sent to the

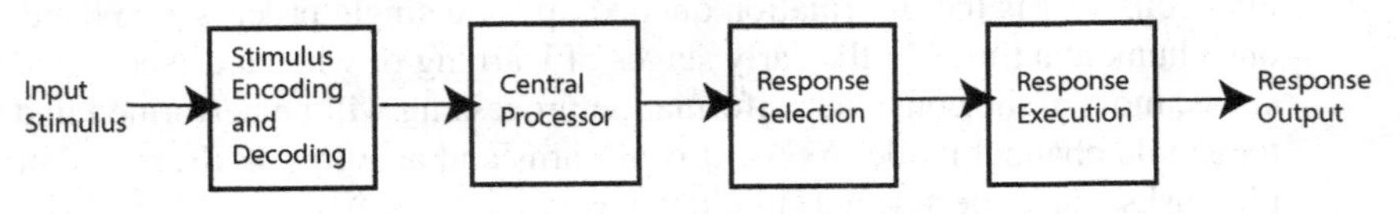

FIGURE 6.3 Four stage serial information processing model.

central processor in the brain, which processes the information further to interpret the stimulus (i.e., decode the meaning of red-signal color), and receiver understands the situation (i.e., the "red" traffic signal will require the driver to stop). The information about the situation is further processed in the third stage to decide on the response to be selected (e.g., the driver decides to stop the vehicle instead of going through the intersection). And finally in the fourth stage, the response is executed (i.e., information or instructions are sent to the appropriate muscles to make movements such as moving the right foot from the gas pedal to the brake pedal to stop the vehicle).

The above four stage model shows that the information is processed serially (i.e., the processing in the sequence of four stages must occur before the response output occurs). Further, the processing of information in a preceding stage must be completed before the next stage can begin processing. This model, thus, suggests that the response time to react to the input stimulus (i.e., time taken between the occurrence of the stimulus and when the human operator initiates an output or a response movement) will be equal to the sum of the time taken by each of the four stages described above.

The concept that information is processed in a series of stages was supported by the following two simple experiments reported in the early information processing literature: Donders's Subtraction Principle, and Psychological Refractory Period.

b) Donders's Subtraction Principle

The principle is based on the additive (and subtractive) nature of the components of reaction time. In order to prove that information is processed in a series of steps (or in a serial manner), Donders, a Dutch physiologist, conducted three reaction time experiments (Kantowitz and Sorkin, 1983). The three experiments, called Experiment A, B and C, are illustrated in Figure 6.4.

In Experiment A, Donders provided a stimulus light (S_1) and a response key (R_1). He asked his subjects to press the response key R_1 as soon as the stimulus light S_1 was turned on. The response time measured, between when the stimulus S_1 came on

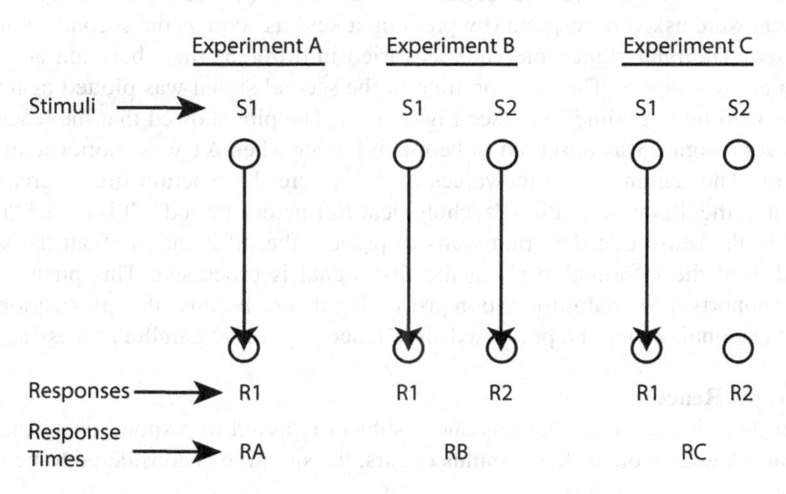

FIGURE 6.4 Donders' three experiments.

and when the response R_1 key was pressed, was called "RA". The situation in this experiment is the simplest (i.e., there is only one stimulus and one response). Thus, the reaction time in such a situation is called the "simple reaction time".

In Experiment B, he provided two stimulus lights (S_1 and S_2) and two response keys (R_1 and R_2). He asked his subjects to respond as quickly as possible, and press the response key $R_{1,}$ if the stimulus light S_1 came on, and press the key $R_{2,}$ if the stimulus light S_2 came on. The response time measured between the occurrence of a stimulus (S_1 or S_2) and the response (R_1 or R_2) was called "RB". The situation in this experiment is more complex as after a stimulus occurs, the subject needs to detect and recognize which stimulus has occurred and also needs to select which key to press. Thus, the reaction time "RB" will be larger than "RA" because the subject makes two additional decisions.

In Experiment C, he provided two stimulus lights (S_1 and S_2) and two response keys ($R_1$and R_2). He asked his subjects to respond as quickly as possible and press the response key R_1 only if the stimulus light S_1 came on; and do nothing if the light S_2 came on. The response time measured between when the occurrence of a stimulus and the correct response was called "RC". In this situation, the subject needed to recognize the stimulus and press the key R_1 if the S_1 came on.

From subsequent analysis of the reaction-time data, Donders found that: RB> RC> RA. This result proved that information processing occurred in a sequential manner because of the following additive nature of the information processing time.

RA = Simple Reaction Time
RB = RA + Stimulus Recognition Time + Response Selection Time
RC = RA + Stimulus Recognition Time

c) Psychological Refractory Period

In another set of reaction-time experiments (Welford, 1952; Fitts and Posner, 1967), subjects were asked to observe two successive signals which were presented within a short period (called the inter-signal interval = ΔT) between the two signals. The subjects were asked to respond (by pressing a key) as soon as the second signal was detected. The inter-signal interval was varied in different trials between about 100 ms to a few seconds. The reaction time to the second signal was plotted against the inter-signal interval time (ΔT) (see Figure 6.5). The plot showed that the reaction to the second signal was observed to be much longer when ΔT was shorter than about 500 ms. The region of shorter values of ΔT, where the reaction time increased, is known in the literature as the "Psychological Refractory Period". It is called "refractory" in the sense that the brain waits to process the information about the second signal until the information about the first signal is processed. This phenomenon, thus, supports the serial information processing theory because the information from both the signals cannot be processed simultaneously (or by parallel processing).

d) Simple Reaction Time

When there is only one stimulus and a subject is asked to respond using only one response key as soon as that stimulus occurs, the situation is considered to be simple

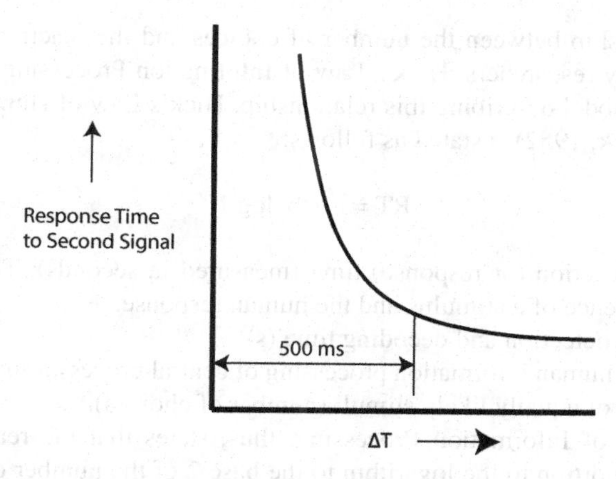

FIGURE 6.5 Reaction time to second signal as a function inter-signal interval (ΔT).

(like Experiment A above). The durations of simple reaction time typically range between about 100 to 300 ms for fully alerted subjects.

Simple reaction time (minimal reaction time) includes the following

a) Receptor delay of about 1 to 38 ms (depending on type of receptor, e.g., mechanical receptors in the inner ear respond more quickly than receptors in the retina as they work on photochemical reactions).
b) Afferent transmission delay of about 2 to 100 ms (The delay time depends upon the signal transmission distance between the sensor and the brain.)
c) Central processor delay of about 70 to 100 ms
d) Efferent transmission delay of about 10 to 20 ms (for signal transmission from the brain to the muscle executing the response)
e) Muscle latency and activation delay of about 30 to 70 ms.

The total of the above times is 113 to 328 ms. If we assume that a driver is traveling at 100 km/h (28 m/s), then during a simple reaction time of about 300 ms, the vehicle will travel about 8.4 m. In addition to the above simple reaction time, more time is generally required to complete a response action, such as moving a hand to press a button or moving a foot to operate a pedal.

e) Choice Reaction Time

When a driver is in a situation where he/she has a choice to select a response among several possible responses and make the selected response, then the situation is called the choice reaction time situation. And the time taken from the instant when the response was requested to the instant when the selected response is made is called the choice reaction time. For example, while approaching an intersection and when the signal light turns yellow, the driver has the following three choices: (a) to decelerate and stop before the intersection; (b) to continue driving without changing speed; or (c) to accelerate. Thus, the time taken by the driver to decide on an action after the signal turned yellow can be called the choice reaction time for the situation.

The relationship between the number of choices and the reaction time has been studied by many researchers. Hick's Law of Information Processing is probably the most popular model describing this relationship. Hick's Law of Human Information Processing (Hick, 1952) is stated as follows:

$$RT = a + b.\ \log_2 N$$

Where, RT = reaction (or response) time (measured in seconds). The time elapsed between occurrence of a stimulus and the human response.

a = stimulus detection and decoding time (s)

1/b = rate of human information processing of central processor measured in bits/s.

N = number of equally likely stimuli (number of choices).

Hick's Law of Information Processing, thus, states that the reaction time will increase in proportion to the logarithm to the base 2 of the number of equally likely choices (N) available to the human operator. Hick developed the logarithm to the base 2 transformation of N based on the concept of information developed by Shannon and Weaver (1949). According to this concept, the information (H) measured in "Bits" is a measure of uncertainty. The value of information is directly related to the amount of uncertainty reduced. Thus, the amount of information (H) in a choice reaction situation that a decision maker processes by associating to an event (choice) among a set of N equally likely events (i.e., each of the N events occur with equal probability of p = 1/N) can be defined as follows:

$$H = \log_2(P_A/P_B)$$

Where,

P_A = Probability at the receiver of the event *after* the information is received
P_B = Probability at the receiver of the event *before* the information is received

In the case when all the events are equally likely, $H = \log_2(1/p) = \log_2 N$

It should be noted that the numerator (i.e., P_A) in the above ratio of probabilities is 1.0. This means that after the information is received, the subject is certain about what has occurred. Thus the probability of what event (or choice) has occurred is 1.0. The probability of the occurrence of the event before the information was received (P_B) was p = 1/N (as the receiver considered the event to be equally likely among the possible N events).

An Excel-based computer program to measure reaction times in choice reaction time situations ranging from N = 1 to 8 is available on the publisher's website (see Appendix 3). The task involves the subject pressing the number key on the computer keyboard corresponding to the number (stimulus) displayed on the screen as soon as the number appears on the screen. The number is randomly selected by the program within a selected set (choices) of numbers (N). The reader can use the program and plot his/her data (reaction times on the y-axis and $\log_2 N$ values on the x-axis) to verify

Hick's law by fitting a straight line to the data. The value of the intercept of the fitted line on the y-axis represents 'a' and 'b' is the slope of the line. The values of the constants 'a' and 'b' in the above Hick's law expression typically vary between 0.1 to 0.5 s and 0.1 to 0.25 s/bit, respectively, for alerted and practiced subjects.

The assumption of all choices being equally likely is applicable to situations like rolling an unbiased six-sided dice or a two-sided coin. In the real world, the possible choices have different probabilities. For example while approaching an intersection a driver may have three choices: do nothing (continue driving at the same speed), brake or accelerate. The probabilities of the three choices will be different and will depend upon other variables such as the distance of the vehicle from the intersection, the traffic signal color and the driver's urgency to reach his intended destination.

Thus, if we assume that there are N events and an ith event occurs with probability p_i such that the sum of all the probabilities, p_1 to p_N is equal to 1.0, then the information in the situation can be computed as follows:

$$H = \sum_{i=1}^{N} p_i \log_2 (1/p_i)$$

Thus, taking into account the unequal probabilities of the N possible choice events, Hick's law was modified by Hyman in 1953. The modified law is known in the literature as Hick-Hyman Law (Hyman, 1953). Hick-Hyman law is stated as follows:

$$RT = a + b. \sum_{i=1}^{N} p_i \log_2 (1/ p_i)$$

In order to get a feel of the amount of information in different situations, Table 6.1 shows calculations of information under two situations: a) Situation 1: when all the events are equally likely, and b) Situation 2: when one of the N events occurs with probability equal to 0.9 (i.e., it occurs 90% of the time) and the remaining N-1 events are equally likely, that is, each of the N-1 events occurs with the probability (1.0-0.9)/(N-1) = 0.1/(N-1). The amount of information in Situations 1 and 2 are called H_1 and H_2, respectively.

Table 6.1 shows the values of H_1 and H_2 for N ranging from 1 to 16. For any row of the table with N>1, the value of H_2 is substantially lower than the value of H_1. Thus, the choice reaction time in Situation 2 will be shorter than in Situation 1 (assuming the same values of the constants "a" and "b" in the above expression). The data in Table 6.1 also illustrates that in most decision making situations, experienced people (experts) will have shorter reaction times than inexperienced people. Because for an experienced individual, all the possible number of choices are not equally likely, and due to his/her experience, he/she will select the most likely choice (i.e., a choice with the highest probability of occurrence) which will result in a lower reaction time than a novice who will treat all choices to be equally likely.

TABLE 6.1
Computation of Information in Two Situations

	Situation 1 using Hicks Law		Situation 2 using Hick-Hyman Law		
N	P = 1/N	$H_1 = \log_2 1/p$	P_1	p_i for i >1	H_2
1	1.000	0.000	1.0	0.000	0.000
2	0.500	1.000	0.9	0.100	0.469
3	0.333	1.585	0.9	0.050	0.569
4	0.250	2.000	0.9	0.033	0.627
5	0.200	2.322	0.9	0.025	0.669
6	0.167	2.585	0.9	0.020	0.701
7	0.143	2.807	0.9	0.017	0.727
8	0.125	3.000	0.9	0.014	0.750
9	0.111	3.170	0.9	0.013	0.769
10	0.100	3.322	0.9	0.011	0.786
11	0.091	3.459	0.9	0.010	0.801
12	0.083	3.585	0.9	0.009	0.815
13	0.077	3.700	0.9	0.008	0.827
14	0.071	3.807	0.9	0.008	0.839
15	0.067	3.907	0.9	0.007	0.850
16	0.063	4.000	0.9	0.007	0.860

Notes:
H_1 = Information when all outcomes are equally likely
H2 = Information when first event occurs with P_i = 0.9 and all other events (i=2 to N) occur with p = (1-.9)/(N-1)
= 0.9 $\log_2(0.9)$ + (N-i) p x log2 *1/p*

f) Factors Affecting Reaction Time

Human reaction time is influenced by many other factors (which also affect the constants "a" and "b" in the above expressions). A partial list including some of the more commonly known factors is given below:

1. Type of sensor or sense modality (e.g., mechanical sensors in human ear have shorter delay times than photochemical sensors in the human eye)
2. Stimulus discriminability or conspicuity with respect to the background or other stimuli (e.g., signal-to-noise ratio or uncluttered background)
3. Number of features, complexity and size of feature elements in the stimuli
4. Amount of search the human operator conducts (size of search set)
5. Amount of information processed (uncertainty, number of choices and their occurrence probability)
6. Amount of memory search
7. Stimulus-response compatibility (e.g., how similar is the mapping or association of the stimuli to the configuration of responses)

8. Alertness of the subject
9. Motivation of the subject
10. Expectancy (how expected – or known from past experience is the event, in terms of when and where it could occur)
11. Mental workload (other tasks that the subject is time-sharing simultaneously)
12. Psychological stress (e.g., emotional state of the subject)
13. Physiological stress (e.g., tired, fatigued, or in environment affecting bodily functions)
14. Practice (how familiar or skilled is the subject to the situation)
15. Subject's age (older subjects are usually slower and more variable)

g) Fitts's Law of Hand Motion

After learning about Hick's law, Fitts developed the law of human movements which is known as Fitts's Law (Fitts, 1954; Fitts and Posner, 1967). Fitts's Law states that the amount of time taken to make a movement to reach a target (usually a hand movement with visual feedback, e.g., to move hand from the steering wheel to touch a button in a radio) is proportional to a variable called the index of difficulty (ID). Fitts's Law is stated as follows:

$$MT = c + [d \times ID]$$

Where, MT = movement time (measured in seconds). It is the time taken to reach to a target of width "W" located at distance "A".

c = muscle preparation time (s)

d = constant (equivalent to the inverse of rate of human information processing of central processor)

ID = Index of difficulty = $\log_2(2A/W)$

A = movement distance

W = target width (denotes accuracy of movement)

Thus, to reduce the time taken to make a movement, the ratio of A/W must be reduced. For example, to reduce the difficulty and time taken to reach to a button (target) on a car radio, the button can be located close to the location from where the driver's right hand is coming from (e.g., 2 o'clock position on the steering wheel) or the button can be made larger or both of the above can be altered to reduce the ratio of A/W.

The Fitts's law can also be easily verified by using targets and hand movement procedure presented in Appendix 4.

HUMAN MEMORY

The human memory system can be considered to have three types of memory storage subsystems. Figure 6.6 shows how the three memory types of work in terms of the flow of information.

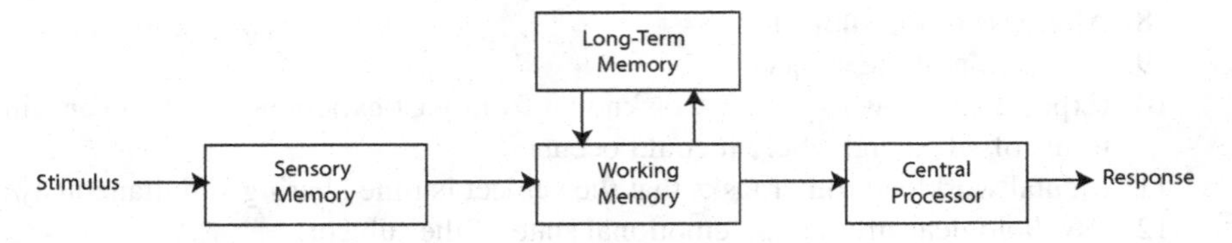

FIGURE 6.6 Three types of human memory subsystems.

Sensory Memory: The human sensory receptors (e.g., sensors in eyes, ears, skin or joints) can store sensed information for a brief period. The sensory memory subsystems that hold information from the eyes and ears are called iconic storage and echoic storage, respectively. The sensed information is stored for a brief time (about 1s for iconic memory and up to 3 to 5 s for the echoic memory – after which it must enter the working memory or be lost (Wickens et al., 1997).

Sensory memory can be experienced as follows. If a subject is shown a 3x3 matrix of nine numbers in a very short period of time (e.g., exposure of less than 100 ms – usually through a tachistoscope), and then asked to recall a number at a particular location in the matrix, the subject can usually recall the number correctly even about a second after the matrix is turned off. This suggests that the image of the matrix is available in the iconic storage and the information from the storage can be retrieved after the original stimulus is withdrawn. Similarly, after a verbal message is announced and when questioned immediately after the message, most people can retrieve (from their echoic memory) what they just heard within a few seconds after the original message was presented.

Working memory: Working memory is at the center of the human memory system. It is also called the short term memory. The information from the sensory subsystem is sent to the working memory. The information cannot be placed or retrieved from long term memory without passing through working memory.

Working memory has a limited capacity. About five to nine items (the magic number 7 plus or minus 2) can be stored in the working memory (Miller, 1956). Rehearsal is the most common control process to maintain information in the working memory. If rehearsal is stopped, the information can be lost from the working memory. Thus, it takes effort (e.g., continual refreshing) to maintain information in the working memory.

A demonstration of the limited capacity of the working memory can be easily given. For example, read a string of 10 one-digit numbers to a person, one at a time (so that the numbers are not connected or associated as two or more digit numbers), in a random order; and then after all the numbers are read, ask the person to recall the numbers (or write them down). It will be found that most people will not be able to recall all the numbers. But if only 5 one-digit numbers are presented in a random order, most people will correctly recall all the numbers. The working memory of most people can easily store 5 (i.e., 7 minus 2) digits (or items) but cannot store more than 9 (i.e., 7 plus 2) digits (or items) in the working memory.

These limitations of the working memory are especially important in designing equipment where people need to remember a number of items (such as lists, keys, preset buttons, components or functions) for a short period of time in their working memory while performing a task. If the working memory is overloaded, people will make errors or they may not be able to perform the task without other aids (e.g., labels, written sheets or visual screens).

Long Term Memory: The information in long term memory can be stored for a very long time, usually through the entire life span of an individual. In contrast with the working memory, no effort or capacity is required to maintain information in long term memory. Once the information has been transferred to long term memory from working memory, it is there forever (though it may be difficult to retrieve). No rehearsal is needed to maintain items in long term memory. Most memory models assume that information in the long term memory is coded (i.e., stored with specific flags or cues) according to certain "meaning". On occasions people find that it is difficult to remember where an item was stored, but once they remember the context (or the flag) under which the information was stored, the information can be retrieved very quickly even after many years of inactivity.

GENERIC MODEL OF HUMAN INFORMATION PROCESSING WITH THE THREE MEMORY SYSTEMS

An updated model of human information processing was provided by Wickens et al. (1998) (see Figure 6.7). This model includes the four-stage model of information processing along with the working and long-term memory used between the middle

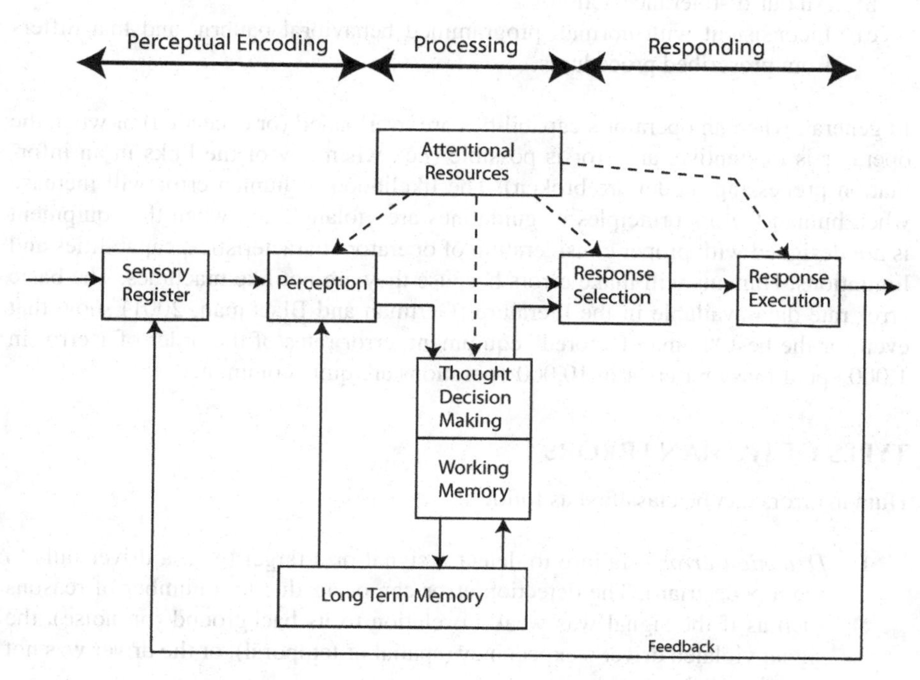

FIGURE 6.7 A more complete model of human information processing model.

Source: Redrawn from Wickens et al., 1998.

two stages involving perception and response selection. It also shows that attention is required for the last three stages, namely, perception, response selection and response execution. The model also includes a feedback loop to receive information about the executed response into the sensory register.

HUMAN ERRORS

Human operators will make errors in operating equipment due to many causes. We are interested here in understanding how a vehicle can be designed to minimize the possibility of driver errors. Most human errors occur due to information processing failures. Treat (1980) studied over 13,500 police-reported accidents and found that in over 70% of the accidents, human errors were identified as the definite causes of accidents. The definite causes were defined as the accidents in which the in-depth accident investigating multi-disciplinary team members were over 95 percent confident that the accidents were caused by human factors (e.g., recognition errors, decision errors, and commission errors).

DEFINITION OF AN ERROR

a) An act involving an unintentional deviation from the truth or accuracy (*Webster's Dictionary*, 1980).
b) An out-of-tolerance action.
c) Inconsistent with normal, programmed behavioral pattern, and that differs from prescribed procedures.

In general, when an operator's capabilities are overloaded (or exceeded) or when the operator is inattentive, an error is possible (i.e., when any of the links in an information processing model are broken). The likelihood of human error will increase when human factors principles or guidelines are violated, i.e., when the equipment is not designed with proper consideration of operator characteristics, capabilities and limitations. Humans will make errors because they are not like machines. The basic error rate data available in the literature (Gertman and Blackman, 2001) show that even for the best "human factored" equipment, error rates of the order of 1 error in 1,000 operations or 1 error in 10,000 operations are quite common.

TYPES OF HUMAN ERRORS

Human errors can be classified as follows:

1. *Detection error* – failure to detect a signal or a target (e.g., a driver fails to see a pedestrian). The detection error can occur due to a number of reasons such as if the signal was weak in relation to its background (on noise), the signal violated driver's expectancy (spatial or temporal), or the driver was not attentive or distracted.

2. *Discrimination error* – failure to discriminate between signals (e.g., stop lamps were perceived as tail lamps).
3. *Interpretation error* – failure to recognize a situation, a signal, a hazard, or a scale (e.g., a tachometer reading was interpreted as the speed reading).
4. *Omission error* – failure (or forgetting) to perform a required action (e.g., forgetting to look in the side view mirror before changing lane).
5. *Commission error* – performing a function that should not have been performed. A commission error can involve addition of an extraneous act, performing a task or a step out of sequence or not being able to perform a task in allotted time. The three commission errors are as follows:
   a) *Extraneous act error* – introducing a step or a task that should not have been performed (e.g., changed radio station while turning on the windshield defrost function; an inadvertent switch actuation due to hitting of body parts).
   b) *Sequential error* – performing a step or a task out of sequence (e.g., changed radio station band after selecting a radio preset station).
   c) *Time error* – failing to perform within an allotted time or excessive time spent (e.g., long looks, more than a single look while using a control or a display).
6. *Substitution error* – using or substituting another item (control or display) instead of the desired one (e.g., pressed the accelerator pedal instead of the brake pedal).
7. *Reversal error* – reversing the direction of activation or interpreting a displayed signal in opposite direction (e.g., increased temperature instead of decreasing).
8. *Inadequate response error* – error in judgment or estimation of signal magnitude, e.g., distance, speed, or insufficient movement or insufficient force applied during control activation (e.g., insufficient brake pedal movement during stopping).
9. *Legibility error* – error related to not being able to read a display (due to factors such as small font size, insufficient light, excessive glare, and parallax in reading an analog display).
10. *Recovered error* – an error has occurred, but the operator could correct the error after some elapsed time. (Note: Many errors made by human operators are recovered, corrected, and thus undesired events for example, accidents, are avoided).
11. *Unrecovered error* – an error that the operator fails to correct (or will not or cannot correct).

UNDERSTANDING HUMAN ERRORS WITH THE SORE MODEL

Many different models have been developed to predict human errors (Leiden et al., 2001). One simple model called the SORE model is described here. The SORE model stands for Stimulus, Operator, Response and Environment (see Figure 6.8). The model shows that the operator responds after detecting a stimulus and the response is

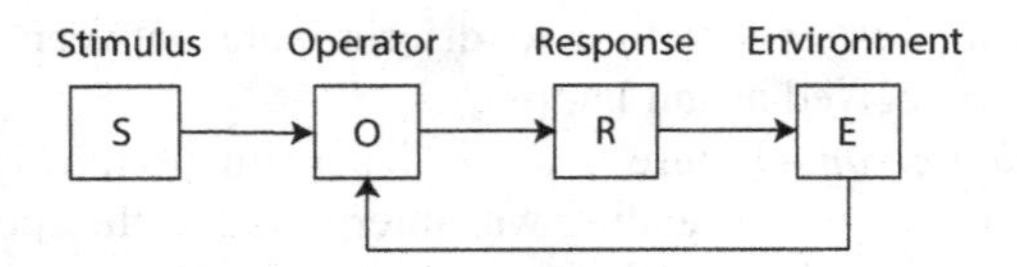

FIGURE 6.8 The SORE model for human error analysis.

seen by the operator in the environment as feedback. In general, when an operator's capabilities are overloaded or exceeded, an error is possible (i.e., when any of the links in the SORE information processing model are broken). For example, an error could occur if: a) S-O link is broken (i.e., stimulus is not detected by the operator), b) O-R link is broken (e.g., the operator does not understand the stimulus thus cannot make a response), c) R-E link is broken (e.g., the response is not seen in the environment), or d) E-O link is broken (e.g., the operator does not receive feedback from the activation of a control).

PSYCHOPHYSICS

Psychophysics is a science that relates the physical characteristics of a stimulus (e.g., brightness of a visual stimulus) to psychological response (i.e., how a person perceives and reacts to the stimulus). Examples of psychophysical problems are: a) detection of a signal (e.g., visibility of a visual target or a change in the pointer position in a gauge), b) just noticeable difference between two stimuli (e.g., difference in brightness between the two tail lamps on the rear end of a vehicle), c) equality (when the two signals are perceptually equal in magnitude, e.g., determining if two stop lamps on the rear end of a car are equally bright), d) magnitude estimation (e.g., estimating speed of a vehicle or distance between two vehicles), e) interval estimation (e.g., determining the time required to reach a particular location from a given vehicle speed), f) production of magnitude of a stimuli (e.g., pushing the brake pedal hard enough to produce required stopping distance), and g) rating on a scale (e.g., providing a rating on ease or difficulty using a 10-point scale, where 1 = very difficult and 10 = very easy).

It should be noted that any one of the above psychophysical problems involves information processing and decision-making. Thus, application of psychophysics will help in improving functionality and customer satisfaction of a product. (Note: Chapter 12 on Automotive Craftsmanship and Volume 2, Chapter 19 on Vehicle Evaluation Methods cover several applications on perception of quality and subjective scaling related to this topic).

VISUAL CAPABILITIES

Psychophysical methods are used to measure different visual capabilities. Some commonly considered visual capabilities of the drivers are:

1. Detection of objects (e.g., visual targets)

2. Perception and identification of colors
3. Differences (or change) in luminance
4. Equality of appearance (e.g., if two trim parts "match", that is, the appearance of the two trim parts is equal by considering their color, brightness and texture)
5. Difference (or change) in colors (color differentiation/tolerance)
6. Recognition of details or object recognition (e.g., reading letters or Snellen visual acuity)
7. Depth perception (i.e., ability to determine if two visual objects are at the same or different viewing distances)
8. Accommodation (i.e., ability of eyes to focus on objects at different viewing distances)

VISUAL CONTRAST THRESHOLDS

One of the basic human visual capabilities (or visual functions) is based on the ability of humans to perceive differences in luminance (physical brightness; measured in fL foot-Lambert or cd/m²). Perception of visual contrast is a basic function that is needed to recognize visual details and objects. For example, a circular object (target) can be seen against a background because of a difference in the luminance of the target against its background.

The visual contrast (C) of a target against its background is defined here as:

$$C = (L_t - L_b)/L_b$$

Where, L_t = Luminance of the target, and L_b = Luminance of the background

Figure 6.9 presents threshold contrast curves as functions of the adaptation luminance and target size. To compute if a target is detectable (i.e., visible), values of the following variables are needed: a) contrast (C) of the target against its background as shown above, b) adaptation luminance (the luminance level to which the observer's eyes are adapted), and c) target size (the angle subtended by the target at the observer's eyes computed in minutes). The target is considered to be visible if the value of the contrast (C) at the abscissa value of the adaptation luminance falls above a threshold contrast curve corresponding to the computed target size (interpolated curve for the target size) while looking for the target.

The visual contrast threshold curves developed originally by Blackwell (1952) and modeled by Bhise et al. (1977) are presented in Figure 6.9. The log values used in the abscissa and ordinate of the graph are to the base value of 10. It should be noted that the contrast threshold curves presented in Figure 6.9 were obtained by Blackwell under 1/30th of a second exposure. The 1/30th of second exposure duration is much smaller than the typical eye fixation durations (about 1/3 s). However, since Bhise et al. (1977) found that the higher contrast thresholds for 1/30th second (as compared to at 1/3rd s exposure) predicted visibility distances to stand-up targets more accurately under the more difficult actual dynamic driving conditions. Thus, Blackwell's contrast thresholds obtained using 1/30th of second were used for modeling the driver's night visibility of targets on the roadway and legibility of displays.

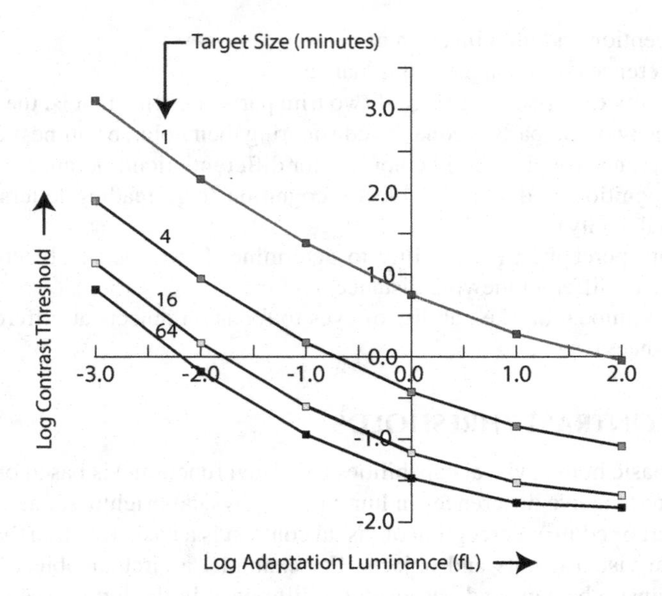

FIGURE 6.9 Visual contrast threshold curves for 1/30th sec target exposure used for driver visibility and legibility modeling.

The visibility and legibility computations using the above thresholds are covered in Chapter 14.

VISUAL ACUITY

Visual acuity is the ability of a person to recognize a visual detail. The most common reference to visual acuity is based on the ability to see and recognize letters. Visual acuity scores are described by ratios such as 20/20, 20/40, and 20/200. The ratio is based on two distances. The numerator denotes the distance in feet from which the smallest size of letters on a visual acuity chart can be correctly read by the subject. The denominator denotes the maximum distance in feet from which the same row of letters can be correctly read by a person with normal vision. Thus, 20/40 means that the subject could read the smallest letters from 20 feet or 6.1 m (this reference distance is set for far vision test in an optometrist's office) which a person with normal vision could have read correctly from 40 feet (12.2 m). The 20/20 means "normal vision", that is, the subject's visual acuity is as good as the person with normal vision. Whereas a score of 20/400 generally means "legally blind" as the person could read the letters from 20 feet (6.1 m) which a person with normal vision could read from a distance of 400 feet (122 m).

"Near visual acuity" is measured at a reference distance of 14 inches (0.36 m) (the shorter distance from which printed text is read by most people under most reading situations). For near acuity, the scores are also presented in ratios 20/20, 20/40, and so forth, like the far acuity scores but the numbers do not indicate actual viewing distances. However, the ratios do indicate the ability of the reader to read the letters

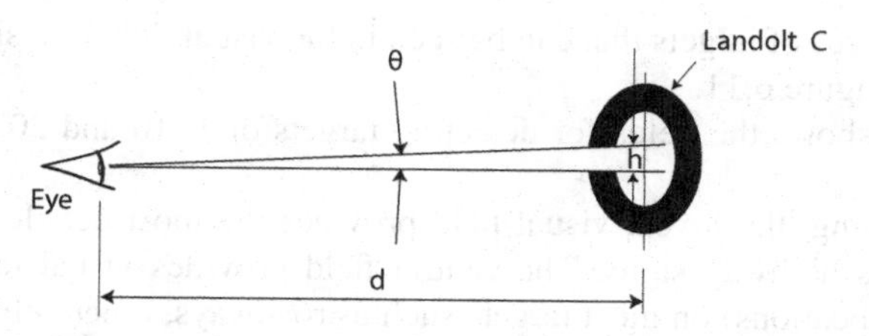

FIGURE 6.10 Visual acuity measurement.

Note: Defined by 1/angle measured in minutes subtended at the gap in the Landolt 'C' of width h =1/ θ at the observer's eye.

in relation to the distance from which a person with normal vision can read the letters located at near distances.

In a standard visual acuity test, six visual acuity tests are conducted under combinations of two distances and three viewing conditions. The two distances are: 14 inches (near) and 20 feet (far or 6 meters if using a metric test); and the three viewing conditions are: left eye only (right eye occluded), right eye only (left eye occluded), and both eyes open.

The minimum visual requirements for a driver's license in most states in the United States are as follows: a) minimum uncorrected (driving without glasses or contacts) far visual acuity score of 20/40 to qualify for an unrestricted driver's license, and b) minimum corrected (with glasses or contact lenses) far visual acuity of 20/50 to qualify for a restricted license (i.e., allowed to drive with corrective lenses).

For visual acuity tests, different targets have been used, for example, letters, checkerboard patterns, gratings, Landolt 'C' rings at various orientations, and alignment of two lines (Vernier acuity). Figure 6.10 illustrates a visual acuity test using the Landolt "C" as the test object where the critical visual detail is assumed to be the gap in the ring (of width "h"). In this test, the subject is shown Landolt rings (or letters) of different sizes (at near or far viewing distances). The visual acuity is defined by the smallest Landolt ring orientation or letter that the subject can correctly identify. The visual acuity is measured by the inverse of the angular size of the smallest visual detail correctly identified by the subject (shown as angle θ measured in minutes) in Figure 6.10. The visual acuity is, thus, equal to $1/([\tan^{-1} h/d)]/60)$. [Note: $\tan^{-1} h/d)$ is measured in degrees].

DRIVER'S VISUAL FIELDS

The normal human visual field with both eyes is called the ambinocular field and it extends about 95 degrees to the left and 95 degrees to the right, about 60 degrees upwards and about 75 degrees downward with respect to the forward line of sight (i.e., axes of both eyes pointed straight ahead). Due to the distribution of the cones (refer to Figure 6.2), the vision is most detailed in the foveal vision (about 1 deg radius). The visual field outside the fovea is called the extra-foveal vision. It can be approximately categorized as the central visual field (up to about 30 degrees eccentricity angle) and the peripheral visual field (beyond 30 degrees eccentricity angle).

The smallest sizes of targets that can be seen in the visual field of a single (right) eye are shown in Figure 6.11.

The figure shows the fields for detecting targets of 1, 10 and 20 minutes for the right eye.

During driving, the foveal visual field provides the most detailed vision to read details such as highway signs. The central field provides visual information (i.e., presence and locations) on most targets such as roadways, other vehicles, and traffic control devices. The peripheral vision provides awareness of larger targets (e.g., vehicles in the adjacent lanes) and provides information on moving targets and motion cues.

The minimum visual field requirements for a driver's license in most states in the United States are that a person with two functional eyes must have a field vision of 140 degrees (horizontal). A person with one functional eye must have a field of vision of 105 degrees (horizontal). Figure 6.12 shows the extent of the field of perception of different colors in the right eye. The size of the field varies depending on the color. It is largest for "Yellow" and "Blue" and smallest for "Green".

Figure 6.13 shows that, for both males and females, the total available visual field narrows with age. Thus, older drivers will not be able to detect targets in far peripheral fields as far outward as compared to younger drivers.

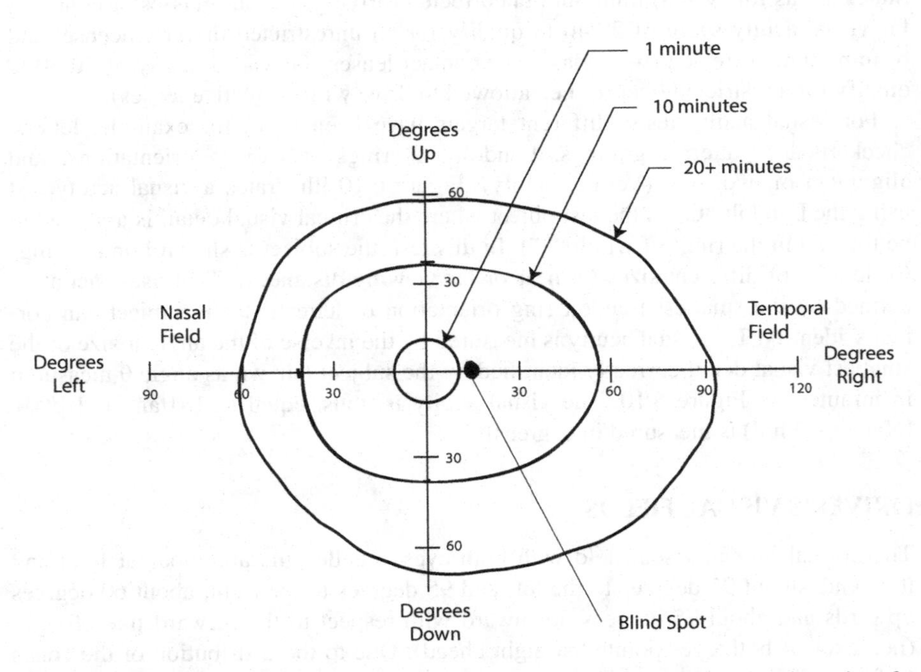

FIGURE 6.11 The normal achromatic monocular visual field of targets of various sizes for right eye.

Source: Redrawn from: Boff and Lincoln, 1988.

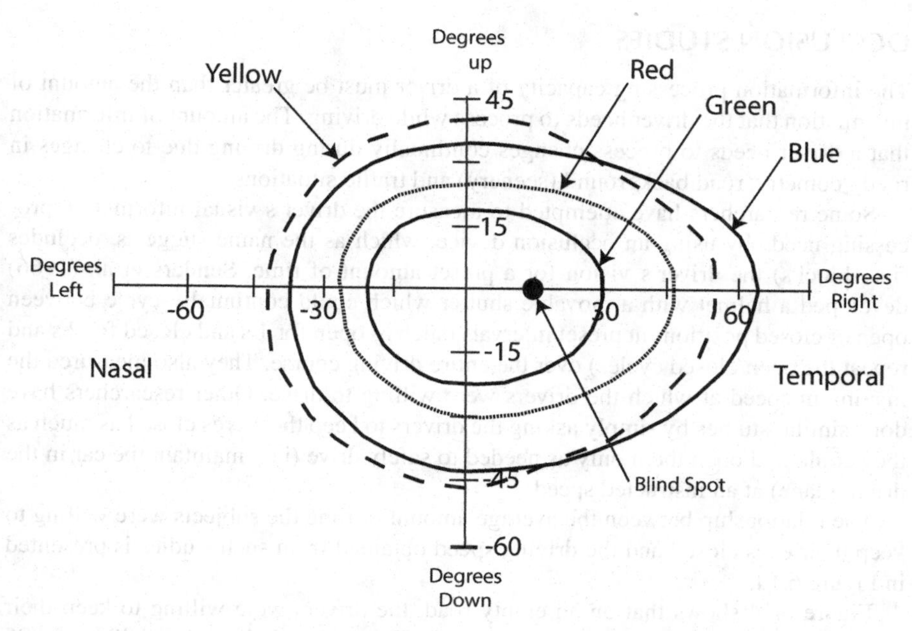

FIGURE 6.12 Right eye's field showing the extent of fields for different colors.
Source: Redrawn from: Boff and Lincoln, 1988.

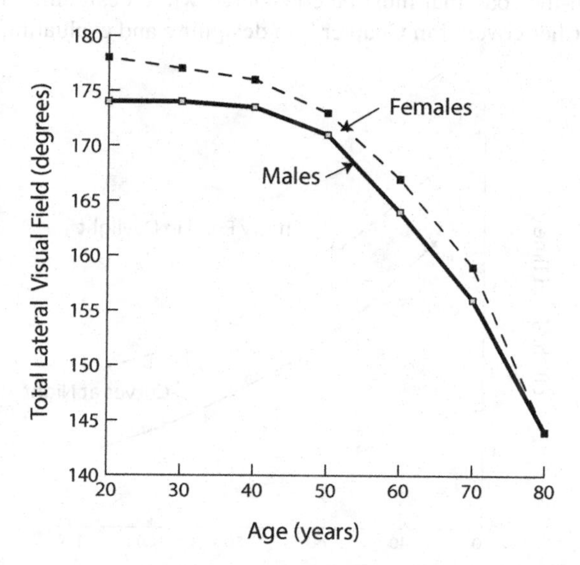

FIGURE 6.13 Effect of age on size of lateral visual field.

Source: Redrawn from: Boff and Lincoln, 1988.

OCCLUSION STUDIES

The information processing capacity of a driver must be greater than the amount of information that the driver needs to process while driving. The amount of information that a driver needs to process changes continually during driving due to changes in road geometry, road background (scenery) and traffic situations.

Some researchers have attempted to measure the driver's visual information processing needs by using an occlusion device, which as the name suggests, occludes (i.e., blocks) the driver's vision for a preset amount of time. Senders et al. (1966) developed a helmet with a movable shutter which could continually cycle between open or closed positions at preset intervals (such as open for 1 s and closed for 2 s and repeat the open/closed cycles) over the entire driving course. They also measured the maximum speed at which the drivers were willing to drive. Other researchers have done similar studies by simply asking the drivers to keep their eyes closed as much as they could and open them only as needed to safely drive (i.e., maintain the car in the driving lane) at an instructed speed.

The relationship between the average amount of time the subjects were willing to keep their eyes closed and the driving speed obtained from such studies is presented in Figure 6.14.

Figure 6.14 shows that on an empty road, the drivers were willing to keep their eyes closed about 2 s on average at 113 km/h (70 mph) and about 4 s at 40 km/h (25 mph). However, when the road environment was more demanding, such as driving on a winding curvy road at night, they closed their eyes on average for only 1 s at 113 km/h (70 mph). The data presented in Figure 6.14 thus provide us with limits on time-away-from-the-road that must be considered while designing driver interfaces. This issue is further covered in Chapter 7 in designing and evaluating driver controls and displays.

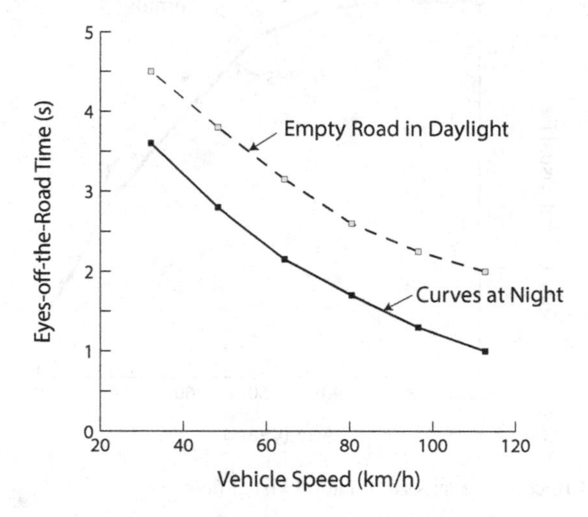

FIGURE 6.14 Relationship between eyes-off-the-road time (occlusion time) as a function of vehicle speed.

INFORMATION ACQUIRED THROUGH OTHER SENSORY MODALITIES

While driving, the driver also obtains information from other sensory modalities. The information obtained through perception of sound, vibrations, vehicle movements, touch and smell also can provide useful information in performing many in-vehicle tasks. The information can also have negative effects such as annoyance, fatigue, and interfering or masking of other useful information. For example, engine and road noises provide the driver with information on vehicle speed. However, too many or too loud noises can induce fatigue and disrupt conversations with other occupants or reception of audio programs.

Human factors engineers work with other engineers in specialized functions such as acoustics and sound engineering, noise, vibrations and harshness (NVH), and interior trim and materials to make sure that the vehicle provides the necessary cues of the right type and magnitude and enhances the pleasing perceptions (e.g., sound of the exhaust that convey engine power, sound of door closing that provides the feeling of "solidness" or "solid build quality", tactile feedback received from "crisp" detent feel while operating electrical switches, and smell of "genuine" leather from the seats).

HUMAN AUDITION AND SOUND MEASUREMENTS

An understanding of human auditory capabilities and limitations provides knowledge for auditory/sound design and evaluation issues related to interior quietness, auditory display designs (e.g., beeps, buzzers, and auditory warnings), voice controls and perception of sound quality.

Important basic information and issues in consideration of auditory systems are given below.

1. The human auditory capabilities are specified by measuring loudness thresholds as a function of the sound frequency (measured in Hertz [Hz], equal to cycles/second). The loudness is measured in decibels (dB) and defined as follows:

$$\text{Loudness (dB)} = -10 \text{ Log}_{10} (P/P_o)^2$$

Where, P = Pressure in sound pressure wave (measured in N/m^2)
P_0 = Lowest perceptible sound pressure

2. The human ear can sense sounds of frequencies between about 20–20,000 Hz. The sound in the human ear is sensed by hair-like sensors inside the cochlea (inner ear). There are about 30,000 hair-like sensors in each inner ear (Konz and Johnson, 2004).
3. The human hearing thresholds vary with sound frequency. The region of best sensitivity (or the lowest loudness thresholds) is within the frequency range of about 500 to 2000 Hz.

4. The human hearing thresholds are measured by instruments called audiometers, which can measure the loudness of a just-perceptible signal as a function of the signal frequency. The sound levels in work environments are measured by instruments called sound level meters. Since the perception of loudness is affected by the sound frequency, most sound level meters use a weighting filter, which converts the incoming sounds (with a combination of different frequencies) and provides a weighted loudness value that is equivalent to the loudness perceived by a listener with normal hearing. The A-weighted filter, used most commonly, is applicable to sound levels around 70 dB. Sound loudness values measured using the A-weighting filter are designated with a unit label called dBA.
5. Human hearing thresholds increase as people get older. The hearing threshold increases are more pronounced in the higher frequency region (over 5000 Hz) (Van Cott and Kinkade, 1972; Sanders and McCormick, 1993).
6. The "phon" is another measure of loudness. It is based on the loudness of the 1000 Hz reference signal that will be perceived to be equivalent in loudness to the loudness of the signal being measured.
7. The just-noticeable differences in loudness and just-noticeable differences in frequencies of humans decrease with an increase in loudness level at which the listener's ears are adapted (Van Cott and Kinkade, 1972). Thus, most people will increase the volume of music to enjoy thorough perception of small differences (or changes – increases or decreases) in sound loudness or sound frequencies (or pitch).
8. In general, the loudness of a sound source experienced by a human decreases according to the inverse square law (i.e., the sound loudness decreases proportional to an inverse of the square of the distance between the source and the listener). The higher frequency sounds attenuate faster than the low frequency sounds as compared to the attenuation predicted by the inverse square law (Van Cott and Kinkade, 1972). Thus, to improve detection of a signal over farther distances, the designer of sound alarm systems should select sound signals of frequencies below 1000 Hz.
9. The sound loudness in modern passenger vehicles with windows closed ranges between about 65–70 dBA at 98 km/hr (60 mph).
10. For a warning signal to be heard inside the passenger compartment by most occupants, its loudness should be at least about 20 dB more than the overall interior noise level. The detection probability of a sound signal will also improve as the frequency of the signal differs from the frequencies of the noise.
11. Noises of higher frequencies (over 2000 Hz) are perceived as "tinny" (cheap and harsh like a tin can) as compared to noise at lower frequencies (below 1000 Hz). Thus, to provide perception of "solidness" from interior components (e.g., sounds from switches, latches, and door closing), the designers should reduce or remove higher frequency sounds created during their operations.

OTHER SENSORY INFORMATION

While driving, a driver also experiences vibrations. The vibrations in vehicles are perceived through three types of sensory systems with overlapping frequencies. The three sensitivity ranges to vibrations are, 1) 0 to 1–2 Hz sensed through vestibular system in the brain (as motion sickness and whole body vibrations), 2) 2 to 20–30 Hz sensed through biomechanical body resonances, and 3) >20 Hz sensed through somesthetic mechanoreceptors such as proprioceptive receptors located in muscles, tendons and exteroceptive receptors located in cutaneous tissues. The human perception and response to vibrations depend upon frequency, amplitude, direction of vibration, duration of vibration, mass of the body segments, posture level of muscle contraction and fatigue and presence of body supports (Chaffin et al., 1999). Vehicle vibrations over 20 Hz are generally perceived as sound or noise and frequencies below 20 Hz are perceived as vibrations. The vehicle suspension and seat design engineers evaluate ride and vibrations experienced by the occupants. Some vibrations are also perceived through a driver's contacts with the vehicle controls such as the steering wheel, pedals and gear shift knobs. Excessive and prolonged vibrations in vehicles are generally associated with higher complaints of driver discomfort and fatigue (especially in heavy truck products). The perception of quality and craftsmanship is also associated with other sensory perceptions related to vibrations, tactile feel of interior touch areas (e.g., smoothness, softness) and smells from emissions from interior materials. Some of these issues are covered in Chapter 12.

APPLICATIONS OF INFORMATION PROCESSING FOR VEHICLE DESIGN

Some Design Guidelines Based on Driver Information Processing Capabilities

Based on the information presented in this chapter and other studies reported in the literature related to the information processing abilities of the drivers, many design guidelines can be developed to reduce the driver's time needed to use displays and operate controls. Important basic guidelines used in designing controls and displays are provided below.

1. Design controls and displays such that they can be used in short (1 s) glances. Minimize or eliminate driver control/display interface configurations that will require multiple glances (or operational steps) away from the forward road scene. Avoid functions buried in menus or hidden features that will require multiple control actions and glances to operate.
2. Place important and frequently used displays close to the driver's normal line of sight while looking forward. Place controls and displays in expected locations and at locations that will involve minimal head, torso and hand/wrist movements. (Note: SAE J1138 provides recommended locations for primary and secondary vehicle controls (SAE, 2009).)

3. Avoid locating controls and displays in areas that are not visible to drivers from their normal eye locations (see driver eyellipses in SAE J941 and in Chapter 5). The steering wheel, stalks, and levers and the driver's hands (hand positions just prior to usage of the control or the display) can cause visual obstructions. Also, avoid obscurations or masking due to veiling glare, specular glare, or reflections into display surfaces or lenses due to the sunlight or other external lights.
4. All displays must be legible to drivers (especially older drivers) from the farthest eye locations under day, dusk (with and without headlamp/panel lights activation) and in night driving conditions (see Chapter 14 for legibility prediction).
5. In developing controls and display layouts, apply principles related to a) importance of the function, b) frequency of use, c) sequence of use, d) functional grouping, and e) controls and display association (see Chapter 7). Further, prioritization within the above principles is generally needed to avoid crowding of controls and displays in regions close to the steering wheel.
6. Place controls at locations that are comfortable to reach, grasp and operate (see Chapters 5 and 7 for maximum reach envelopes). Also, avoid locating controls that require awkward hand/wrist deviations. Controls located too close to the torso generally require severe wrist deviations (e.g., chicken-winging) – especially if the driver cannot move his/her elbows away from the torso (see minimum reach zones in Chapter 7).
7. Tiny controls that require more precise hand movements are difficult to locate and reach. (Note: Fitts's Law covered earlier in this chapter directly relates hand movement time to index of difficulty.)
8. The direction of movements of controls and displays should conform to population stereotypes (see SAE J1139 in SAE, 2009).
9. All controls should provide distinct, perceptible feedback (visual, auditory and/or tactile) to inform the driver about the completion of control activations.
10. All controls and displays must be labeled (provide identification and/or setting label) by using accepted wording or symbols. (Refer to Federal Motor Vehicle Safety Standard 101 (NHTSA, 2010); SAE J1138 (SAE, 2009)).
11. Avoid similar-looking controls or displays. Provide cues/coding (e.g., by incorporating changes in shape, size, color, texture, sound, and force feel) to discriminate between different controls or display details to reduce confusion between similar looking items.
12. Some features such as head-up displays, voice displays, and voice activated controls will require redundant (or alternate) methods of use due to their possible unavailability or undesirability under some driving environments (e.g., head-up displays under bright roadway background) and to drivers with certain deficiencies (e.g., speech or hearing impaired drivers cannot use voice controls/displays).

CONCLUDING COMMENTS

The automobile manufacturers are currently facing a difficult challenge of determining what features should be provided for the drivers to use while driving. The rapidly advancing sensor, microprocessor, information, communication, GPS (global positioning system), display and lighting technologies are allowing development of new features/devices that can provide the driver with information quickly at high speeds and in large quantities. These technologies also have the potential to perform many tasks performed normally by drivers. Thus, if these systems are developed such that they provide just the right type of information and/or control in the right amounts and at the right time and place, then there is a great potential to assist the driver by improving driver convenience and safety. Additional information on future in-vehicle devices is provided in Chapters 17, 18 and 25.

Many traffic safety experts and organizations are concerned that these new devices can overload information processing capabilities of drivers and thus distract drivers from the primary driving tasks. The benefits that these new devices can offer are also substantial. For example, while most drivers realize the additional risk of using cellular phones while driving, there are a number of benefits (e.g., keeping in touch with home/office and a feeling of security – knowing that you can call for help immediately) that may well preclude their prohibition. Similar emotional arguments for access to the Internet in cars are also currently being made. Additional information on driver workload measurement is provided in Chapter 16.

The human factors and safety engineering community must, therefore, come up with performance requirements and design guidelines to ensure that useful and safe in-vehicle devices can be developed. Additional issues related to this area are covered in Chapter 7 and Volume 2 of this book.

REFERENCES

Bhise, V. D. 2002. Designing Future Automotive In-Vehicle Devices: Issues and Guidelines. *Proceedings of the Annual Meeting of the Transportation Research Board*, Washington, D.C.

Bhise, V. D., E. I. Farber and P. B. McMahan. 1977. Predicting Target Detection Distance with Headlights. *Transportation Research Record*, No. 611, Washington, D.C.: Transportation Research Board.

Blackwell, H. R. 1952. Brightness Discrimination Data for the Specification of Quantity of Illumination. *Illuminating Engineering*, 47(11): 602–609.

Boff, K. R. and J. E. Lincoln. 1988. *Engineering Data Compendium: Human Perception and Performance*. Vol. 1. Wright-Patterson Air Force Base, Ohio: Harry G. Armstrong Aerospace Medical Research Laboratory.

Chaffin, D. B., G. B. J. Andersson and B. J. Martin. 1999. *Occupational Biomechanics*. New York, NY: John Wiley & Sons, Inc.

Fitts, P. M. 1954. The Information Capacity of the Human motor System in Controlling Amplitude of Movement. *Journal of Experimental Psychology*, 47: 381–391.

Fitts, P. M. and M. I. Posner. 1967. *Human Performance*. Belmont, CA: Books/Cole Publishing Company.

Gertman, D. L. and Blackman, H. S. 2001. *Human Reliability and Safety Analysis Data Handbook*. New York, NY: Wiley.

Hick, W. E. 1952. On the Rate of Gain of Information. *Quarterly Journal of Experimental Psychology*, 4: 11–26.

Hyman, R. 1953. Stimulus Information as a Determinant of Reaction Time. *Journal of Experimental Psychology*, 45: 423–432.

Jackson, D., Murphy, J. and V. D. Bhise. 2002. *An Evaluation of the IVIS-DEMAnD Driver Attention Demand Model*. Paper presented at the Annual Congress of the Society of Automotive Engineers, Inc., Detroit, Michigan.

Kantowitz, B. H. and R. D. Sorkin. 1983. *Human Factors: Understanding People-System Relationships*. New York, NY: John Wiley and Sons.

Konz, S. and S. Johnson. 2004. *Work Design-Industrial Ergonomics*. Sixth Edition. Scottsdale, AZ: Holcomb Hathaway.

Leiden, K., K. R. Laughery, J. Keller, J. French, W. Warwick and S. Wood. 2001. *A Review of Human Performance Models for the Prediction of Human Error*. Report prepared for NASA Ames Research Center, Moffett Field, CA by Micro Analysis & Design, Inc., Boulder, CO.

Miller, G. A. 1956. The Magical Number Seven Plus or Minus Two: Some Limits on Our Capacity for Processing Information. *Psychological Review*, 63: 81–97.

National Highway Transportation Safety Administration. 2010. Federal Motor Vehicle Safety Standards. *Federal Register*, CFR (Code of Federal Regulations), Title 49, Part 571, U. S. Department of Transportation. www.ecfr.gov/current/title-49/subtitle-B/chapter-V/part-571 (accessed: March 10, 2024)

Rockwell, T. H., V. D. Bhise, and Z. A. Nemeth. 1973. Development of a Computer Based Tool for Evaluating Visual Field requirements of Vehicles in Freeway Merging Situations. Report no. VRI-1. Warrendale, PA: Society of Automotive Engineers, Inc.

Sanders, M. S. and E. J. McCormick. 1993. *Human Factors in Engineering Design*. Seventh Edition. New York, NY: McGraw-Hill, Inc.

Senders, J. W., A. B. Kristofferson, W. H. Levison, C. W. Dietrich and J. L. Ward. 1966. The Attentional Demand of Automobile Driving. *Highway Research Record*. Also in the proceedings of the 46th Annual Meeting of the Highway Research Board. Washington, D.C.

Shannon, C. E. and W. Weaver. 1949. *The Mathematical Theory of Communication*. Urbana, IL: University of Illinois Press.

Society of Automotive Engineers, Inc. 2009. *SAE Handbook*. Warrendale, PA.: Society of Automotive Engineers, Inc.

Treat, J. R. 1980. *A Study of Precrash Factors Involved in Traffic Accidents*. Highway Safety Research Institute Report, University of Michigan, Ann Arbor, MI.

Van Cott, H. P. and R. G. Kinkade (Eds.). 1972. *Human Engineering Guide to Equipment Design*. Sponsored by the Joint Army-Navy-Air Force Steering Committee, McGraw-Hill, Inc./U.S. Government Printing Press.

Webster's New Collegiate Dictionary. 1980. Springfield, MA: G. & C. Merriam Company.

Welford, A. T. 1952. The Psychological Refractory Period and Timing of High Speed Performance: A Review and a Theory. *British Journal of Psychology*, 43: 2–19.

Wickens, C., S. E. Gordon and Y. Liu. 1998. *An Introduction to Human Factors Engineering*. Upper Saddle River, NJ: Pearson Prentice Hall (Addison Wesley Longman, Inc.).

Yarbus, A. L. 1967. *Eye Movements and Vision*. New York, NY: Plenum Press.

7 Controls, Displays, and Interior Layouts

INTRODUCTION

The driver obtains information available from different displays (e.g., road views from the windshield and window openings, inside and outside mirrors, speedometer, fuel gauge, radio display, warning lights, warning sounds and labels) and generates outputs (e.g., moves pedals and gear shifter, turns steering wheel, operates push buttons, and turns stalks) to control the vehicle motion and/or change the states of various in-vehicle devices (e.g., change the radio station). To obtain information from displays and operate controls, the driver uses various information processing and control activation capabilities. Depending upon the levels of different capabilities and available resources, the driver may or may not make an appropriate control action. The time taken by the driver to make a control action will also depend upon the amount of information that the driver will need to process and his/her information processing capabilities. If the controls and displays are not designed for ease in performing of these tasks, the driver may not be able to complete the tasks within the available time and could make errors.

The controls and displays design research began with studies of pilot errors (Fitts and Jones, 1961a and 1961b). Soon after World War II, the Air Force launched a systematic study of errors made by pilots in situations where accidents and near-accidents occurred. The pilots were asked to recall incidents where they almost lost an airplane or witnessed a co-pilot make an error in reading aircraft displays or operating controls. From the analyses of the data gathered from these critical incidents, Fitts and Jones found that, regardless of experience or skill, practically all the pilots reported making errors in using cockpit controls and instruments. They also concluded that it should be possible to eliminate or reduce most of these pilot errors by designing equipment in accordance with human requirements. Similarly, driver errors in using displays and controls can be reduced if they are designed in accord with the human engineering criteria.

Chapter 6 provided basic information on driver information processing, driver errors and some information processing-based guidelines for designing controls and displays. This chapter is intended to provide additional information and details on many key considerations related to design and evaluation of the controls and displays.

DOI: 10.1201/9781003485582-7

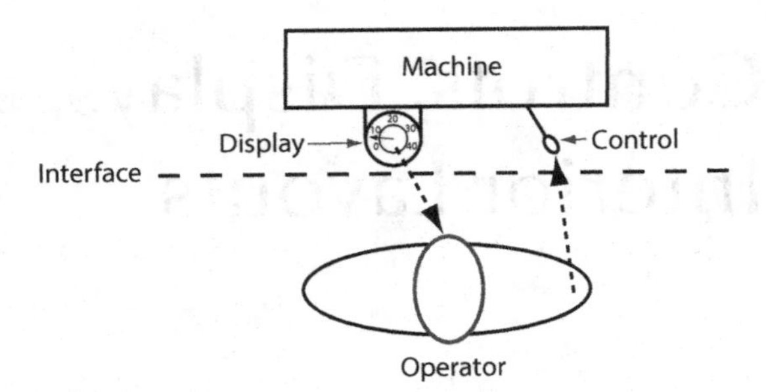

FIGURE 7.1 Controls and displays serve as the interface between the human operator and the machine.

CONTROLS AND DISPLAYS INTERFACE

The controls and displays are the "interface" between the human operator (the driver) and the machine (vehicle). Figure 7.1 illustrates this interface. Thus, the problem of controls and displays design is regarded as a problem of "Human–Machine Interface" (HMI) design.

This figure provides a sketch of human–machine interface. It shows that controls and displays are the "interface" between the human operator and the machine. In automotive design, the human is the driver, and the machine is the vehicle with its controls (e.g., steering wheel, pedals and hand controls) and displays (e.g., speedometer, fuel gage and mirrors) as the interface.

In designing the driver interface, the vehicle designer should always keep in mind the following basic considerations:

- Drivers will prefer to minimize their mental and physical efforts while using controls and displays.
- People will prefer not to use what they don't understand.
- Consider the user population and the variability among the users in the population. Characteristics of users (such as their age, familiarity with the equipment and situations, expectations, eye and hand positions, and visual characteristics) must be considered.
- Consider the use conditions (i.e., the variables and situational complexities involved when displays and controls are used). These conditions can involve informational needs, time constraints, environmental, road and traffic situations, presence of other tasks, and so forth.

CHARACTERISTICS OF A "GOOD CONTROL"

1. A driver should be able to operate a control quickly with minimal mental and physical effort.
2. The operation of a control should be accomplished by the driver by using a minimal number of eye glances (e.g., most turn signal activations are performed without looking [with zero glances] at the turn signal stalk).

3. Any control activation should require minimal hand/finger movements (e.g., operating a control without moving hands from the steering wheel). Note that hands-free operations or controls activated by voice commands can reduce or eliminate hand/finger movements.

CHARACTERISTICS OF A "GOOD VISUAL DISPLAY"

1. The driver should be able to read and understand the display (i.e., obtain the required information) quickly with minimal mental and physical efforts.
2. The driver should be able to acquire the necessary information from a visual display in a few short eye glances. (One short glance [of less than 1 sec] to read a visual display is preferable.)
3. The driver should not require any gross body movements (such as leaning over involving excessive head and torso movements) to obtain needed information. (Note that auditory displays generally do not require the driver to make any eye, head or torso movements.)

TYPES OF CONTROLS AND DISPLAYS

There are many different types of controls and displays. A designer has to select the right types of controls and displays such that drivers can understand and associate them with their functions. The controls and displays must be designed so that they work as a system. Therefore, their layout (locations and orientations of both the displays and controls) is very important. They must be placed in expected locations and in obscuration-free areas, should be well-labeled and should move in directions expected by most drivers (i.e., they meet direction-of-motion stereotypes). They must be associated with each other in terms of characteristics such as physical (e.g., size, shape and tactile coding), visual (i.e., appearance) and/or functional grouping. Many controls and displays are combined together. For example, many controls have displays such as identification and setting labels. And some displays have controls within them, e.g., a touch screen (see Figure 7.2).

IN-VEHICLE CONTROLS

The controls can be classified as follows:

1. *Continuous vs. Discrete*: Continuous controls allow for the setting of a controlled parameter at any point within its control or movement range (or scale). Typical controls used for this application are rotary controls (e.g., volume control in audio products), slide switches, joy sticks, and levers. Discrete controls have detent (or preset) positions that only allow a user to set at one of their detents. Typical examples of such controls are rotary fan speed controls (with "Off", "Low", "Mid" and "High" settings), detented slide controls, rocker switches, toggle switches, and gear shifters.

2. *Push Buttons*: Push button switches require the simplest kind of grasp called the "contact grasp", which merely involves an extended finger to touch and apply force (usually about 1.8 to 5.3 N in low current switches) to activate a micro-switch contact. They do not require the operator's fingers to grasp the control surface (as compared to a control with a knob that requires bending of the fingers and grasping before activating). Since many push or touch buttons can be activated with very little force, the drivers will prefer to have a finger or palm rest area where the operating hand (or other fingers) can be supported close (within the hand/finger reach distance) to the button.

   It should be noted that with the advances in electronics (low current circuits), the push button is probably the cheapest control to produce, and many small buttons can be mounted within small areas on the control panels and arranged in different patterns (e.g., keyboards). The operator's ability to precisely locate and activate a small button, especially following a large hand movement, can become a source of user complaint. Fitts's law of hand motion discussed in Chapter 6 is important to reduce difficulty and movement time in reaching to a push-button.
3. *Touch Screens*: The touch screen displays have a control surface overlaid on the top of the display and, thus, it can be operated by finger touch without any extra input control device (see Figure 7.2). The touch areas are displayed visually, and they act as controls. The touch controls are the most direct form of the control interface as the information display and controls are on one surface. Thus, they have the potential to be intuitive, expected and natural (like pointing a finger).

   Since the touch control area is accessed by visual information about its location and size, the accuracy of activation will depend on the visibility of

FIGURE 7.2 Touch screen with controls overlaid on the display surface.

the touch button, finger size, application of Fitts's law (to reduce index of difficulty, see Chapter 6) and provision of hand/finger support. The advantages of the touch displays are as follows: a) the input device is also the output device (generally the same display), b) it reduces the hand and eye movements needed to find and touch/grasp the control, and c) it eliminates finger bending and grasping motions. Thus, it generally reduces activation time as compared to a control with a grasp area, and d) no extra input device or packaging space is needed.

Some problems of the touch displays are: a) obstructions of the touch areas due to the operating finger and the hand, b) broad finger contact does not allow for fine control movements, c) long finger nails can cause difficulties in orienting a finger and achieving adequate skin contact, d) inadvertent/unintended touching of the control area can activate the control, e) lack of tactile feedback in conventional touch screens, f) sunlight falling on the screen or reflections in the screen will reduce legibility, g) a parallax will affect finger positioning accuracy (especially infrared based touch systems), h) a capacitive touch screen will not work with gloved hand or pen stylus (unless it is conductive), i) finger touch can cause prints/smudge marks on the display surface, and j) touch screen surface can wear and get scratched unless a hard coating or other protective materials are applied.

While touch screens have been available for many years, recent advances in resistive and capacitance technologies have increased their use in automotive applications. The resistive technology uses two layers of material separated by a gap. As pressure is applied to the surface of the film, the two layers touch, and an event is triggered. Thus, the resistive technologies actually require the user to touch and press the screen whereas the capacitive touch technology only requires the properties of capacitive fields broken by the proximity of a finger to trigger an event. The key disadvantage for both of these technologies is that they lack force feedback typical with the conventional mechanical buttons. Since there is no mechanical actuation with the touch switch, it is difficult to determine if and when an event was triggered. However, graphic (visual indication), sound and haptic (touch feel) feedback can be incorporated to reduce this deficiency. The visual feedback can be accomplished by simply changing brightness and/or color of the visual elements in the screen as soon as a finger touch is sensed. The sound feedback can be employed by an internal sound device on the printed circuit board of the device, while a second option is to use the vehicle speaker system. An alternative to an audible indication is a haptic feedback upon a touch event. With this design, an actuator creates perceptible tactile/force feedback. One method to accomplish this feedback is to use a floating structure with specified amplitude versus time vibratory movements or a crisp tactile feeling similar to a conventional switch. In general, users prefer at least sound or haptic feedback in addition to visual feedback.

There are other advantages of touch switch sensors. These include reduced package space due to the thin switch structure and easy to clean surfaces with no edges or gaps. They are generally more durable due to a lack of moving parts or electrical contacts to wear out. They also can achieve a significant weight reduction due to using fewer parts.

4. *Rocker Switches*: Two-position rocker switches that select between two modes (e.g., "On" or "Off") are the most common switches on the automotive instrument panels. The non-protruding portion of the rocker switch indicates the set (currently selected) mode of the switch. On the other hand, the protruding portion gives visual and tactile cues to indicate the mode available for future setting action. SAE standard J1139 provides guidelines on the direction of activation of different automotive switches (including the rockers) when placed in different mounting orientations (SAE, 2009). Some rocker switches have more than two positions. Such rocker switches are difficult to set to a required setting and are prone to setting errors. The difficulty arises because the rocker does not provide sufficient visual cues to determine its selected setting and other available settings unless a visual display is placed next to the switch to provide the setting information.
5. *Rotary Switches*: Most common rotary switches have knobs that are grasped and rotated in the clockwise (to increase or turn-on) or counterclockwise (to decrease or turn-off) directions. The knobs can be designed in different shapes, sizes and with pointers or markings (e.g., rotary headlamp switches, thumb wheels, ring switches) and they can be mounted on different surfaces and in different orientations (e.g., rotary headlamp switch mounted at the end of a turn signal stalk). SAE J1139 standard provides guidelines on the direction of activation of the switches in different mounting orientations (SAE, 2009).

   The rotary controls can be continuous or discrete (detented). The continuous rotary can be set at any position within its range, whereas the discrete rotary can be set at one of its fixed detent positions.

   Figure 7.3 shows a continuous rotary temperature control in a climate control. It incorporates a momentary push button "to turn on and off" the automatic feature of the climate control. The rotary control is also surrounded by five momentary push buttons to set different climate control modes. Each mode switch (except the off) also has a small LED to indicate its status. The center button incorporates a push button for "auto" on and off and a rotary control to set temperature level.
6. *Multi-function Switches*: There are many combination switches created to allow activation of many functions (e.g., a rotary switch that can be pulled or pushed, a rotary switch that can also be moved like a joystick (see Figure 7.4), a stalk that can have rotary and slide switches, push buttons arranged in different groups, layouts and orientations, etc.). To indicate various available functions, visual labels and additional cues (e.g., shape, texture, color, and orientation) are generally necessary for successful implementation of such controls.
7. *Haptics Controls*: These are essentially programmable switches that can change their functions and tactile feel characteristics (e.g., force-deflection profiles, number of detents, feel of detents [shallow versus deep, fine versus

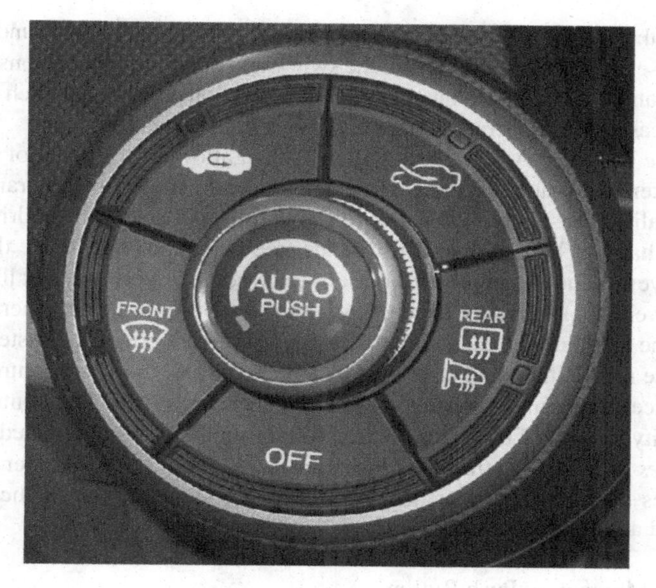

FIGURE 7.3 Rotary temperature switch with a push button in the center (to set "Auto" climate control) and surrounded by five push buttons to select modes for climate control.

FIGURE 7.4 A multifunction rotary switch (in the middle) which can be rotated as well as moved in different directions like a joystick.

large], gains and activation directions) depending up on the selected mode. To understand their movement characteristics and "present" setting, the driver must get immediate tactile feedback or information through its associated visual or auditory display.

8. *Programmable/Reconfigurable Switches*: The programmable, or reconfigurable, switch can change its controlled function depending upon the selected mode. For example, the BMW's i-Drive uses a multi-function rotary control mounted with its axis normal to the horizontal surface of the center console. The reconfigurable switch must provide the driver with clear information on the selected function

and the available settings. In general, since the driver needs to understand the "present" (or selected) mode of the switch and other possible options, the task of operating such a switch involves more information processing which in-turn may increase control operation times and errors in its activations.

9. *Voice Controls*: A voice recognition system allows recognition of the driver's spoken words and sets system functions depending upon the programmed functionality. Thus, in principle, the voice controls will not require a driver to make any hand or body movements (hands-free operation) to activate the controls. However, due to a number of reasons (e.g., some drivers may not like to talk to the vehicle, some may have a temporary disability in voice generation, noise in the vehicle may reduce the accuracy of voice recognition system, delay in voice recognition, and errors in voice recognition), the voice controls may not be acceptable in all driving conditions. Therefore, redundant controls and the ability to turn off the voice control system should also be provided.
10. *Types of Hand Controls Used in Automotive Products*: A number of different types of controls are used in automotive products. Some of the commonly used automotive controls are:

    a) Momentary Push Button
    b) Latching Push Button
    c) Touch Button and touch movements
    d) Radio Push Buttons (selected function button remains pushed in)
    e) Multi-button Mode Selection
    f) Multi-button Directional
    g) Momentary Rocker
    h) Latching Rocker
    i) Hinged Pull up/Push Down Rocker
    j) Continuous Rotary
    k) Detented Rotary
    l) Continuous Thumbwheel
    m) Detented Thumbwheel
    n) Stalk Rotary Continuous
    o) Stalk Rotary Detented
    p) Toggle Switch
    q) Slider Continuous
    r) Slider Detented
    s) Levers Continuous
    t) Lever Detented
    u) Joystick

IN-VEHICLE DISPLAYS

The displays can be classified as follows:

1. *Static vs. Dynamic*: A static display does not change its displayed information or characteristics with time, that is, the display content is fixed (e.g., a

printed label). A dynamic display will change its characteristics, e.g., to indicate change in the magnitude of the displayed variable, it will show change in the graphics – like the speed indicated by a speedometer.

2. *Quantitative versus Qualitative*: A quantitative display will provide a numeric value or precise magnitude of a displayed parameter (e.g., value of instantaneous fuel economy displayed in km/L or miles per gallon). A qualitative display will show a category or change in category of information but not its precise value (magnitude or quantity). It will only indicate direction or category, e.g., a qualitative temperature gauge which only shows if the temperature is within the "normal" range (or "green" zone) or in abnormal (or "red" danger) zone.
3. *Symbolic, Graphic or Pictorial*: Theses displays provide information by using symbols or graphics that the user can associate after recognizing their meaning, e.g., a low tire pressure warning symbol (a cross section of a tire with an exclamation mark). Such symbols or graphics have the potential to be understood by users from different countries (hence the labels in different languages are not needed). It should be noted that many symbols used in the automotive instrument panels were created by technical personnel; and such symbols show related parts of the vehicle (e.g., the low tire pressure symbol shows the cross section of the tire, check engine symbol shows a cross section or a side view of the engine, the brake malfunction symbol shows a drum brake). Such symbols are initially difficult for many non-technical drivers to understand.
4. *Dedicated versus. Programmable*: A dedicated display is a display that presents only a given (assigned) function, whereas a programmable (or reconfigurable) display can change its function and format depending upon the selected display mode. For example, in an audio display, if the radio mode is selected then the display will show the selected radio frequency. But in the CD mode, the same display will show the selected track being played.
5. *Visual, Auditory, Tactile and Olfactory*: Displays can be presented in different forms (or mechanisms) so that they can be identified by using certain sensory modalities (e.g., vision, hearing, tactile, or olfactory). Visual displays allow presentation of the most complex form of information by use of many features such as color, luminance, highlighting, size, shape, font, thickness (or stroke width), and so forth. However, the disadvantage of a visual display is that the driver must take eyes away from the road scene to look at the display (i.e., the eye axes need to be aimed at the display) and the operator must be attentive or alert to process the information available from the display. On the other hand, auditory information can be presented from a region not visible to the driver; and thus, the auditory display can be used to alert the driver. However, the amount of information that can be provided in an auditory display should be relatively short and simple; but it can be coded by use of presentation mode such as loudness, pitch/tone, voice, and musical notes. The tactile display can provide information by a change in physical characteristics such as surface shape, size of grip, texture, movement or vibrations of the grasp/contact area that can be sensed and perceived by touch. The olfactory display is based on

sensation of odor, or a smelly substance emitted by the display (e.g., the smell added in the natural gas to indicate a gas leak or new car (or leather) scent sprayed in the vehicle interior).

6. *Head-down versus Head-up Visual Displays*: Most visual displays used in the automotive products are mounted in the instrument panel, and thus, require the driver to look down by making eye and head movements (thus, they are head-down displays). The time required to acquire information from a visual display involves the time required to move the eyes and head (if required for sightline changes over about 20 degrees), refocus, transmit the image information to the brain, and process the information. The head-down displays should be placed no more than 30 degrees down from the driver's normal horizontal line of sight to allow monitoring of at least some aspect of the forward scene through peripheral vision.

   The head-up displays (HUDs) are displays that are placed at higher locations such that the driver does not need to tilt his head down to look at the display. The head-up displays are generally projected on the windshield (but focused on distances farther than the front bumper) so that the driver can view the display with little or no refocusing and without looking away from the forward scene. The advantages of a head-up display are a) reduction in eyes-off-the-road time (time involved in eye travel, convergence, accommodation, search to locate and read), b) elimination of refocusing and eye-axes convergence movements (which are needed to view displays located at near distances on the instrument panel). Older drivers who cannot focus on near distances can view the HUD clearly, and c) more time available to be spent on the road scene.

   The disadvantages of a head-up display are a) targets can be masked or hidden behind the HUD image, b) the projected image may not be seen over brighter backgrounds in sunlight and glare (however, newer laser HUDs have higher luminance levels), c) poor quality optics can cause annoyance (e.g., double images), d) attention switching/sharing, distraction and visual clutter, e) possibility of "Cognitive Capture" (i.e., capture driver's attention) under less demanding visual environments, and f) need for additional controls associated with the HUD (e.g., brightness control, location control and on/off control) will increase control operation complexity and time.

7. *Types of Displays Used in Automotive Products*: A variety of different types of displays are used in automotive products. Some of the commonly used automotive displays are:
   a) Static Symbol or Icon (Used for identification)
   b) Static Word (or abbreviation) Label (Also, identification and setting labels on controls)
   c) Analog Display with Scale(s) and Pointer(s)
   d) Analog Display with Bars (Bar Display)
   e) Digital Display
   f) Graphic Display with Pictures
   g) Changeable Message Display

h) Color Change Indicator
i) Programmable or a Reconfigurable Display (Screens with dynamic images)
j) Auditory displays (beeps, tones, buzzers, voice messages)

DESIGN CONSIDERATIONS AND LOCATION PRINCIPLES

Basic Considerations in Designing Controls and Displays

1. Prioritize and provide only limited amounts of information to the driver such that the driver does not have to reduce attention to the basic driving tasks, especially when traffic demands are critical.
2. The time required to operate a control is the sum of time taken to a) find and recognize the control, b) access the control, and c) operate the control.
3. The time required to use a display is the sum of time taken to a) find the display, and b) read and interpret the information presented in the display.
4. The usability of a product (with controls and displays) is dependent upon the number of applicable ergonomic guidelines that the product meets. (Note: The guidelines are presented in the next section). In general, user satisfaction with a product will increase as the product meets a greater percentage of the ergonomic design guidelines. Violation of one or more key ergonomic design guidelines can make a product very difficult to use. For example, a control placed in an unexpected location and obstructed by a vehicle component will be very difficult to find.
5. "What people don't understand, does not exist". If a user does not understand how a product or its features (e.g., controls or displays) are to be used, then he/she will not be able to use the feature, i.e., the user will disregard the feature, or it will not exist as a usage choice in the user's mind.
6. "People inherently seek to conserve their energy and reduce effort". They like to minimize their physical and mental workload. (Note: Most drivers like [thus prefer] features such as remote key fobs, steering wheel mounted redundant controls, power windows, power mirrors, and automatic transmissions – which reduce the number of hand movements and force exertions.)
7. Information that the driver does not need should not be presented. This will avoid unnecessary information clutter that may only increase search and discrimination time.
8. Seamless integration (coordination, prioritization, and consistency) of available features is important with potentially greater number of new features to be offered in future products (see Volume 2, Chapter 25).
9. "Hands-free is not risk-free". Voice displays and voice-activated controls can interfere with other visual tasks, especially because verbal information processing generally requires more conscious attention. However, simple well-designed voice displays, or voice activated controls that do not overburden a driver's working memory are generally superior to visual-manual interfaces.
10. Locate visual displays close to the driver's normal line of sight so that the driver's eye movement time is reduced and peripheral detection (and/

or monitoring) of visual cues related to the primary driving tasks is not compromised. Locations within about 30 to 35 degrees from the normal line of sight are generally convenient due to: a) shorter eye movements and head movements will be needed to view the display, and b) higher retinal sensitivity and higher information transmission speed in the visual field within and closer to the fovea (see Chapter 6).

CONTROL DESIGN CONSIDERATIONS

In designing a hand control, the following issues must be considered:

1. *Location*: Controls should be located such that the drivers can easily find and reach them. The location of any automotive control should be based on driver expectancy (i.e., where most drivers expect the control to be located in the vehicle space). Ideally, the driver should be able to locate a control blindly (without looking at it). But if the control is complex (e.g., it includes displays such as setting labels or it is combined with other controls) and if it is used while driving, then it should be located in a visible area and the eye movements (i.e., magnitude of the angle with which the line of sight needs to be moved to view the control from the straight ahead viewing direction) should be as small as possible (preferably no more than 30 degrees). The location of the control should also be based on its grouping with other controls with similar functions (e.g., grouping of all light controls or climate controls and maintaining associations with locations of other displays and controls related to the same function). SAE standard J1138 recommends locations for various primary and secondary hand controls (SAE, 2009).
2. *Visibility*: Controls should be placed in obscuration-free areas and should not require excessive head or torso movements to view them. Control size, color, luminance, and contrast with the background should aid the driver in quickly finding and recognizing the control. In some cases, it may not be necessary for the entire control to be placed within the visible area. A partially visible control can be found with a slight head movement (smaller than about ±50 mm lateral head movement). Some controls can be found and operated without looking, i.e., by blind positioning (groping without looking) of hands and by providing tactile and/or shape coding of the grasp areas of the controls. SAE J1050 standard provides a procedure to determine obscuration free areas around the rim, spokes and hub area of the steering wheel (SAE, 2009).
3. *Identification*: In general, each control should have an identification label or symbol (except for the controls that ae well-known to the drivers, e.g., steering wheel) and the labels should be placed in a visible location and is in close visual proximity to the control. Some controls can be identified by touch by providing unique shape, texture or tactile coding. When controls are placed in close association or grouping with other controls or displays, they can be identified by the function of the associated items. FMVSS 101 (NHTSA, 2024) and SAE J 1138 (SAE, 2009) provide requirements on identification symbols and labels.

4. *Interpretation*: A control should be designed so that its operation (i.e., how it is operated or moved) is easy to understand and interpret. The configuration, shape, appearance and touch feel (e.g., texture and soft rubber like feel) of the grasp area of the control should provide additional cues about its operability (i.e., direction of activation) with minimal reliance on the user's memory. The shape of the knob should be designed such that it invites certain actions (movements) that are compatible with its shape (e.g., rotary knob with a pointer needs to be rotated, flat ends of stalks need to be pulled or pushed, and knobs can be pulled and pushed).
5. *Control Size*: The grasp area of control should provide sufficient surface to grasp, and clearances should be provided for hand/finger access (see Table 4.1 for finger dimensions). The target size of the control should be designed to minimize the hand movement time by considering Fitts's law of hand motions (see Chapter 6 in Volume 1).
6. *Operability*: The direction of motion of a control should meet the population stereotypes specified in SAE J1139 standard (SAE, 2009). The control must provide feedback (visual, tactile or auditory) to convey completion of the control activation movement. The control effort (torque or force needed to operate) should be less than 20 percent of the 5th percentile female maximum strength of the muscles producing the control movement. Such a low effort level not only ensures that most users can operate the control, but it also makes the control operation experience acceptable (i.e., pleasing). The feeling of smoothness during control movements, crispness of detent feels and reduction in free-play or "slop" are also important attributes in improving the perception of quality feel in automotive switches.
7. *Error-free Operation*: The control should be designed to minimize the possibility of errors during operation (see Chapter 6 for types of human errors). For example, similar looking controls should not be placed in close proximity to avoid substitution errors.
8. *Inadvertent Operation*: Important controls such as those who control vehicle motion or driver visibility (e.g., gear shifter, light switch) should be designed to ensure that their settings will not be inadvertently changed during normal and accidental hitting (movements) of an operator's hands and body parts (e.g., driver's knee or elbow bumping against a switch on the door trim panel or the instrument panel). In such cases, additional clearances, recessing or shields around the grasp area of the control should be considered.

VISUAL DISPLAY DESIGN CONSIDERATIONS

In designing any visual display, the following issues must be considered:

1. *Findability and Location*: The display should be located such that the driver can easily find it with a minimum search-and-recognition time and without any body movements (e.g. head or torso movements). The location of any

automotive display should be based on driver expectancy (i.e., where most drivers expect the display to be located in the vehicle space), eye movements (i.e., magnitude of the angle with which the sightline needs to be moved from the straight ahead viewing direction; preferably less than 30 degrees), and locations of other associated displays and controls. SAE standard J1138 recommends locations of various primary and secondary hand controls and embedded or associated displays (SAE, 2009).

2. *Visibility*: The display should be placed in an obstruction-free area and should not require excessive head or head and torso movements to view it. The display size, color, luminance and contrast with the background should aid the driver in searching and quickly finding the display. SAE J1050 standard provides the procedure to determine obscuration free areas around the rim, spokes and hub area of the steering wheel (SAE, 2009).

Figure 7.5 shows a view of an instrument cluster through the steering wheel. The speedometer is the most frequently used display. If we assume that a driver looks at the speedometer about 3 times per km and drives about 20,000 km per year, the speedometer usage will be 60,000 times/year. Care must be taken to locate the speedometer such that at least 95 percent or more drivers will see the speedometer and the instrument cluster area without any obscurations by the steering wheel rim, spokes, the hub area and the stalk controls. To maximize the visibility of the instrument cluster, the steering wheel, the instrument panel and the stalks should be designed as a system by continuous evaluation of the visibility during the design process.

3. *Identification*: The appearance and content of the display should allow the driver to identify its function and display information. Placement of a closely associated identification label or a symbol will help the driver in identifying the display and thus, reduce any unnecessary time required in interpreting the function of the display. For some displays, the setting labels (e.g., "A/C" for the climate control) or units label (e.g., "MPH" for the speedometer) can

FIGURE 7.5 Instrument cluster located within obscuration free area between the steering wheel rim, spokes, hub and the turn signal and the wiper control stalks.

provide sufficient information to identify the display. Thus, additional identification labels for some displays would not be necessary.

4. *Legibility*: All letters, numerals and graphic elements in the displays should be legible under day, night and dawn/dusk conditions. The viewing distance, letter size, font, height-to-stroke- width ratio, width-to-height ratio, luminance contrast, background luminance, glare illumination and angle should be considered to ensure that the display is legible to at least 65-year-old drivers. The legibility can be predicted by using available models (Bhise and R. Hammoudeh, 2004) (Also, see Volume 2, Chapter 14).
5. *Interpretability*: The content of the display should be evaluated to ensure that its displayed information can be correctly interpreted (not confused) and understood by most drivers. The appropriate use of display type, layout, scales/pointers, direction of motion stereotypes, use of colors, coding, frame of reference, number of similar looking displays in close proximity should be evaluated to ensure interpretability of the display (see checklist in Table 7.2).
6. *Reading Performance*: The driver should be able to read the needed information very quickly (preferably in a short single glance). The information acquisition time (e.g., reading time) and reading errors should be evaluated.

CONTROL AND DISPLAY LOCATION PRINCIPLES

In determining locations of the hand controls and displays and their associations, the following principles must be considered:

1. *Sequence of Use Principle*: The controls and displays should be located in the order of the sequence of use to reduce eye and hand movements. The driver's eye fixation location and location of the hand (to be used for control operation) prior to the use of the control and display should be considered for possible reduction of eye and hand movements.
2. *Location Expectancy Principle*: Controls and displays should be located based on the driver's expectancy of location of the controls and displays. To establish controls and displays expectancy, high volume vehicles in the market segment (e.g., the segment leaders) for which the vehicle is planned must be studied to determine the most common locations of primary and frequently used secondary controls and displays.
3. *Importance Principles*: Controls that are perceived by the drivers to be important should be located close to the steering wheel. Important displays should also be located close to the driver's forward line of sight.
4. *Frequency of Usage Principles*: Locate frequently used controls close to the steering wheel. Locate frequently used displays close to the driver's forward line of sight.
5. *Functional Grouping Principle*: Controls and displays associated with a similar function (e.g., light controls, engine controls, climate controls, and audio controls) should be grouped and located together for ease in finding and operating.

FIGURE 7.6 Functional grouping of controls in the center stack.

Figure 7.6 shows a center stack unit with over fifty different controls. To reduce driver workload and confusion, the controls are grouped in seven rows. The lower two rows have climate controls and seat-temperature controls. The audio controls are grouped in the middle rows. The top row includes buttons for the trip and fuel consumption display. The rows of controls are separated by spaces which are covered by a bezel with a different appearance (due to differences in material, texture and color) than the appearance of the push buttons and the rotary knobs. The continuous rotary controls are used to adjust audio volume, radio tuning and temperature controls with the "clockwise to increase" direction of motion convention. The frequently used on/off switches for the audio and climate control and fan controls are placed closer to the driver on the left hand side. The least frequently used heated-seat controls are placed in the bottom of the center stack. The hazard switch is placed in the easy-to-reach top row. The display is also placed higher in the center stack, well within a 30 degrees down-angle from the horizon.

6. *Time Pressure Principle*: Controls should be located close to the steering wheel or in a prominent area if they are to be used quickly and cannot be used under the driver's discretion due to demand from external situations (e.g., sudden fogging of the windshield – which requires quick operation of the windshield defrost switch, unexpected or erratic maneuver by other vehicles – which may require quick use of the horn switch, the high beam switch, or the hazard switch).

METHODS TO EVALUATE CONTROLS AND DISPLAYS

Given a list of functions to be incorporated in the interior of a new vehicle, different design alternatives can create a large number of possible layouts and configurations using different types of controls and displays. Realizing that many current luxury vehicles have over a hundred different controls and displays located in the instrument panel, console and door trim areas, literally thousands of different layouts can be generated to meet different styling concepts. In general, creating a new design for the sake of change or innovation alone does not produce a superior design, because the changes must be made to support functional improvements. Thus, it is important for the ergonomics engineer to evaluate alternate designs and select the few that are ergonomically superior.

The possible methods that can be used to arrive at one or more superior designs generally involve a combination of the following methods or approaches:

1. Apply methods, tools, models and design guidelines available in various ergonomic standards (e.g., SAE (SAE, 2009), company practices, other regulatory requirements (e.g., FMVSS (NHTSA, 2024), ECE (EC, 2000), lessons learned from customer feedback and complaints, warranty experience).
2. Develop and apply ergonomic checklists and summarize the results of the checklists (see Tables 7.1 and 7.2).
3. Conduct a task analysis on selected operational tasks to uncover potential driver errors and to suggest product improvements (see Chapters 10 and 11 for task analysis).
4. Conduct quick-react studies to evaluate driver/user performance and preferences on selected product issues using field tests, laboratory or driving simulator studies (see Volume 2, Chapters 15).
5. Conduct systematic drive evaluations using representative subjects. In these tests, the subject should be asked to perform a set of different tasks (e.g., turn on the radio and find a FM station of your preference, set climate control to reduce heat) that will require the driver to use all the controls and displays and provide usability ratings. The subjects could also be observed (or video recorded) to measure the time taken to read the displays and operate the controls. Any errors made using the controls and displays can also be observed (e.g., pushed a wrong button, looked at the buttons for a long time by making many long glances). After the test, the subjects could be debriefed and asked to describe problems encountered and "likes and dislikes" in using the control and displays (see Volume 2, Chapter 19 for additional information on vehicle evaluation methods).
6. Inclusion of other leading competitors (benchmarking vehicles) in the above methods is also very useful in establishing the superiority of candidate designs.

SPACE AVAILABLE TO LOCATE CONTROLS AND DISPLAYS

After the driver position is established in the vehicle package by locating the SgRP, AHP, BOF/PRP and the steering wheel (see Chapter 5), the vehicle package engineer should create zones to place the controls and displays. The controls and displays zones are bounded by: (a) the maximum reach zone; (b) the minimum reach zone; (c) the visible zone through the steering wheel; and (d) the 35 degree down angle zone. Figure 7.7 presents a side view drawing showing the sections of the zones considered for locating controls and displays.

Maximum Reach Zone: The maximum reach zone is developed by applying SAE J287 procedure which will define the space (in front of the driver) that 95 percent of drivers can reach (SAE, 2009). The left hand reach and right hand reach boundaries are illustrated in Figure 5.21. To establish the reach zones, the members of the SAE Human Accommodation Committee conducted studies by measuring how far forward drivers seated in different vehicle bucks could reach forward (in the X-direction) for different Y and Z locations, using a 3-finger pinch grip holding a 25 mm diameter rotary knob. The reach distance data (in X-direction) from a vertical plane (called the HR plane, placed perpendicular to the X-axis) at different Y and Z locations are provided in tabular form in SAE J287 standard. The standard provides tables for different combinations of following three variables (a) restrained or unrestrained reach; (b) male-to-female ratios of drivers; and (c) G-factor values that define how the package is set up (The G-factor values range from -1.3 for a sports car with low chair height package to +1.3 for a heavy truck package with a high chair height). The reach distances provided in the tables are increased by 50 mm to obtain extended finger reach (to operate a push button) and decreased by 50 mm to obtain full grasp reach (like grasping a floor mounted shifter knob with all fingers turned inward).

Once the left and right hand maximum reach surfaces (or envelopes) are placed in the drawing (or 3D CAD model), they set the forward boundary for locating all the controls that are operated during driving to ensure that 95% of the drivers can reach the controls.

Minimum Reach Zones: The minimum reach zones define the closest distance (with respect to the driver's body) for locating controls. Figure 7.7 illustrates the

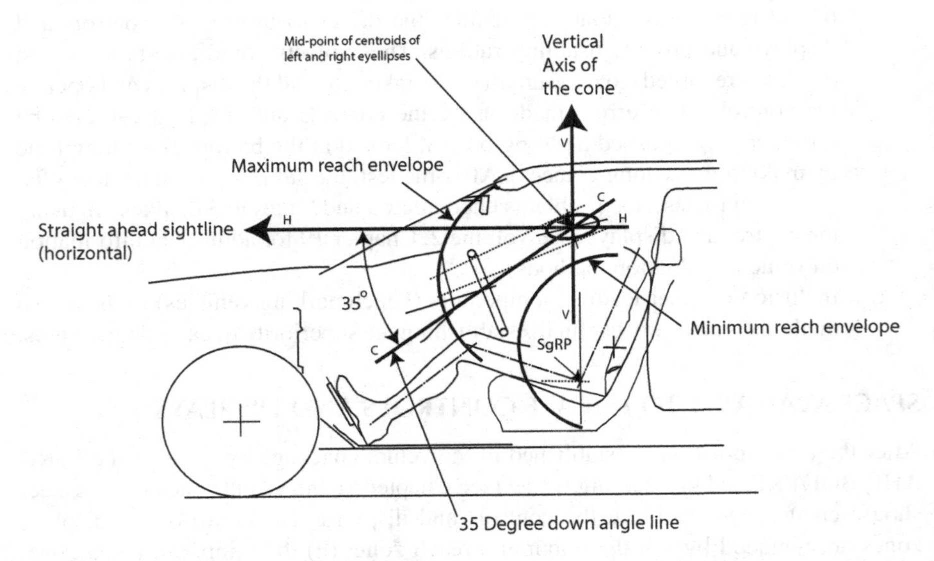

FIGURE 7.7 Maximum reach, minimum reach, 35 down angle and visibility through the steering wheel (The side view is shown in the vertical plane passing through the SgRP.).

minimum reach. The minimum reach zones are hemi-spherical in shape with centers at the left and right elbow locations of the short driver who sits forward at the 5th percentile location in the seat track (i.e., at the 5th percentile location of the H-point defined using SAE J1517 or SAE 4004).

Minimum reach zones are not specified in any SAE standard. (Different vehicle manufacturers have their own internal guidelines). They define the shortest distance (closest location to the driver) at which a control can be located without inconveniencing most drivers. Thus, controls should be located forward of the minimum reach zones.

A typical procedure for determining the minimum reach zones is as follows: (a) consider a short "female" driver sitting in the forward-most seating position; (b) find her elbow points just touching the seat back (upper arms hanging down and elbows touching the bolsters of the seat back); (c) determine her (5th percentile female) elbow-to-knuckle length (i.e., lower arm length, which is about 400 mm). Set this dimension as the radius (R) of the hemi-spherical minimum reach envelopes and (d) create two hemi-spherical zones of radius R with centers at the elbow points.

Figure 7.7 shows a side view of both the maximum and minimum reach zones positioned in a vehicle package. The space between the maximum and the minimum reach zones is thus the space available to locate controls that are easily reached by most drivers.

Zone defining the visibility through the steering wheel: Figure 7.7 shows the sightlines (dotted lines) of drivers from the 95th percentile eyellipses touching the steering wheel (inside of the rim and spokes and hub areas) and projecting on to a plane of the instrument cluster (which includes gages such as the speedometer, fuel gauge, etc.). SAE J1050, Appendix D (SAE, 2009) provides a drafting procedure for determination of the boundary of the zone on the instrument cluster plane which will be visible to most drivers (i.e., below the obscurations caused by the top side of the steering wheel rim and above the obscurations caused by the upper spokes and the hub area of the steering wheel). It is the area on the cluster plane that will be visible to at least one eye of all drivers whose sightlines from the left and right eyes are tangents to the 95th percentile left and right eyellipses, respectively.

35 Degree Down Angle Cone: The 35 degree down angle zone defines the lower boundary for locating controls and displays such that the driver will not require eye movements larger than 35 degrees down in the vertical (elevation) direction (see Figure 7.7). The limit of useful peripheral/central vision is considered to be about 30–35 degrees for detecting stop lamps of lead vehicles while simultaneously looking down to view a display or a control.

Figure 7.7 shows the construction details to create a conical shaped zone. The primary displays and controls (that cannot be blindly reached and operated) should be placed above this conical zone and close to the forward (straight ahead) line of sight. The procedure for generation of the cone is as follows: a) find the mid-point (or cyclopean centroid) of the left and right eyellipse centroids, b) construct 35 (or 30) degree down line (shown as line CC in Figure 7.7) from the straight ahead sightline (shown as line HH in Figure 7.7) passing through the cyclopean centroid, c) rotate the 35 deg down line to form a cone around the vertical axis (shown as line VV in Figure 7.7) passing through the mid-point of the eyellipse centroids, and d) locate displays and controls above the cone.

Space Available to Locate Controls and Displays: The space available to locate control and displays is thus bounded by (a) space rearward of the maximum reach; (b) space forward of the minimum reach; (c) space above the 35 degree cone; and (d) visible regions seen through and around the steering wheel and the stalks.

CHECKLISTS FOR EVALUATION OF CONTROLS AND DISPLAYS

Since there are so many issues and principles to consider while designing and evaluating each control or a display, using a set of comprehensive checklists is an efficient approach used by ergonomists. Table 7.1 provides an example of such a checklist for evaluation a control. The questions in the checklist are grouped according to steps related to finding, identifying, interpreting, reaching, grasping and operating a control. Scoring schemes can also be developed by placing appropriate weight on each question in each group for quantitative comparisons of ergonomic qualities of different controls. Scoring weights can be determined based on attributes such as importance of the control (urgency of usage), frequency of usage, and consequence of errors in not finding or using the control incorrectly.

Table 7.2 similarly provides a checklist for evaluation of visual displays.

TABLE 7.1
Checklist for Evaluation of a Control

	No	Question
Findability	1	Can this control be easily found?
	2	Is the control located in the expected region?
	3	Is the control visible from the normal operating posture?
	4	Are head or head and torso movements required to see the control?
	5	Is the control visible at night from the normal operating posture?
Identification	6	Is the control logically placed and/or grouped to facilitate its identification?
	7	Is the control properly labeled?
	8	Is the label visible?
	9	Can the label be read (legible?) from the normal operating posture?
	10	Is the label illuminated at night?
	11	Can the label be read (legible?) at night from normal operating posture?
	12	Can the control be identified by touch?
	13	Can the control be discriminated from other controls located close to it?

TABLE 7.1 (Continued)
Checklist for Evaluation of a Control

	No	Question
Interpretability	14	Can the control be confused with other controls or functions?
	15	Can an unfamiliar operator guess the operation of the control?
	16	Does the shape of the control convey/suggest activation directions?
	17	Does the control work like most other controls of that control type?
	18	Is the control grouped logically?
	19	Is the control placed within a group of controls that control the same basic function?
	20	Are there other controls within 2–3 inches that have similar visual appearance or tactile feel?
Control Location, Reach and Grasp	21	Is the control located within maximum comfortable reach distance?
	22	Can the control be reached without excessive bending/turning of operator's wrist?
	23	Is the target area of the control large enough to reach the control quickly ?
	24	Can the control be reached without complex/compound hand/ foot motions?
	25	Can the control be reached without torso lean?
	26	Can the control be grasped comfortably without awkward finger/ hand orientations?
	27	Is there sufficient clearance while grasping the control?
	28	Is there sufficient clearance for a person with long (15 mm) fingernails?
	29	Is there sufficient clearance space for operator's hands/knuckles?
	30	Is there sufficient clearance to grasp the control with winter gloves?
	31	Is there sufficient foot clearance (if foot operated control)?
	32	Is the control located at "just about right" location?
	33	Is the control located too high?
	34	Is the control located too low?
	35	Is the control located too far?
	36	Is the control located too close to the driver?
	37	Is the control located too much to the left?
	38	Is the control located too much to the right?
	39	Is the control oriented to facilitate its operation?
	40	Is the control combined or integrated with other controls?
	41	Can the location of the control be changed when setting of any other control is changed? Findability Identification Interpretability Control Location, Reach and Grasp

(continued)

TABLE 7.1 (Continued)
Checklist for Evaluation of a Control

	No	Question
Operability	42	Can the control be operated quickly?
	43	Can the control be operated blindly or with one short glance?
	44	Is the operation of the control part of a sequence of control operations?
	45	Can the control be operated without reading more than 2 words or labels?
	46	Can the control be operated without looking at a display screen?
	47	Can the control be operated easily without excessive force/torque/effort?
	48	Does the control provide visual, tactile or sound feedback on completion of the control action?
	49	Does the control provide immediate feedback (without excessive time lag)?
	50	Does the control move without excessive dead space, backlash or lag?
	51	Is sufficient clearance space is provided for operating hand/foot as the control is moved through its operating movement?
	52	Is regrasp required during operation of the control?
	53	Can the control be moved without excessive inertia or damping?
	54	Does the control direction of motion meet the direction of motion stereotypes?
	55	Are more than one simultaneous movements required to operate this control?
	56	Is the direction of screen/display movement related to the control movement compatible?
	57	Is the magnitude of displayed movement related to the control movement "about right" ?
	58	Can the control be activated easily with gloved hand?
	59	Can the control be operated easily by a person with long finger nails?
	60	Does the surface texture/feel of the control facilitate its operation?
	61	Can the operation of the control be performed with little memory capacity (5 or less items)?
	62	Are surfaces on the control rounded to reduce sharp corners and grasping discomfort during its operation?

TABLE 7.2
Checklist for Evaluation of a Visual Display

	No.	Question
Findability	1	Can this display be easily found?
	2	Is the display located in the expected region?
	3	Is the display visible from the normal operating posture?
	4	Are head or head and torso movements required to see the display?
	5	Is the display illuminated and visible at night from normal operating posture?
Identification	6	Is the display logically placed and/or grouped to facilitate its identification?
	7	is the display properly labeled? (e.g. units shown)
	8	Is the label visible? (not obstructed or not obscured by glare/ reflections)
	9	Can the label be read (legible) from the normal operating posture?
	10	Is the label illuminated at night?
	11	Can the label be read (legible) at night from normal operating posture?
	12	Can the display be identified by its appearance? (e.g., clock)
	13	Can the display be discriminated from other displays located close to it?
Interpretability	14	Can the display be confused with other displays? (e.g., similar in appearance)
	15	Can an unfamiliar operator guess the functionality of the display?
	16	Does the association of the display with a control convey its function?
	17	Does the display work like most other displays of that display type?
	18	Is the display grouped logically?
	19	Is the display placed within a group of displays or controls that have similar functions?
	20	Are there other displays within 2–3 inches that have similar visual appearance?
	21	Are any coding methods (color, shape, outlines, etc.) used to improve its comprehension?
Display Location	22	Is the display located at comfortable viewing distance?
	23	Is the display located close to the driver's primary line of sight (above the 35 degree down angle cone)?
	24	Is the display area large enough to accommodate displayed information?
	25	Does the display appear cluttered?
	26	Is the display located at "just about right" location?
	27	Is the display located too high?
	28	Is the display located too low?
	29	Is the display located too far?
	30	Is the display located too close to the driver?
	31	Is the display located too much to the left?
	32	Is the display located too much to the right?
	33	Is the display oriented to facilitate its viewing?

(continued)

TABLE 7.2 (Continued)
Checklist for Evaluation of a Visual Display

	No.	Question
Usability	34	Can the display be read quickly?
	35	If the display contains scales: Are the numerals, the scale(s) and pointer(s) easy to read? (Consider: End Points, Progression, Placement, Orientation, Size and Font of the Numerals. Scale Markings: Major/Minor, Size. Pointer Length/Width. Obscuration of Numerals by Pointer, etc.)
	36	Can the display be read on bright sunny days with sun rays directed at the display?
	37	Is the display required? (Does it serve useful function?)
	38	Is the reading of the display part of a sequence of display readings steps? (e.g., menus)
	39	Does the display provide other than visual (e.g. sound, vibrations, tactile) cues related to the displayed the information?
	40	Does the display provide immediate feedback (without excessive time lag) of a control action or change in status?
	41	Does the display change too slowly or fails to display quickly (inaction, damping or lag)?
	42	Is the display too sensitive to small changes in displayed function?
	43	Does the display direction of motion meet the direction of motion stereotypes?
	44	Are more than one simultaneous control movements are required to access this display?
	45	Is the direction of screen/display movement compatible to its related control movement?
	46	Is the magnitude of screen/display movement related to the control movement "about right"?
	47	Can the display be easily read by an older person at night?
	48	Does the background surface/texture/color of the display facilitate its readability?
	49	Can the displayed information be understood with little memory capacity (5 or less items)?
	50	Are surfaces on or close to the display provide bright discomforting reflections of external or internal sources?

ERGONOMICS SUMMARY CHART

Table 7.3 provides an example of an ergonomics summary chart (called the "Smiley faces" chart). The chart lists each control and display in different interior regions of the vehicle on the left hand side of the table. The evaluation criteria are grouped into nine columns located in the middle of the table. The nine criteria groups are labeled as follows:

1. Visibility, Obscurations and Reflections
2. Forward Vision Down Angle

TABLE 7.3
Ergonomics Summary Chart

Ergonomics Evaluation : Vehicle X

Key:	⊘ Not Appl	● Low	O Mid	☺ High
Rating:		1-2	3	4-5

Region / Location	No.	Interior Items: Controls, Displays and Handles	Control and Display Evaluation Criteria: Visibility, Obscurations & Reflections	Forward Vision Down Angle	Grouping, Association & Expected Loc.	Identification Labeling	Graphics Legibility and Illumination	Understandability/ Interpretability	Max.& Min. Comfortable Reach Distance	Control Area, Clearance & Grasping	Control Movements, Efforts & Operability	Comments: Specific Problems and Suggestions
Door Mounted Controls	1	Inside Door Handle	☺	⊘	☺	☺	☺	☺	☺	☺	☺	
	2	Door Pull Handle	☺	⊘	☺	☺	☺	☺	O	O	⊘	Requires some chicken-winging-- should be moved forward 25-50 mm. Cannot use full power grasp.
	3	Door Lock	☺	☺	☺	☺	☺	☺	☺	O	☺	Difficult to push -- touch area moves under bezel surface.
	4	Window Controls	☺	☺	☺	☺	☺	☺	☺	☺	☺	
	5	Window Lock	☺	☺	☺	☺	☺	☺	☺	☺	☺	
	6	Mirror Control	☺	☺	☺	☺	☺	☺	☺	☺	☺	
Column/S/W	7	Turn Signal Stalk	☺	☺	☺	☺	☺	☺	☺	☺	☺	
	8	Wiper Switch (right Stalk)	☺	☺	☺	☺	☺	☺	☺	☺	☺	
	9	Ignition Switch	O	⊘	☺	☺	☺	☺	☺	☺	☺	Requires head movement to see the control during key insertion.
	10	Cruise Controls	☺	☺	☺	☺	☺	☺	☺	☺	☺	
	11	Shifter	☺	⊘	☺	☺	☺	☺	☺	☺	☺	Located on console.
Left I/P	12	Light Switch (on left stalk)	☺	☺	☺	☺	☺	☺	☺	☺	☺	
	13	Panel Dim	☺	☺	☺	☺	☺	☺	☺	☺	☺	On left side of I/P.
	14	Parking Brake (on console)	☺	☺	☺	☺	☺	☺	☺	☺	☺	
	15	Hood Release	O	⊘	☺	●	☺	☺	☺	☺	☺	Difficult to see from the driver's seat. Not labeled.
Cluster	16	Tachometer	☺	☺	☺	☺	☺	☺	⊘	⊘	⊘	
	17	Speedometer	☺	☺	☺	☺	☺	☺	☺	☺	☺	
	18	Temperature	☺	☺	☺	☺	☺	☺	⊘	⊘	⊘	
	19	Fuel Gauge	☺	☺	☺	☺	☺	☺	⊘	⊘	⊘	
	20	Oil Pressure							⊘	⊘	⊘	Not on this vehicle.
	21	PRNDL Display	☺	☺	☺	☺	☺	☺	⊘	⊘	⊘	
Center of Instrument Panel / Console	22	Radio	O	O	☺	☺	☺	☺	☺	☺	☺	1/2 - 2/3 of the radio controls below 35 deg. Silver buttons on silver background difficult to locate quickly.
	23	Climate Control	☺	O	☺	☺	O	☺	☺	☺	☺	Mode selector symbols difficult to read on silver background. Low location & disassociated display.
	24	Clock (top in center stack)	☺	☺	☺	☺	☺	☺	☺	☺	☺	
	25	Backlight Defrost	☺	☺	☺	☺	O	☺	☺	☺	☺	Symbol difficult to read on silver background.
	26	Windshield Defrost	☺	☺	☺	☺	O	☺	☺	☺	☺	Symbol difficult to read on silver background.
	27	Traction Control										Not on this vehicle.
	28	Parking Brake	☺	☺	☺	☺	☺	☺	☺	☺	☺	On console.
	29	Hazard Switch	☺	☺	☺	☺	☺	☺	☺	☺	☺	On center above the CD slot.
	30	Ash Tray	☺	●	☺	☺	☺	☺	☺	☺	☺	Located too low.
	31	Cigarette Lighter	☺	●	☺	☺	☺	☺	☺	O	☺	Located too low. Not enough clearance for finger grasp.
	32	Cupholder	☺	●	☺	☺	☺	☺	●	☺	☺	Requires chicken-winging. Located too close and too low.
	33	Shifter	☺	☺	☺	☺	☺	☺	☺	☺	☺	
Other	34	Glove Box Latch	☺	⊘	☺	☺	☺	☺	☺	☺	☺	
	35	Cruise Control On/Off	☺	☺	☺	☺	☺	☺	☺	☺	☺	On right spoke of S/W.
	36	Trunk Release	O	☺	☺	☺	☺	☺	☺	☺	☺	Difficult to see from the driver's eyepoint.
	37	Fuel Fill Door Release	☺	☺	☺	☺	☺	☺	☺	☺	☺	On floor-- left side.

3. Grouping, Association and Expected Locations
4. Identification Labeling
5. Graphics Legibility and Illumination
6. Understandability / Interpretability
7. Maximum and Minimum Reach Distance
8. Control Area, Clearance and Grasping
9. Control Movements, Efforts and Operability

A 5-point rating scale (with 5=Highest score and 1= Lowest score) is used to evaluate the ergonomic guidelines in each of the above nine groups. The ratings are usually obtained from trained ergonomists (based on data obtained from 3D-CAD model of occupant package, sitting in available interior bucks and results from applicable design tools and models) and graphically displayed by using a graphic scale of "smiley" faces for each of the above nine groups, for each item listed in each row. The chart provides an easy-to-view format that can be used to provide an overall ergonomics status of a vehicle interior and was found by the author to be a useful tool in various design and management review meetings. The objective of the ergonomics engineer is to convince the design team during the design review meetings to remove as many "black dots and black donuts" from the charts and increase the number of "smiley faces" by making the necessary design changes.

SOME EXAMPLES OF CONTROL AND DISPLAY DESIGN ISSUES

1. *Speedometer Graphics Design*
Figure 7.8 shows four analog speedometer graphic designs that were created to work with the same speedometer hardware. An ergonomics engineer was asked to evaluate the four graphic designs and recommend the best design. The top left design has the speed numerals inside the scale (tic marks) and the numerals are progression of 10's, which are easier to interpolate and read precise speed as compared to the top right and lower left designs. The top right design looks less cluttered due to the numerals presented in progression in 20's; whereas the lower left design has numerals ending in 5's, which the designer claimed may be "easier to use because many of the speed limits on the roads are in multiples of 5 such as 25, 35, 45, 55 and 65 MPH". The lower right design was created by placing the speed numerals outside the scale such that the pointer cannot block any of the numerals.

To evaluate the designs, the ergonomics engineer created photographic slides of the four speedometers with the pointer positioned at different random angular locations and presented the speedometers to the subjects using a 0.5 s exposure tachistoscopically. It was found that when the scales were presented in the numeric progression of 10's, the subjects could read the speeds with lower error rates, and they also preferred using the speedometers (top left and bottom right). Additional information on scale design issues and progression of numbers in analog gauges can be found in Van Cott and Kinkade (1972) and Sanders and McCormick (1993).

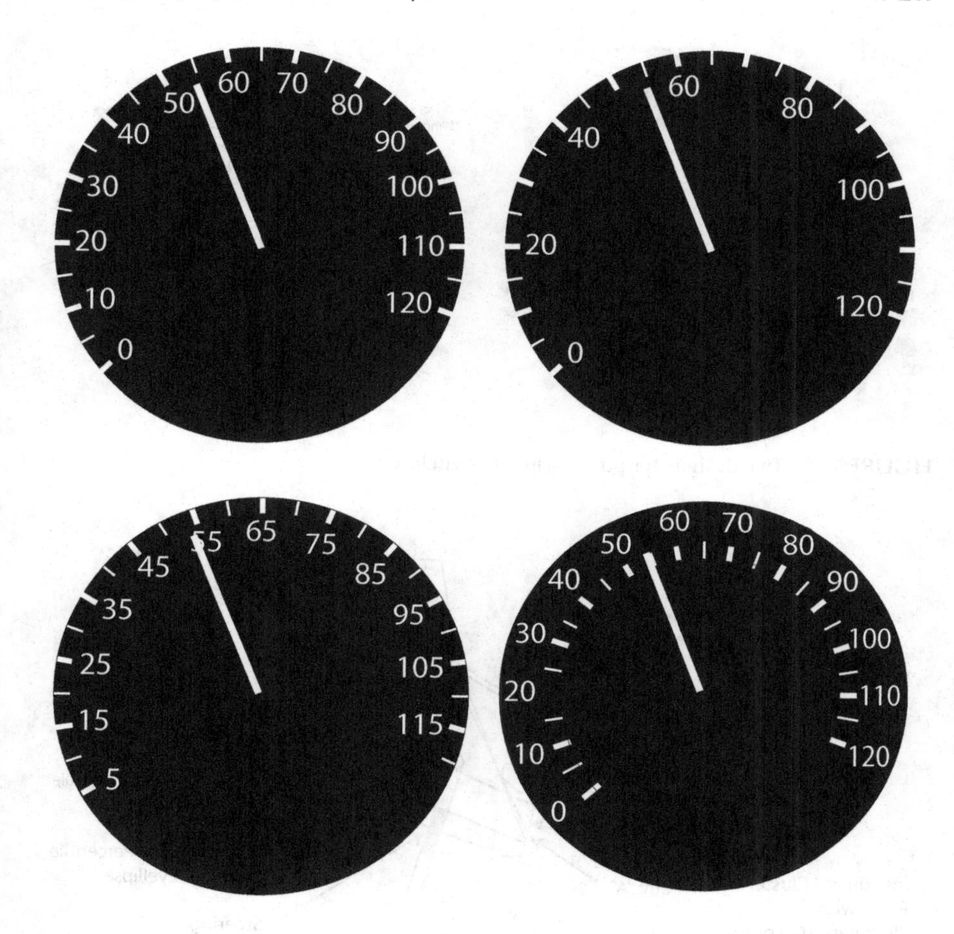

FIGURE 7.8 Four different speedometer designs.

2. *Power Window Location*

Figure 7.9 presents two different locations of power window switches. The figure on the left shows the power window rocker switches on the center console; whereas the figure on the right shows the power window switches grouped and located on the driver's door. When drivers in an evaluation study were asked to drive the vehicles with the above two power window switch configurations and asked to operate the switches at different times while driving, they said that the switch configuration located on the driver's door was more convenient to operate because the window switches "belong on the door" (because of their association) and they were placed higher (required only about 20–25 degrees down-eye movements to look at them from the forward line of sight) as compared to the console mounted window switches which required over 50 degrees of eye movements which the drivers said were "buried way down in the console".

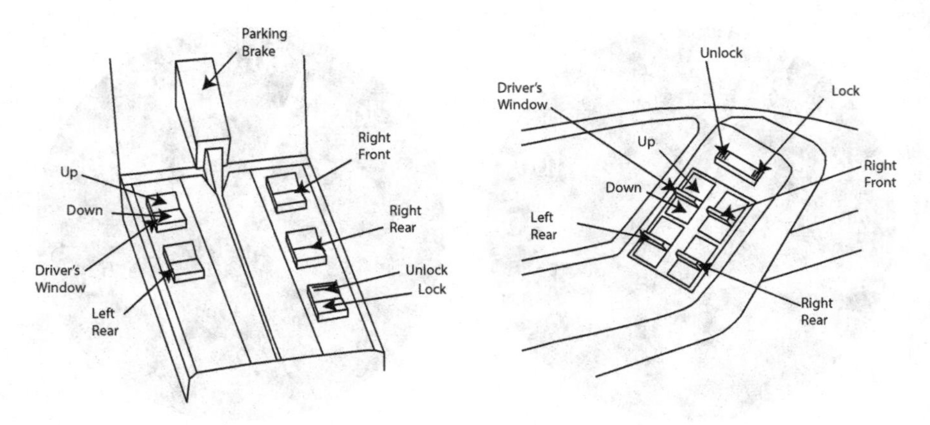

FIGURE 7.9 Two designs for power window switches.

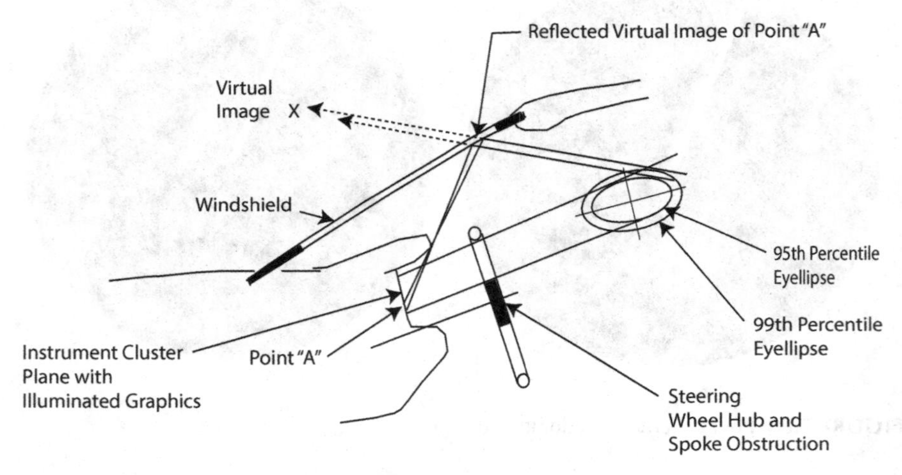

FIGURE 7.10 Sightlines analysis to predict locations of reflected images of illuminated graphics in the windshield.

3. *Annoying Reflections*

Figure 7.10 shows a drawing created to analyze reflections of lighted setting labels on switches mounted on the lower part of the instrument panel. The drivers, especially those with taller eye locations, complained that at night they were able to see reflections of lighted switch labels in the windshield every time they looked at the inside rear view mirror. The drivers complained that the reflections were "annoying" during night driving. The drawing in Figure 7.10 shows that rays from the lower part of the instrument cluster and the switches, when reflected into the windshield could pass through the upper portions of the 95th and 99th percentile eyellipses. Thus, the tall drivers whose eyes will be located in that upper part of the eyellipses would be able to see the reflection of the lighted graphics (as virtual image, see Figure 7.10) from the lower side the instrument panel via the windshield.

Other typical reflection problems of interior lighted components are, a) Reflections of the lighted graphics of headlamp switches mounted on the left lower side of the instrument panel into the driver's side glass. These reflections in some vehicles occur along the driver's line of sight to the left outside mirror. Thus, every time the driver uses the left outside mirror, the reflections of the lighted headlamp switch can be seen. The reflections can mask some part of the driver's mirror field. b) Reflections of the lighted instrument panel in the backlite (back window behind the driver) of pickup trucks: In some pickup trucks, the drivers can see reflections of the lighted instrument panel in the backlite when the driver uses his inside mirror to view the rear field. Thus, an ergonomics engineer must perform reflection analyses to ensure that any lighted components should not cause annoying reflections in the driver's field of view (see Chapter 8 for field of view evaluations).

4. *"Hard to Read" Labels and Radio "Difficult to Operate"*
Figure 7.11 shows a sketch of a center stack with a high mounted radio display. The display was separated from its controls because of the three circular air registers that were placed between them. Further the radio buttons and their surrounding background had a silver (brushed nickel) appearance. The alphanumeric labels on the radio buttons were printed in black, and thus, they appeared black during daytime. But when the panel lights were turned on, the labels looked red. Because of the separation of the radio display from its controls and the difficult to read buttons on the radio during daytime (low visual contrast between the black printed labels and silver button background), the drivers during a drive test complained saying that they liked the "high" radio display location but did not like the locations and the legibility of the radio buttons (see Chapter 14 for information on legibility evaluations).

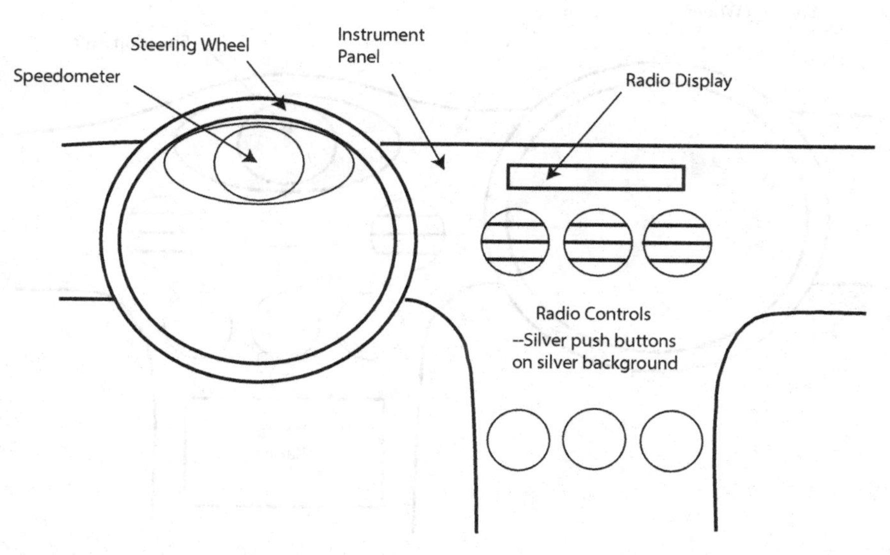

FIGURE 7.11 Disassociated radio controls with silver background.

5. *Center Speedometer and Low Radio Location*
Figure 7.12 shows a picture of an instrument panel in which the instrument cluster was located high and at the center of the vehicle. The radio was located well below the 35 degree down angle location. The initial impression of the drivers to the instrument panel was that they thought that the centrally located speedometer was very "unusual", but after they drove the vehicle on a short trip, they found that they could easily look at the speedometer and monitor the road through the front windshield. The down angle (with respect to the horizontal) to the speedometer was only about 12–15 degrees as compared to the traditional, slightly lower-mounted speedometers (at about 20 degrees down angle) which are viewed through the steering wheel. The radio was mounted well below the 35 degrees down angle and the drivers found it difficult to operate the radio because while looking at the radio they could not see any of the forward road view in their peripheral vision. Additional information on the speedometer location study can be found in Bhise and Dowd (2004).

This figure shows a sketch of a center stack showing a high mounted radio display. The display was separated from its controls because of the three circular air registers that were placed between them. The labels on the radio buttons were hard to read because of their silver surrounding background and when the lights were turned on, the labels looked red.

6. *Door Trim Panel Layout*
Figure 7.13 shows two door-trim panels on the driver-side door. Most drivers found the left door was very difficult to open from the inside because the inside door opening

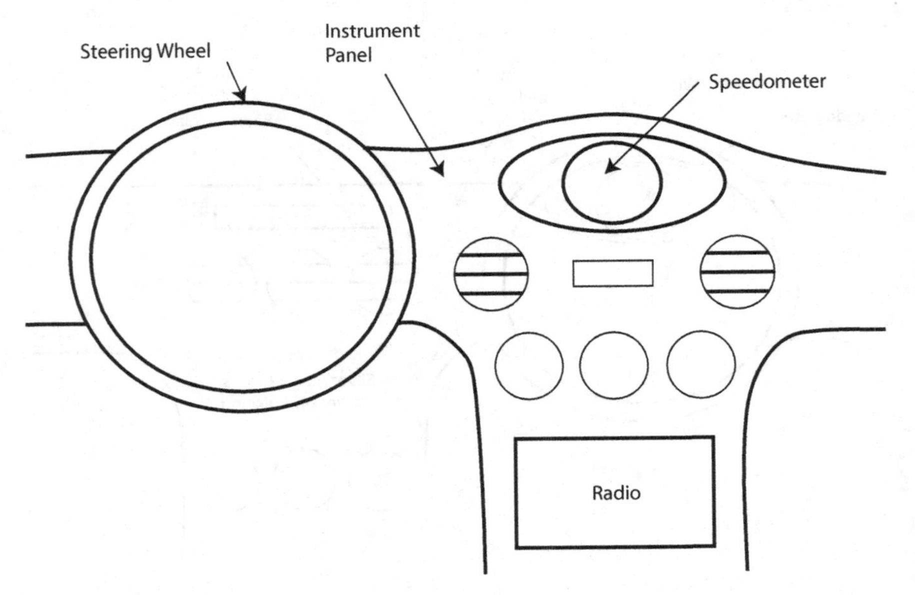

FIGURE 7.12 High center mounted instrument cluster and low mounted radio.

FIGURE 7.13 Layout of two door trim panels.

handle was located too low. The door was also found to be difficult to close because the door pull cup was located too far rearward in the armrest, as it was located too close to the driver, that is, inside the minimum reach zone. The door trim panel in the right picture was easier to use because of the high-mounted inside door opening handle and the forward-mounted angled pull grip handle, which was located well within the maximum comfortable reach zone, and it was also easier to reach when the door was open.

QUALITY FUNCTION DEPLOYMENT

Quality function deployment (QFD) is a technique used to understand customer needs (voice of the customer) and to transform the customer needs into engineering characteristics of products or processes in terms of functional or design requirements. It relates "what" (what are the customer needs?) to "how" (how should an engineer design to meet the customer needs?), "how much" (the magnitude of design variables, i.e., their target values), and it also provides competitive benchmarking information, all in one diagram. The description, procedure for conducting the QFD analysis and an example of its application is provided in Chapter 3.

CONCLUDING REMARKS

A number of basic considerations in designing controls and displays were covered in this chapter. The basic considerations such as minimization of mental and physical efforts, visibility of displays, legibility and identification, and reach to hand control are applicable to designing driver interfaces with hand controls and visual displays. With the advances in technologies related to driver information systems and entertainment systems, the content of the driver interface has been steadily increasing over time. Recent advances in reconfigurable displays, touch screens, multifunction and steering wheel mounted controls, voice controls, data storage devices (Bhise, 2006), and Bluetooth communications have allowed even more choices in incorporating more features and their controls and displays. With these technologies, the driver's expectations and demands on new in-vehicle features have also increased substantially. The increases in functionality and features have led to additional issues such as interactions between different systems, priorities in displaying the status of different systems and operating their controls, increases in driver workload and driver distractions. Volume 2 of this edition covers additional issues such as prediction of visibility and legibility, driver performance measurement, driver workload evaluation, new technology implementation to aid the ergonomics engineer in designing future vehicles.

REFERENCES

Bhise, V. D. 2006. *Incorporating Hard Disks in Vehicles – Usages and Challenges*. SAE Paper no. 2006-01-0814. Presented at the SAE 2006 World Congress, Detroit, Michigan.

Bhise, V. D. and J. Dowd. 2004. Driving with the Traditional Viewed-Through-The-Steering Wheel Cluster vs. the Forward-Center Mounted Instrument Cluster Proceedings of the Human Factors and Ergonomics Society 48th Annual Meeting in New Orleans, Louisiana.

Bhise, V. D. and R. Hammoudeh. 2004. *A PC Based Model for Prediction of Visibility and Legibility for a Human Factors Engineer's Toolbox*. Proceedings of the Human Factors and Ergonomics Society 48th Annual Meeting in New Orleans, Louisiana.

European Commission. 2000. *The European Statements of Principles on Human Interface for Safe and Efficient In-Vehicle Information and Communication System*. In the Official Journal of European Communities, Document no. C(1999) 4786.

Fitts, P. M. and R. E Jones. 1961a. Analysis of Factors Contributing to 460 "Pilot Errors" Experiences in Operating Aircraft Controls, Memorandum Report TSEAA-694-12, Aero Medical Laboratory, Air Materiel Command, Wright-Patterson Air Force Base, Dayton, Ohio, July 1947. In *Selected Papers on Human Factors in the Design and Use of Control Systems*, ed. H. Wallace Sanaiko. New York, NY: Dover Publications, Inc.

Fitts, P. M. and R. E. Jones. 1961b. Psychological Aspects of Instrument Display. I: Analysis of Factors Contributing to 270 "Pilot Errors" Experiences in Reading and Interpreting Aircraft Instruments, Memorandum Report TSEAA-694-12A, Aero Medical Laboratory, Air Materiel Command, Wright-Patterson Air Force Base, Dayton, Ohio, July 1947. In *Selected Papers on Human Factors in the Design and Use of Control Systems*, ed. H. Wallace Sanaiko. New York, NY: Dover Publications, Inc.

Kantowitz, B. H. and R. D. Sorkin. 1983. *Human Factors: Understanding People-System Relationships*. New York, NY: John Wiley and Sons.

National Highway Transportation Safety Administration. 2024. Federal Motor Vehicle Safety Standards. *Federal Register*, CFR (Code of Federal Regulations), Title 49, Part 571, U. S. Department of Transportation. www.ecfr.gov/current/title-49/subtitle-B/chapter-V/part-571 (Accessed: March 11, 2024)

Sanders, M. S. and E. J. McCormick. 1993. *Human Factors in Engineering Design*. Seventh Edition. Washington, DC: McGraw-Hill, Inc.

Society of Automotive Engineers, Inc. 2009. *The SAE Handbook*. Warrendale, PA: SAE.

Van Cott, H. P. and R. G. Kinkade (Eds.). 1972. *Human Engineering Guide to Equipment Design*. Sponsored by the Joint Army-Navy-Air Force Steering Committee, McGraw-Hill, Inc./Washington, D.C.:U.S. Government Printing Press.

8 Field of View from Automotive Vehicles

INTRODUCTION TO FIELD OF VIEW

The objective of this chapter is to provide a background into ergonomic issues related to designing the daylight openings (called the DLOs or the window openings) and other fields of view providing devices such as mirrors and cameras to ensure that drivers can view the necessary visual details and objects in the roadway environment. This chapter will present methods used in the industry to locate various eye points in the vehicle space and draw sightlines used to measure and evaluate fields of view.

LINKING VEHICLE INTERIOR TO EXTERIOR

The field of view analyses link the vehicle's interior design to its exterior design. The interior package provides the driver's eye locations and interior mirror. The vehicle exterior defines daylight openings and exterior mirrors. Thus, the interior and exterior designs must be developed in close coordination to ensure that drivers can see all the needed fields to drive their vehicles safely.

WHAT IS FIELD OF VIEW?

The field of view is the extent to which the driver can see 360 degrees around the vehicle in terms of left and right (horizontal or azimuth) angles and up and down (vertical or elevation) angles of the driver's line of sight to different objects outside the vehicle. (The interior field of view issues, primarily related to the visibility of controls and displays, are covered in Chapter 7). Some parts of the driver's visual field are obstructed due to the vehicle structure and components such as pillars, mirrors, instrument panel, steering wheel, hood, lower edges of the window openings (called the beltline), and headrests.

Thus, what the driver can see while seated in a vehicle depends upon the characteristics of (a) the driver; (b) the vehicle; (c) the targets (e.g., pedestrians, signs, signals, lane lines, etc.); and (d) the environment (i.e., road geometry, weather, and day/night light levels). Important variables associated with the above items are described below.

DOI: 10.1201/9781003485582-8

Driver Characteristics: The amount of visual information that a driver can obtain will depend upon driver eye locations defined by the eyellipses in SAE J941 standard (SAE, 2009), visual capabilities (e.g., visual contrast thresholds, visual acuity, color perception and discrimination, visual fields), visual sampling behavior (e.g., eye movements), head turning abilities (e.g., range of comfortable head turn angles), head movements (e.g., leaning forward or sideways), information processing capabilities, and the driver's age.

Vehicle Characteristics: The vehicle characteristics related to the driver's field of view are window openings and glazing materials (e.g., optical and installation characteristics of the glass), size and locations of vehicle components that can reduce visibility due to obscurations, glare and/or reflections of brighter objects (e.g., external light sources, high reflectance or glossy materials on vehicle surfaces), indirect vision devices (e.g., mirrors, sensors, cameras and displays), wiping and defrosting systems, vehicle lighting and marking systems (e.g., headlamp beam patterns, signal lamps and reflectors).

Targets: The sizes, locations and photometric characteristics of different targets and their backgrounds will affect the amount of visual information the driver can acquire. The targets include: the roadway and traffic control devices (road geometry, lane markers, signs, and signals), other vehicles (e.g., their color and reflectance, their visibility due to exterior lamps and reflectors at night), pedestrians (their size, location, movements, and clothing reflectance), animals and other roadside objects.

Environment: This will include the visual conditions due to illumination: day, night, dawn/dusk, weather (fog, snow, rain), other sources of illumination and glare (from sun, on-coming headlamps, street lighting), reflections of interior and exterior light sources, and the roadway.

ORIGINS OF DATA TO SUPPORT REQUIRED FIELDS OF VIEW

The field of view analysis is based on information about the size of the visual fields that are needed to drive a vehicle safely. Information can be obtained from several sources. These include the following.

- *Targets that must be seen*: Spatial information (i.e., location related) about objects/targets relevant to the driving tasks (e.g., lane keeping, collision avoidance and route guidance information/knowledge related to driving lane, road geometry, presence of other vehicles, objects on collision course, maneuvering space, pedestrians, and signs and signals).
- *Driver Capabilities*: Knowledge about drivers' capabilities in perceiving vehicle heading, distances to targets, speeds, acceleration/deceleration, and distances with respect to other objects in the visual scene, driver response capabilities and vehicle maneuvering (control) capabilities.
- *Driver Feedback on Vehicle Design Features Related to Visibility*: Driver/owner complaints received by the vehicle manufacturers on problems related to visibility (e.g., obscurations due to vehicle pillars, smaller mirror fields, obscurations due to large seatbacks and headrests, height of steering wheel,

mirrors, instrument panel, hood, cowl and deck points, and higher beltlines) are useful in determining fields of view needed to satisfy the customers.

- *Accident Experience*: Analyses of accident databases may show higher accident rates for some vehicle designs under certain visibility situations (e.g., wider A-pillar obscurations have been implicated to cause higher accidents in left-turning situations).
- *Past Research Studies*: Results from field studies involving measurements of driver behavior, performance and preferences while using different driver vision systems under different vehicle use and traffic situations (e.g., eye glance behavior while using different mirror systems, effect of forward visibility on vehicle speed, distance estimation while using different mirrors, effect of plane and convex side view mirrors on accident rates and effect of hood visibility on parking performance).

TYPES OF FIELDS OF VIEW

The driver's fields of view can be classified based on a) direct versus indirect fields and b) field coverage with each eye separately, either eye or both eyes.

The direct field consists of the views that the driver sees directly by moving his/her eyes and head. These include a) forward view (through the windshield), b) rear view (directly looking back through the backlite [rear window]), and c) side views (directly looking through the left and right side windows). The indirect field is what the driver views indirectly by use of imaging devices such as the inside mirror, the outside mirrors, or display screens showing camera views or locations of objects detected by other sensors (e.g., blind area detection systems, back up sensors).

The monocular field is the view obtained by only one eye. Figure 8.1 shows a plan view of the human head showing the horizontal fields of view from the left and right eyes. The field of the left eye shown as "L" is the monocular field of the left eye. Similarly, "R" denotes the monocular field of the right eye. The ambinocular field is

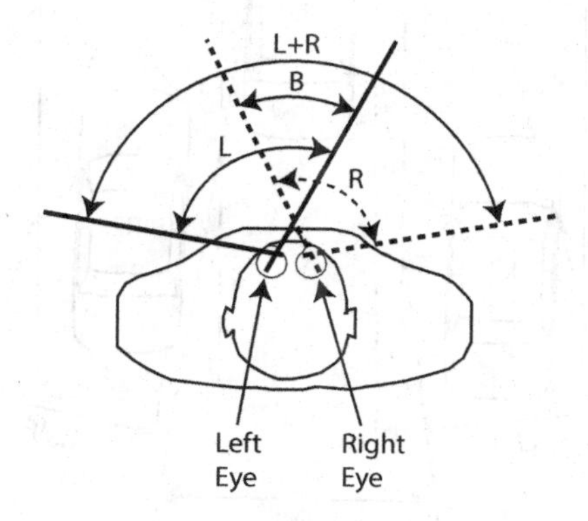

FIGURE 8.1 Left monocular (L), right monocular (R), binocular (B) and ambinocular (L+R) fields. The picture represents a plan view of the driver and horizontal fields of the left (L) and the right (R) eyes.

the sum of the fields obtained from the left and the right eye (L + R). The binocular field (B) is the common (i.e. only the overlapping portions) field seen by both eyes.

SYSTEMS CONSIDERATION OF 360 DEGREES VISIBILITY

The direct and indirect fields that a driver can obtain while seated in a vehicle should be designed such that the driver can always obtain 360-degree visibility around his vehicle. Figure 8.2 shows that the driver in the subject vehicle (labeled in the figure as

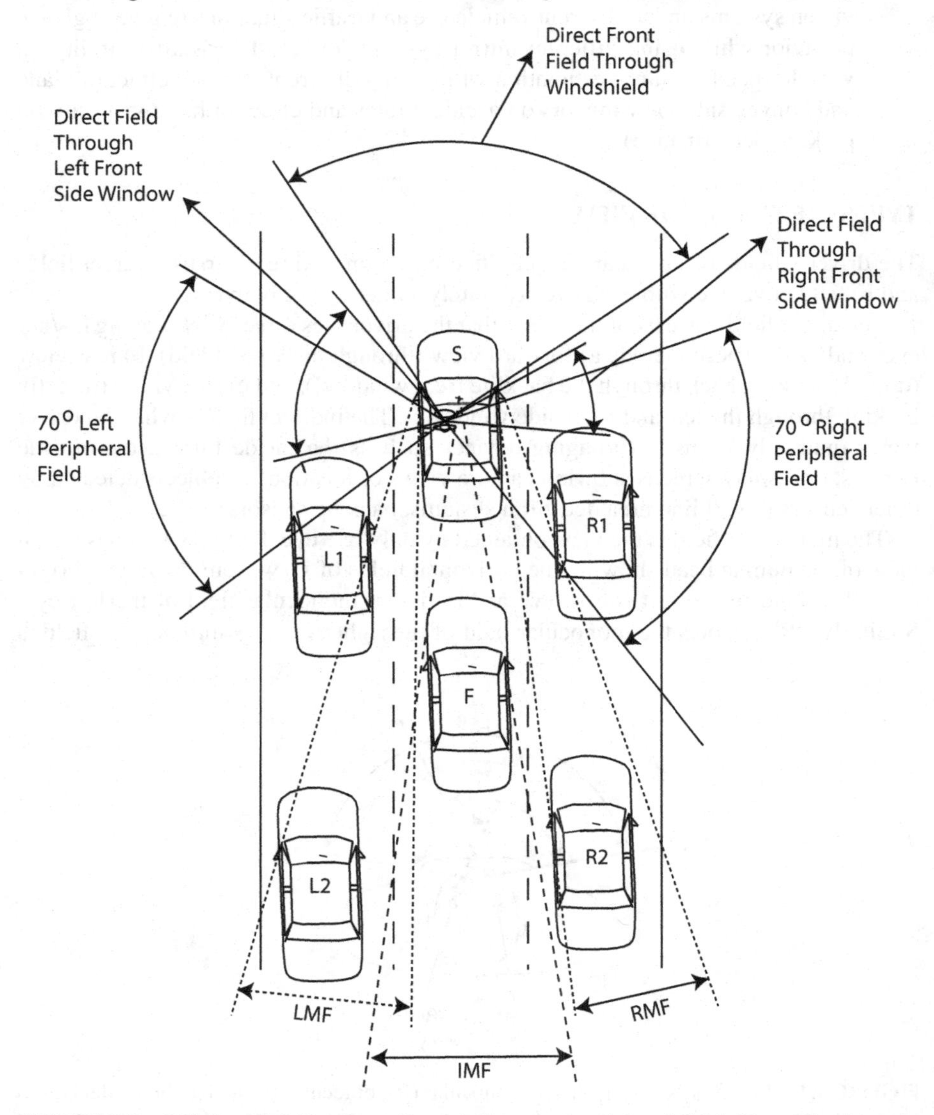

FIGURE 8.2 360 deg. visibility from direct, indirect and peripheral fields.Note: The driver in vehicle "S" can see at least a portion of any of the vehicles on his sides and rear.

"S"), shown in the middle lane of the three lane highway, can see 360-degrees around his vehicle through direct, indirect and peripheral visual fields. The fields that the driver can see directly are what he can see from his windshield and side windows by turning his eyes and head. The driver can also see objects in his indirect mirror fields shown in the figure as "LMF", "IMF" and "RMF" through the use of left outside, inside and right outside mirrors, respectively. In addition, the figure shows peripheral direct fields (labeled as 70^0 left peripheral and 70^0 right peripheral fields) that the driver can see when looking at the left outside or the right outside mirrors. Thus, the figure shows that the driver can see at least a part of any of the surrounding vehicles (labeled as vehicles "L1" and "L2" in the left adjacent lanes, "F" as the following vehicle and "R1" and "R2" as the vehicles in the right adjacent lanes) in his direct, indirect or peripheral fields under both daytime and night driving situations. Under night driving situations, the driver can see at least one headlamp or side marker lamp of each vehicle in the left and right side adjacent lanes, and at least one headlamp of the following vehicle in one of the three mirror fields or in one of his peripheral fields. Thus, all other vehicles around the subject vehicle can be seen either in the direct field of view, or in a direct peripheral field, which was found to extend peripherally about 70 degrees from the line of sight while viewing into the left or right outside mirror (Ford Motor Company, 1973).

The vehicle designer, thus, must ensure that the direct and indirect fields (with proper aiming of the three mirrors) can provide to most drivers 360-degree visibility of surrounding vehicles (i.e., the driver can see at least a part of each of the surrounding vehicles). This means that the locations of the three mirrors, their sizes, the vehicle green-house (pillars and window openings) and the driver eye locations should be all designed as a system to ensure that the driver can see any of the vehicles in the left and right adjacent lanes and directly behind and ahead in the driving lane. The vehicle package engineers and ergonomics engineers use CAD models of the vehicles during the concept development phase to verify that the vehicle meets this 360-degree visibility requirement.

MONOCULAR, AMBINOCULAR, AND BINOCULAR VISION

The views obtained by any one eye, sum of the fields of both eyes and only the field common to both eyes are called monocular, ambinocular and binocular fields, respectively. Figures 8.3 and 8.4 present photographs taken from the same vehicle to illustrate what the driver sees from his left and right eyes respectively, while looking toward a left outside mirror. Since the driver receives information available from both the eyes and the brain fuses these images together from both the eyes, Figure 8.5 shows superimposed views of both the monocular views.

To understand the differences between the two images and what the driver sees from the fused images, Figure 8.6 shows the outlines of the left A-pillar and the left outside mirror in Figure 8.5 from the two eyes. Figure 8.6 shows that the right eye's view (shown in the dotted lines) and the left eye's view (shown in the solid lines) are different, that is, the corresponding lines of A-pillar and mirror obstructions do not fall on top of each other. Figure 8.7 shows the binocular obscuration (shaded areas)

FIGURE 8.3 Monocular view from left eye.

FIGURE 8.4 Monocular view from right eye.

caused by the A-pillar and the outside mirror. The binocular obscuration is observed to be smaller than the obscuration in either the left or right monocular views. This is because some portions of obscuration in the left eye are visible to the right eye, and vice versa. Thus, the binocular obscuration (i.e., the field not visible to both the eyes simultaneously) is always smaller than any of the monocular obscurations.

Figure 8.8 shows another interesting effect. It shows the ambinocular outside mirror field (shaded area), which is what the driver will see as the total mirror field from the left and the right eyes together (assuming that the mirror surface is the same as the shroud that surrounds the mirror). Thus, for a two-eyed driver, the ambinocular mirror field is larger than what he can see with either eye. The binocular field through the mirror, where the driver can see the same field with both eyes, is smaller than the field seen by either eye. Since the objects in the binocular mirror field are viewed by

FIGURE 8.5 Superimposed view of views from left and right eyes.

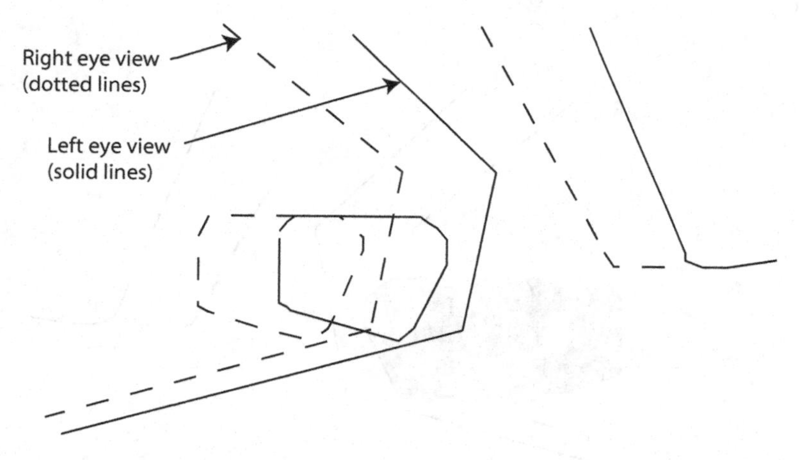

FIGURE 8.6 Superimposed views from both eyes of outlines of obstructions caused by the A-pillar and the side view mirror.

both eyes (and images of the object seen by each eye are slightly different), the driver gets additional information, which improves the perception of depth (or distances) of the objects seen in the mirror field. The portions of the ambinocular field that do not contain the binocular field are only seen by one eye. Any object seen monocularly (i.e. only with one eye) does not get the additional cues (from the other eye), and therefore, the driver's judgment of depth and location of the object seen only in the monocular field are less precise than the judgments made when the object is seen in the binocular field.

This is also an important issue in realizing that if a single camera (non-steroscopic) is mounted outside the vehicle to replace an outside mirror, the camera view provided to the driver on a display screen mounted inside the vehicle can only provide a

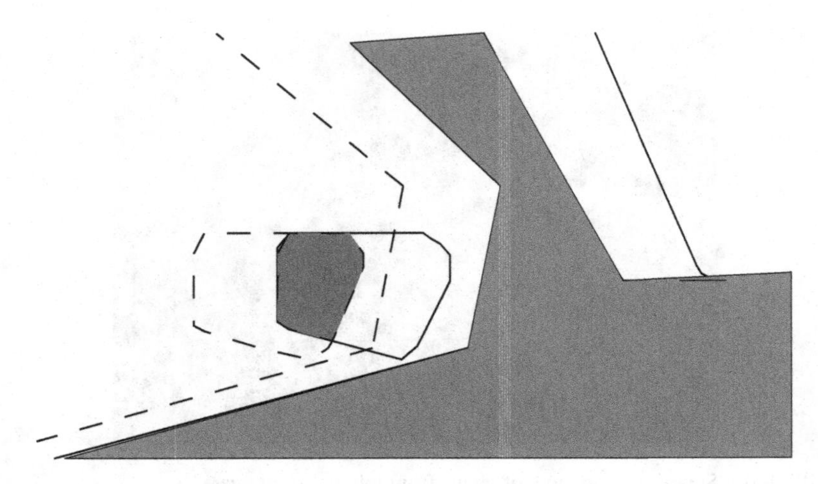

FIGURE 8.7 Binocular obstructions (dark areas) caused by the A-pillar and the side view mirror.

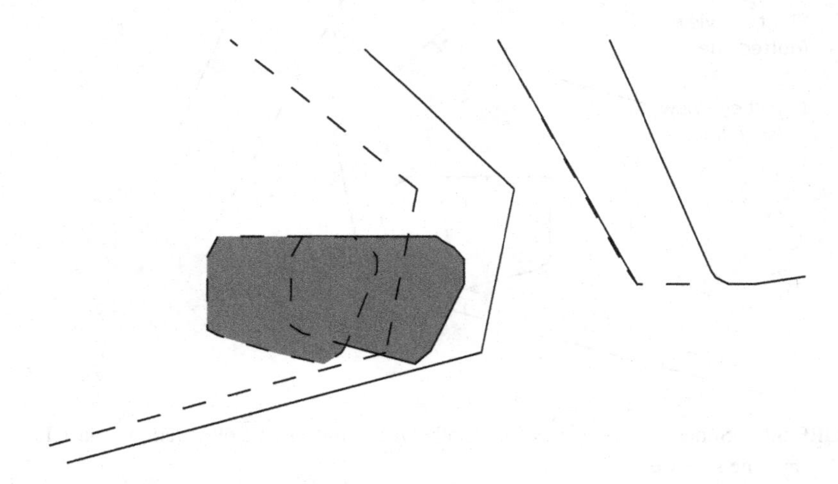

FIGURE 8.8 Ambinocular mirror field from left and right eyes.

"monocular view". The camera view, thus, lacks the binocular cues obtained by the driver using a plane outside mirror.

The field of view determination procedures, therefore, must measure the monocular, ambinoular and binocular views of the driver and evaluate a) locations and sizes of binocular obstructions, and b) the sizes of ambinocular fields.

FORWARD FIELD OF VIEW EVALUATIONS

To determine if a vehicle design will provide satisfactory forward field of view, the vehicle design team must conduct many analyses to evaluate different driver needs and requirements. Important issues related to the forward field of view are presented below.

Up and Down Angle Evaluations

The up angle (A60-1) from the tall driver's eye points (95th percentile eye location) and down angle (A61-1) from the short driver's eye points (5th percentile eye location) should be determined by drawing tangent sightlines (in the side view) from the 95th percentile eyellipse to the top and bottom edges of the windshield (daylight openings considering the black-out paint) as shown in Figure 8.9 below (refer to SAE Standard J1100 for definitions of the angles). The above angles are measured in the vertical plane passing through the driver centerline (i.e. through the driver's SgRP) and using a mid-eyellipse. The angles are measured with respect to the horizontal. A smaller up angle (A60-1) indicates that a tall driver will experience difficulty in looking at high-mounted targets (e.g., the tall driver may have to duck his head down to view a high mounted traffic signal while waiting at an intersection). Whereas a smaller down angle (A61-1) indicates insufficient visibility for the short drivers over the cowl area.

The visibility of short drivers should be also evaluated by drawing tangent sightlines over the top of the steering wheel, top of the instrument panel binnacle (most upward protruding parts) and over the hood from the short (2.5 or 5th percentile female) driver eye's locations.

Visibility of, and over, the Hood

1. The visibility of the road surface (i.e., the closest longitudinal forward distance from the front bumper at which the road surface is visible, also called the

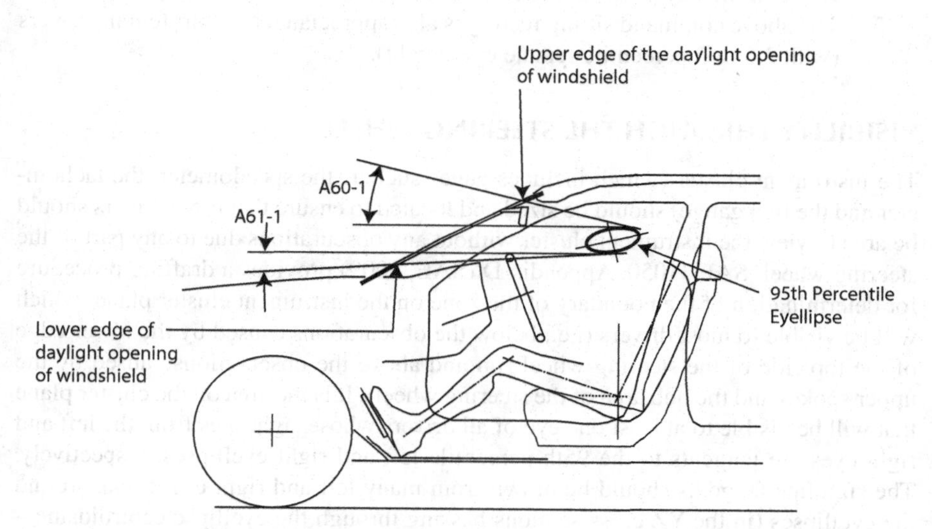

FIGURE 8.9 Tall driver up angle (A60-1) and short driver down angle (A61-1).

ground intercept distance) is of critical concern to many drivers. The problem is worse for short drivers. In general, most drivers want and like to see the end of the hood, the vehicle corners (extremities) and the road at a close distance. As more aerodynamic vehicle designs with low front ends were introduced in the United States after the mid-1980s, many drivers who were accustomed to the long hoods with their visible front corners complained about not able to see the ends of their hoods.

2. The view of the hood provides better perception of the vehicle heading with respect to the roadway, providing a feeling of ease in lane maintenance and while parking. (Note: Racing cars have a wide painted strip over the hood at the driver centerline to provide highly visible vehicle-heading cues in their peripheral vision).
3. Heavy trucks with long hoods experience a larger obstruction of the road due to the hood. The problem can be severe if the obstruction is large enough to hide a small vehicle (e.g., bike rider, sports car) located in front of the hood. The problem often occurs when the truck is behind a small vehicle while waiting for a signal change at an intersection.

Command Seating Position

1. The "Command Seating Position" provides the feeling of "sitting high" in the vehicle. It is opposite to the feeling of "sitting in a well" or "sitting too low" in the vehicle.
2. For the command seating position, provide a) higher SgRP location from the ground, b) low cowl point, c) low beltline, d) adequate visibility of the hood and d) greater visibility of the roadway (shorter ground intercept distance from the front bumper). It should be noted that the command sitting position is one of the key positive attributes of an SUV.
3. The above command sitting feeling is also appreciated by short female drivers (with 2.5 or 5 percentile female eye-height).

VISIBILITY THROUGH THE STEERING WHEEL

The instrument cluster (which includes gages such as the speedometer, the tachometer and the fuel gauge) should be sized and located to ensure that most drivers should be able to view the instrument cluster without any obscurations due to any part of the steering wheel. SAE J1050, Appendix D (SAE, 2009) provides a drafting procedure for determination of the boundary of the zone on the instrument cluster plane, which will be visible to most drivers (i.e., below the obscurations caused by the lower edge of the top side of the steering wheel rim and above the obscurations caused by the upper spokes and the hub area of the steering wheel). It is the area on the cluster plane that will be visible to at least one eye of all drivers whose sightlines from the left and right eyes are tangents to the 95th percentile left and right eyellipses, respectively. The sightline tangents should be drawn from many left and right eye points around the eyellipses (in the YZ cross-sections passing through the eyellipse centroidsinc – luding the eyepoints of tall drivers, drivers leaning to the right, short drivers and drivers leaning to the left) are considered to determine visible region that would be common from all eyepoints on the cluster plane while viewing under the top rim of the steering wheel and over the steering wheel hub and upper spokes.

SHORT DRIVER PROBLEMS

Short drivers are drivers with shorter (5th percentile and below) sitting eye-heights and/or shorter (5th percentile and below) leg lengths. The visibility problems encountered by such short drivers are as follows: 1. Obstruction of the road by the steering wheel (top part of the rim) and instrument panel (or cluster binnacle; causing smaller down angle [A61-1]).

1. Obstruction of the road by the steering wheel (top part of the rim) and instrument panel (or cluster binnacle; causing smaller down angle [A61-1]).
2. Inability to see any part of the hood – no visibility of the end of the hood). (Note: providing a raised hood ornament near the front of the hood can provide useful information in maintaining vehicle heading. Similarly, providing visibility of the corners of the hood or ends of the front fenders, via placement of "flag poles" as provided on some trucks with longer hoods, can improve ease in parking and lane maintenance).
3. The side view mirrors may obscure the forward direct view. The upper edge of the side view mirrors should be placed at least 20 mm below the 5th percentile female's eye point.
4. The closest distance at which a driver can see the road (over the hood) is much longer for the short driver than the visible distances for other drivers.
5. The shorter drivers will experience reduced rear visibility during reversing or backing up, especially with a higher deck point and taller rear headrests. (Note: One check that many package engineers consider is whether a short driver can see a 1-m high target, simulating a toddler, in the rear view while backing-up in the direct rear view with the driver's head turned rearward and also while looking in the inside mirror.)
6. Since short drivers (with shorter leg lengths) sit more forward in the seat track, the driver's side A-pillar will create larger obscurations in the forward field of view for short drivers as compared to tall drivers.
7. Short drivers require larger head-turn angles to view side view mirrors due to their more-forward seating position as compared to the taller drivers. This problem is more severe for short drivers (typically older short females) who have arthritis, because they may have a shorter range of head-turn angles.

TALL DRIVER PROBLEMS

The tall drivers are those with greater (95th percentile and above) sitting eye heights and/or longer (95th percentile and above) leg lengths. The visibility problems encountered by such tall drivers are as follows:

1. External objects placed at higher locations (placed above the upper sightline at up angle (A60-1 in Figure 8.9) may be obstructed from the view of tall drivers (e.g., a tall driver may have to duck his head down to view overhead traffic signals at intersections). The visibility near the top portion of the windshield is further limited by the shade bands and/or the black-out paint applied to the windshields.
2. The inside rearview mirror may block the tall driver's direct forward field. Therefore, the lower edge of the inside mirror should be placed at least 20 mm above the 95th percentile eye points.
3. The tall driver also sits farther from the mirrors due to a more rearward sitting position. Thus, the mirrors provide smaller fields of view to the tall drivers as compared to mirror fields of other drivers.
4. The tall driver may have more side visibility problems because of: a) more forward B-pillar obscurations in direct side viewing (because the tall driver sits more rearward in the seat track than the shorter drivers), and b) more forward peripheral awareness zones (i.e., the peripheral zones [see Figure 8.2] do not extend as much rearward as for the shorter drivers) while using the side view mirrors.

SUN VISOR DESIGN ISSUES

1. The sun visor drop-down height and its length should be designed to prevent the incidences of direct sunlight on the driver's eyes from different sun angles and sun glare from the windshield and driver's side window.
2. The sun visor dropped-down position should be adjustable, and it should be capable of dropping down to accommodate the short driver's needs.
3. If the sun visor hinge mechanism becomes loose, the sun visor may accidentally swing and drop down and cause obstruction in the forward field of view. This obstruction will be more severe for the taller drivers.

WIPER AND DEFROSTER REQUIREMENTS

The SAE standards J902 and J903 (SAE, 2009) also FMVSS 103 and 104; NHTSA, 2024, provide requirements on how to establish areas in the driver's forward field that must be defogged (or defrosted) and wiped by the wipers, respectively. The requirements specify the sizes of these areas and percentages of each area to be covered (cleaned) by the defoggers and wipers. The areas are specified by establishing four tangent planes to the 95th percentile eyellipses.

The areas to be covered by the wiper sweep pattern are defined as areas A, B and C (SAE J903; SAE, 2009). The areas A, B, and C are defined by drawing up, down, left and right tangent planes to the eyellipses as shown in Figure 8.10. The area A is bounded by the upper tangent plane at 10 degrees up angle, the bottom tangent plane at 5 degrees down angle (see side view in Figure 8.10), the left tangent plane at 18 degrees angle to the left eye ellipse and the right tangent plane at 56 degrees angle to the right eyellipse (see plan view in Figure 8.10). Similarly, the angles defining area

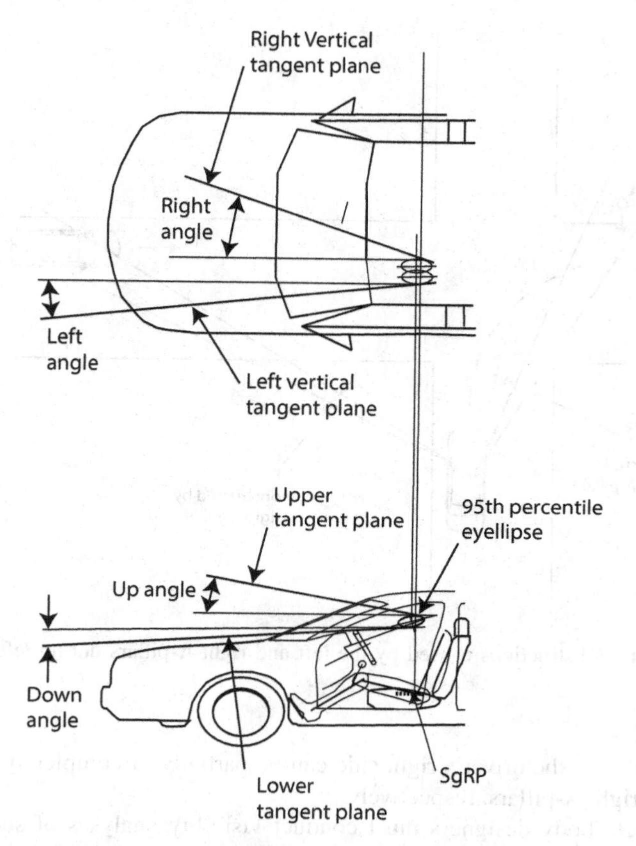

FIGURE 8.10 Plan and side views showing the four tangent planes that define the wiping areas to be covered by the wipers.

B are 5 degrees up, 3 degrees down, 14 degrees left, and 53 degrees right. The angles defining area C are 5 degrees up, 1 degree down, 10 degrees left and 15 degrees right. The wiper and defroster/defogging systems must be designed so that at least 80 percent of area A, at least 94 percent of area B, and at least 99 percent of area C must be cleared wiped by the wiping/defrosting systems.

OBSCURATIONS CAUSED BY A-PILLARS

The left and right front roof pillars (called the A-pillars), depending upon their size and shape of their cross-sections at different heights with respect to the driver's eye locations, can cause binocular obstructions in the driver's direct forward field of view. The obstructions can hide targets such as pedestrians and other vehicles during certain situations. Figure 8.11 shows that during an approach and left turn through an intersection, a pedestrian crossing the street on the driver's left side and vehicles

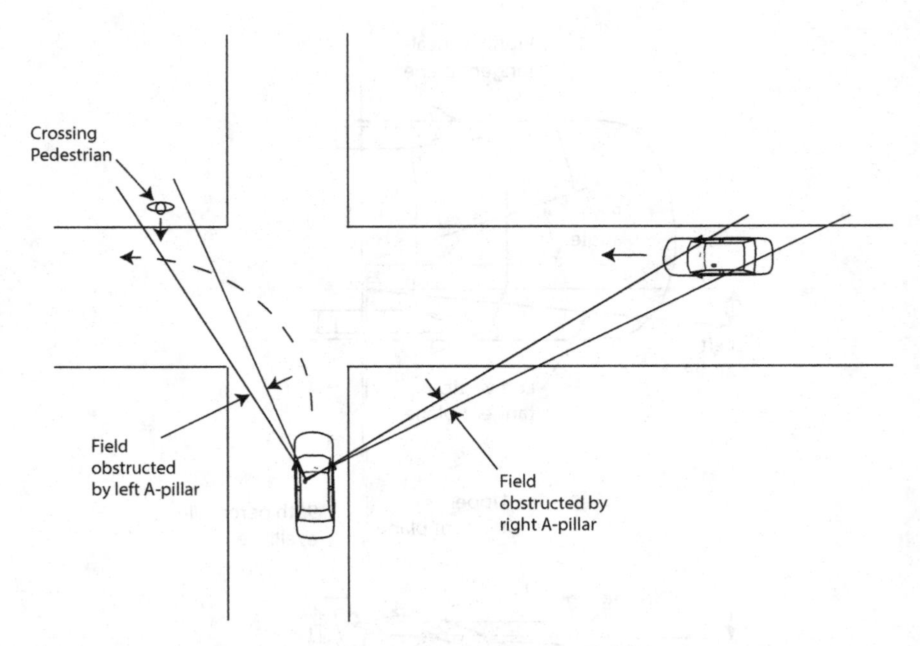

FIGURE 8.11 Obstructions caused by the left and right A-pillars during left-turning at an intersection.

approaching from the driver's right side can be partially or completely obscured by the left and right A-pillars, respectively.

The vehicle body designers must conduct visibility analyses of such situations and minimize the obstructions caused by the pillars. The body engineers make trade-offs between a) increasing the cross section of the pillars to meet the roof-crush requirements in FMVSS 216 (NHTSA, 2024), b) applying padding to reduce head impact injuries, and c) reducing the binocular visual obstructions simultaneously for left and right side viewing . As the vehicle design progresses, there are other issues such as: a) addition or increase in the thickness of the rubber seals used to secure the windshield to the pillars, b) the black-out paint applied to the glass to hide the joints and improve appearance from the exterior, and c) manufacturing variations in pillar cross-sectional areas with respect to the SgRP. These three issues tend to increase the obstructions caused by the pillars.

The SAE standard J1050 in its Appendix C (SAE, 2009) provides a procedure to measure the visual obstruction caused by the A-pillar. Figure 8.12 below illustrates the binocular obstruction angle obtained by drawing sightlines tangent to the cross section of the left and right pillars. The obstruction angles, β_L and β_R, are determined by using the following steps:

1. The obstruction is measured in the horizontal plane passing through the eyellipse centroids. Thus, Figure 8.12(a) shows the left A-pillar cross section

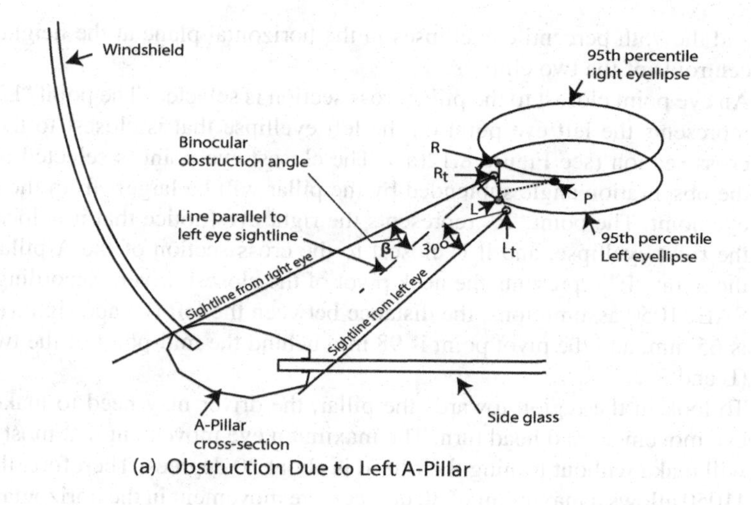

(a) Obstruction Due to Left A-Pillar

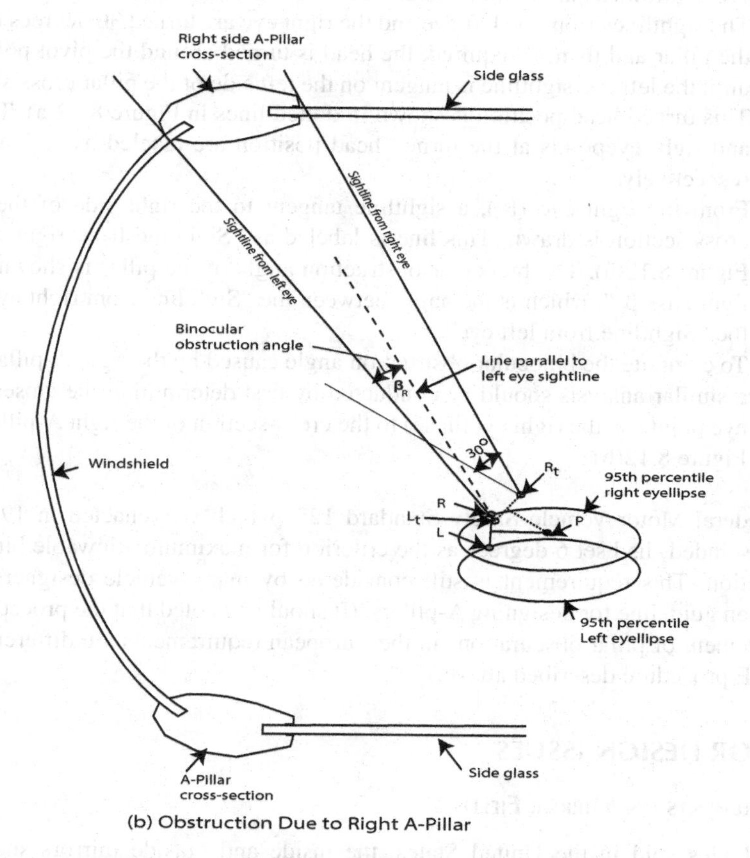

(b) Obstruction Due to Right A-Pillar

FIGURE 8.12 Determination of binocular obscuration angle due to left and right A-pillars.

and the 95th percentile eyellipses in the horizontal plane at the height of the centroids of the two ellipses.

2. An eye point closest to the pillar cross section is selected. The point "L", thus, represents the left eye point on the left eyellipse that is closest to the pillar cross-section (see Figure 8.12(a)). The closest eye point is selected because the obscuration angle subtended by the pillar will be largest from the closest eye point. The point "R" represents the right eye (notice that it is located on the right eyellipse, and it is closest to the cross-section of the A-pillar), and the point "P" represents the neck pivot of the closest driver. According to the SAE J1050 assumptions, the distance between the left (L) and right (R) eyes is 65 mm, and the pivot point is 98 mm behind the mid-point of the two eyes (L and R).
3. To look in the region towards the pillar, the driver may need to make some eye movement and head turn. The maximum eye movement that most people will make without turning their head is about 30 degrees. Therefore, the SAE J1050 allows a maximum of 30 degrees eye movement in the horizontal plane. The sightlines from the left eye and the right eye are turned 30 degrees toward the pillar and then, if required, the head is turned around the pivot point "P" until the left eye sightline is tangent on the left side of the pillar cross-section. This turned head position is shown in dotted lines in Figure 8.12(a). The left and right eyepoints at the turned head position are labeled as "L_t" and R_t" respectively.
4. From the right eye (R_t), a sightline tangent to the right side of the pillar cross-section is drawn. This line is labeled as "Sightline from right eye" in Figure 8.12(a). The binocular obstruction angle of the pillar is shown in the figure as "β_L", which is the angle between the "Sightline from right eye" and the "Sightline from left eye".
5. To compute the binocular obstruction angle caused by the right A-pillar "β_R", a similar analysis should be conducted by first determining the closest right eye point (on the right eyellipse) to the cross-section of the right A-pillar (See Figure 8.12(b)).

The Federal Motor Vehicle Safety Standard 128 (which was enacted in 1978 and later rescinded) had set 6 degrees as the criterion for maximum allowable binocular obstruction. This requirement is still considered by many vehicle designers as an unwritten guideline for designing A-pillars. (It should be noted that the procedure for measurement of pillar obscurations in the European requirements are different from the SAE procedure described above.)

MIRROR DESIGN ISSUES

REQUIREMENTS ON MIRROR FIELDS

For vehicles sold in the United States, the inside and outside mirrors should be designed to meet the field of view requirements specified in FMVSS 111 (NHTSA,

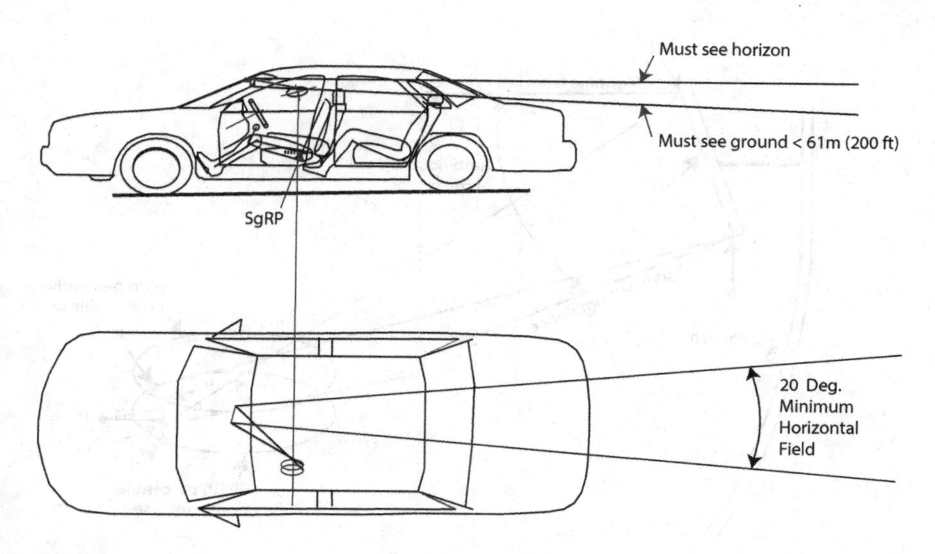

FIGURE 8.13 Inside mirror field required for passenger cars.

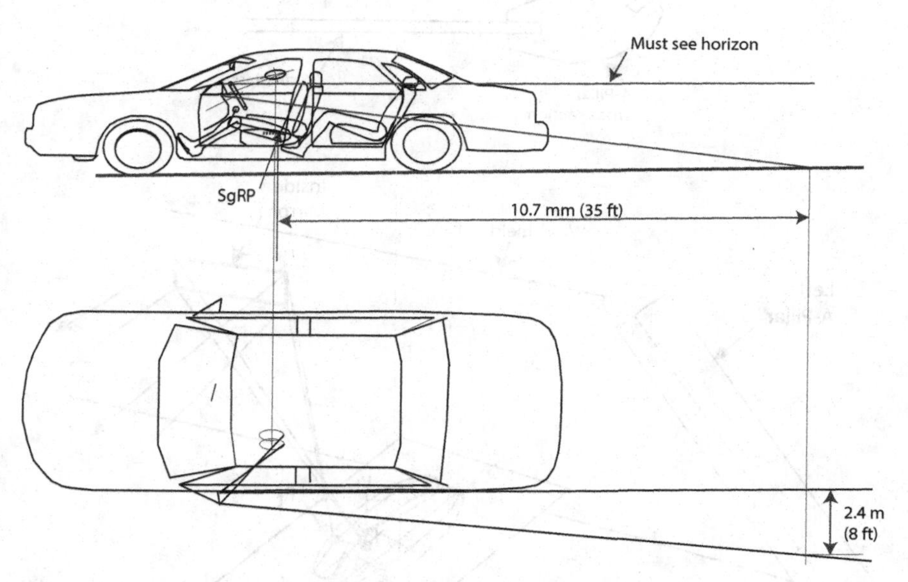

FIGURE 8.14 Driver's side mirror field required for passenger cars.

2023). Figures 8.13 and 8.14 show the minimum required fields for inside and driver's side outside mirrors, respectively, for passenger cars.

The inside plane mirror should provide at least a 20 degrees horizontal field and the vertical field should intersect the ground plane at 61 m (200 ft.) or closer from the driver's SgRP to the horizon (see Figure 8.13). A procedure for determination of the mirror field of view is covered in a later section (see Figure 8.15).

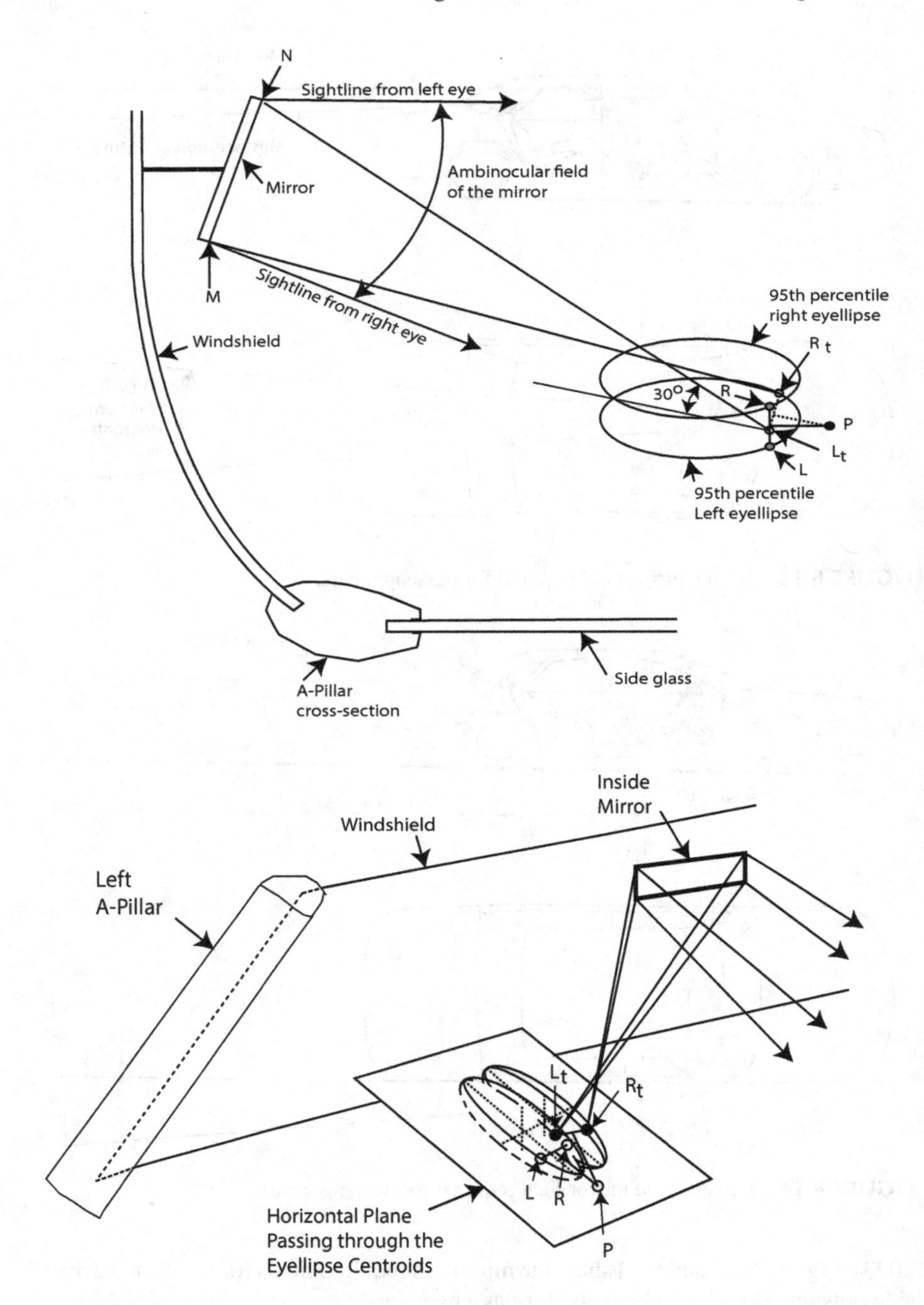

FIGURE 8.15 Determination of inside mirror field of view.

The driver's side outside plane mirror should provide (as specified in the FMVSS 111 [NHTSA, 2023]) a horizontal field of 2.4 m (8 ft) width at 10.7 m (35 ft) behind the driver at ground level and the vertical field should cover the field from the ground line at 10.7 m (35 ft) to the horizon (see Figure 8.14).

Mirror Locations

Inside Mirror Location

The inside mirror should be located with the following design considerations:

a) The mirror should be placed within 95th percentile maximum reach envelope with full hand grasp using SAE J287 procedure. This ensures that at least 95 percent of the drivers should be able to re-aim the mirror by grasping with their right hand without leaning forward.
b) The lower edge of the mirror should be located at least 20 mm above the 95th percentile driver eye height. This ensures that the mirror will not cause obstruction in the forward direct field of view for at least 95 percent of drivers.
c) The mirror should be placed outside the head swing area (during a frontal crash) of the driver and the front passenger (refer to FMVSS 201, NHTSA, 2023).

Outside Mirror Locations

The driver's side outside mirrors should be located with the following design considerations:

a) The driver's side outside mirror should be located such that a short driver who sits at the forward-most location on the seat track should not require a head turn angle of more than 60 degrees from the forward line of sight.
b) The upper edge of the mirror should be placed at least 20 mm below the 5th percentile driver eye location to avoid obscuration in the direct side view of at least 95 percent of the drivers.
c) The mirror aiming mechanism should allow for a horizontal aim range large enough for a short driver to see a part of his vehicle and tall driver to aim outward to reduce the blind area in the adjacent lane.
d) In addition, to improve the aerodynamic drag and wind noise, the mirror housing design needs (reduced frontal area) should be considered along with the reduction in obscuration caused by the mirror and the left A-pillar in the driver's direct field of view (see Figure 8.7).

The passenger's side outside mirror is located symmetrically to the driver's side outside mirror. FMVSS 111 does not require an outside passenger mirror on passenger cars or a truck if the inside mirror meets its required field of view. However, if the passenger side outside mirror is provided, FMVSS 111 requires it to be a convex mirror with a radius of curvature not less than 889 mm and not more than 1651 mm (NHTSA, 2023). FMVSS 111 also provides alternate requirements for trucks and multi-purpose vehicles that cannot provide any useful field from their inside mirrors due to blockage by cargo or passenger areas.

Procedure for Determining Driver's Field of View through Mirrors

SAE Standard J1050 presents a procedure to determine the field of view through a mirror (SAE, 2009). To determine the horizontal field of view that at least 95 percent of drivers can view, one begins with a given mirror with its known location and size and the location of the 95th percentile eyellipses in the vehicle space. Figure 8.15 (see top figure) presents a plan view showing the inside mirror (on the right side of the driver) and the 95th percentile eyellipses.

The horizontal mirror field of view (from a plane mirror of unit magnification) is determined by using the following steps:

1. The mirror surface is shown in Figure 8.15 (top figure) as "MN" and the left and right eyellipses are drawn. The aim of the mirror is determined iteratively to ensure that the ambinocular mirror field defined by the reflected sight lines from the left and the right eyes is contained within the backlite.
2. First locate the eyepoint that is farthest from the mirror. The farthest eyepoint is selected because it provides the smallest mirror field. The left eyepoint labeled as "L" on the left eyellipse is the farthest eyepoint from point "N" (which is the farthest point on the mirror from the eyellipses).
3. For the farthest left eyepoint "L", the corresponding right eye (65 mm apart) on the right eyellipse is identified as "R". The neck pivot point for head turn corresponding to the eyepoints "L" and "R" is "P". The pivot point "P" is located 98 mm behind the two eye points.
4. From the left eye "L", the line of sight at 30 degrees to the right is drawn to show maximum eye turn. The head is turned around the pivot point "P" until the 30 degree turned line of sight from the left eye (L) passes through point "N". The left eye at the turned head position is shown as "L_t".
5. The right eye at the turned head position is indicated as "R_t". The line of sight from R_t is connected to the left point on the mirror "M".
6. The mirror ambinocular field of view is the angle between the sightlines that are reflected (shown in the Figure as "Sight line from right eye" and "Sight line from left eye") at points "M" (i.e., reflection of sight line R_tM) and "N" (reflection of sight line L_tN), respectively.

The horizontal field of the left outside mirror can be computed by using a similar procedure. The analysis for the left side mirror will be flipped, that is, the farthest eye point from the left mirror will be the farthest right eye point.

To compute the vertical ambinocular mirror field, the above analysis is conducted in three dimensions where, instead of just reflecting the sight lines from points "N" and "M" as described above, the sight lines from R_t and L_t to all the four corners (top and bottom points on the left and right sides) of a rectangular mirror are constructed (see lower figure in Figure 8.15). The L_t and R_t turned head eye points can also be raised if the driver requires to tilt head upwards to maintain up angles of the sight lines up to 45 degrees maximum (see SAE J1050, SAE 2009 for more details).

CONVEX AND ASPHERICAL MIRRORS

If a convex mirror is used instead of a plane mirror, the driver's field of view through the mirror will be larger. The radius of convex mirrors used in vehicles usually ranges between about 1016 to 1524 mm (40 to 60 inches). FMVSS 111 requires the average convex mirror radius to be between 889 mm (35 inches) to 1651 mm (65 inches) for convex mirrors mounted on the passenger side. The field of view covered by a convex mirror increases as the mirror radius is decreased. However, the size of the images of objects in the convex mirrors will decrease (be minified) as the mirror radius is decreased. Convex mirrors with radii below about 889–1016 mm (3540 inches) are not recommended because the images of the objects (e.g., other vehicles in the rear field) will be too small to get an accurate estimation of their distances and speeds relative to the subject vehicle; and radii greater than 1524–1651 mm (60–65 inches) are not recommended as the images of the objects cannot be well discriminated with images seen in a plane mirror (with radius equal to infinity).

Furthermore, the images of objects in a convex mirror are located at a much closer distance to the driver's eyes than the images of the objects in the plane mirror. (The image of an object in the convex mirror is located near its focal point [which is located at half the radius of the convex mirror] behind the mirror surface.) Thus, it is more difficult for the drivers to estimate distances of objects seen in convex mirrors. Because of the minified images, drivers will tend to overestimate distances of objects (such as other vehicles viewed in the mirror field). Further, since older drivers cannot focus on closer distances, they will have difficulty in using convex mirrors placed on the driver's side. The difficulty is increased due to the appearance of double images for some drivers, especially when the convex mirrors are placed at closer viewing distances. Since the convex mirrors placed on the passenger side door are at a larger distance from the driver, viewing the minified images becomes the primary problem in estimating distances of objects in the mirror.

Aspherical mirrors: An aspherical mirror has a continuously changing radius in the lateral (horizontal) direction, usually changing from a large radius (at the edge closer to the driver) to a small radius (at the edge farthest from the driver with a radius of 735 mm [30 inches] or more). (Note: the plane mirror has a radius equal to infinity). The advantage of aspherical mirrors is that they provide a larger mirror fields than a plane mirror of the same size. However, many drivers experience visual strain because a) the images of views seen by the driver's two eyes have different minification and b) the distance-estimation difficulties of drivers with the minified images. Thus, the drivers need considerable practice to get used to the aspherical mirrors. Older drivers typically have more difficulty in getting used to aspherical mirrors than younger drivers, due to the inability of older drivers to focus on closer distances and the visual strain associated with double images.

FMVSS 111 currently does not require aspherical mirrors on vehicles used in the United States.

METHODS TO MEASURE FIELDS OF VIEW

The field of view issues described above should be analyzed to ensure that a vehicle being designed will not cause visual problems when it is used in different driving situations by drivers with differing visual characteristics within the target population.

During the early design phases, when the vehicle greenhouse is being defined and the driver's eye locations have been established in the vehicle space, vehicle package engineers and ergonomics engineers should conduct a number of field of view analyses. The field of view analysis methods are generally incorporated in the CAD systems used for digital representation and visualization of the vehicle. The methods involve projecting the driver's sightlines to different components (such as pillars, window openings, mirrors, instrument panels, hoods and deck surfaces) on to different projection planes such as ground planes, vertical target planes, spherical surface (for polar plots) and instrument panel surfaces (for visibility of displays and controls). Physical devices (e.g., sighting devices, light sources, lasers and cameras) have also been used to conduct evaluations of physical properties (e.g., bucks, production or prototype vehicles). However, the positioning of such devices in vehicle space with high precision is time consuming and costly.

Early feedback on vehicle designs that reduce driver visibility by increasing obstructions (e.g., due to wider pillars, bigger headrests, higher beltlines, or smaller mirrors) should be investigated fully by creating full-size bucks or even drivable mock-ups for market research clinics or human factors field tests. Such problems, if not fixed early, would be extremely time consuming and expensive to change during the later stages of vehicle development.

Ground Shadow Plot

A ground shadow plot is analogous to creating a shadow of the entire vehicle's greenhouse by placing two light sources at the driver's left and right eye points on the horizontal ground plane of the road surface. The dark areas in the ground shadow plot represent areas that are not seen from the driver's both eye locations (i.e., binocular obscurations caused by pillars, outside mirrors, hood, instrument panel and beltline). Conversely, the lighted areas in the ground shadow plot present areas that are visible to the driver. The ground shadow plots are generated by creating sightlines from the driver's eye points to the daylight openings and outside mirror structures using a CAD system. The ground shadow plots of different vehicles can be compared by superimposing them by using common ground plane and SgRP projected location in the plan view. Figure 18.16 presents an example of a ground shadow plot. The test vehicle is placed in the middle plane of the 3-lane highway. The 45-degree hatched areas in this figure represent the obscurations on the ground plane by the pillars, hood, beltline and the outside mirrors from the eye points of the driver in the test vehicle.

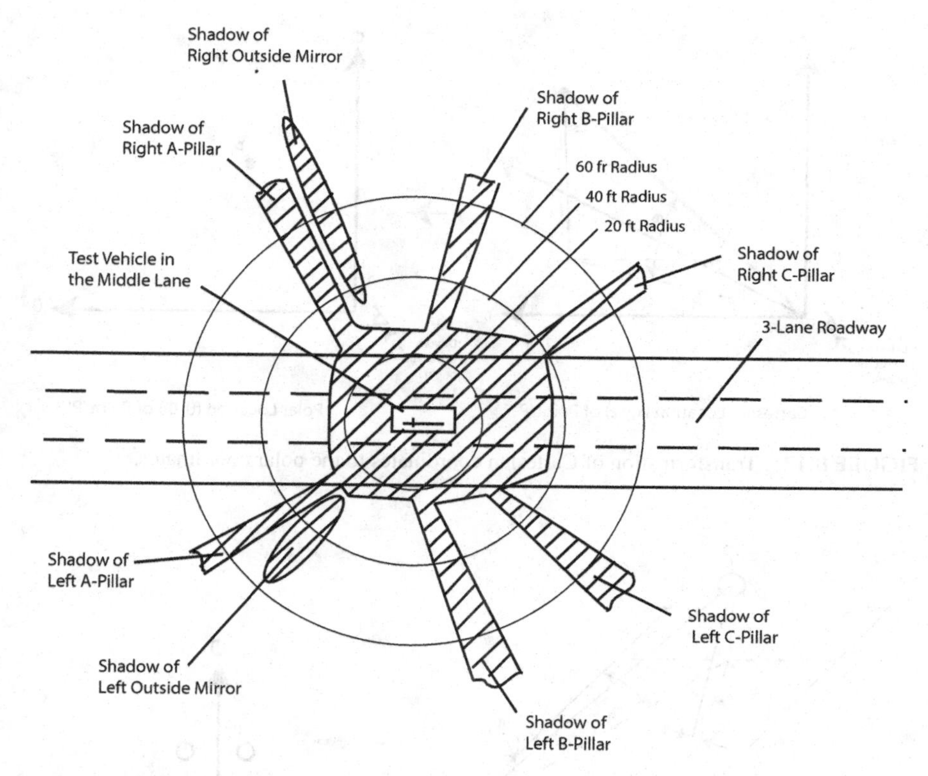

FIGURE 8.16 An illustration of a ground shadow plot.

Polar Plot

Creating a series of polar plots to conduct different field of view analyses is a very effective method for visualizing and measuring fields of view issues (McIssac and Bhise, 1995). A polar plot is especially useful for ergonomic analyses, as it allows direct measurements of angular fields, angular location of different objects, angular sizes of different objects, and angular amplitudes of eye movements and head movements required to view different objects. It also allows incorporation of views from both eyes, and thus, facilitates the evaluation of monocular, ambinocular and binocular fields and obscurations. It also simplifies the three-dimensional analysis by reducing it into a two-dimension analysis.

A polar plot involves plotting the visual field from the driver's (one or both) eye points in angular coordinates. It is equivalent to projecting the driver's view on a spherical surface with the driver's eyes at the center of the sphere. The driver's eye point is considered as the origin from which sight lines are originated. Each sightline aimed at a target point can be located by determining its azimuth angle in degrees (θ) and elevation angle in degrees (Φ) with respect to the eye point (as the origin) of a coordinate system. If point P is defined by (x,y,z) as its Cartesian coordinates (with the eye point as the origin), then its polar (angular) coordinates (θ, Φ) can

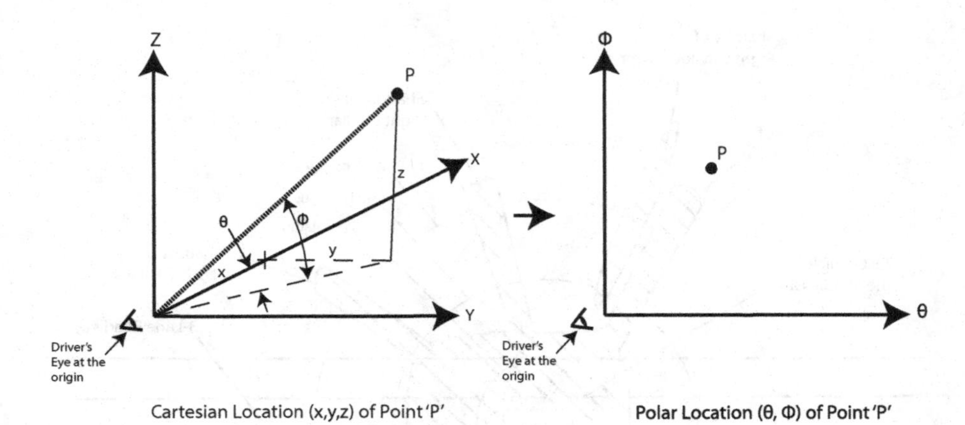

FIGURE 8.17 Transformation of Cartesian coordinates to the polar coordinates.

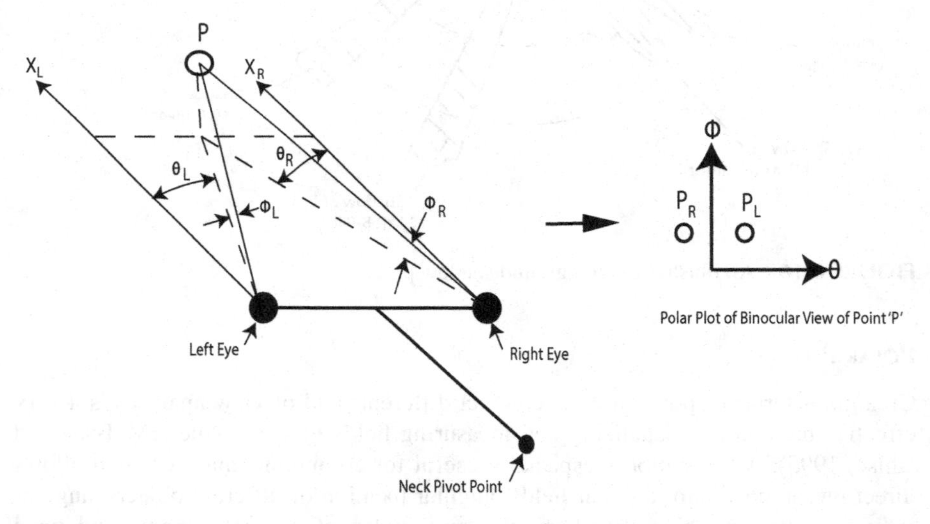

FIGURE 8.18 Illustration of viewing of point 'P" by both eyes and its polar plot.

be computed as follows: $\theta = \tan^{-1}(y/x)$ and $\Phi = \tan^{-1}[z/(x^2 + y^2)^{0.5}]$. It should be noted that this polar plotting method does not use the distances from an eye point to any object point for the analysis. Figure 8.17 illustrates the transformation of the Cartesian location of point P to its polar location.

Figure 8.18 illustrates the polar representation of a point as it is viewed by both eyes. The figure shows that 'P' is located at angular positions of (θ_L, Φ_L) and (θ_R, Φ_R), respectively, from the left and right eyes of an observer. Thus, the images of point 'P' will be at different locations on the retinas of the observer's two eyes. However, the two images will be fused and perceived as a single image by the observer's brain. The

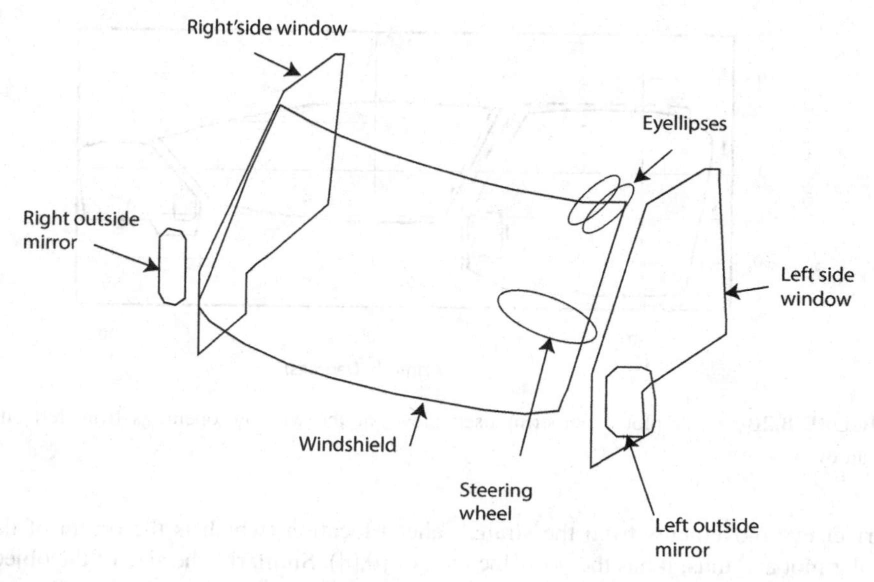

FIGURE 8.19 Data inputs for polar plotting.

figure on the right, therefore, shows locations of both the image points, namely (θ_L, Φ_L) and (θ_R, Φ_R), on a common origin. This technique of plotting polar locations of objects seen by each eye on a common origin point is very useful because it provides information on what the driver sees with each eye and also what can be obscured binocularly.

The above described concepts are used in evaluating the driver's field of view as follows. First the Cartesian coordinates of all relevant objects in the driver's field of view are measured. Figure 8.19 shows a view of the daylight openings of a truck-type vehicle with outside mirrors and eyellipses. The inputs required to create a plot would be the Cartesian coordinates of points along all edges of the window openings shown in Figure 8.19 and the left and right eye points represented by the centroids of the left and right eye ellipses, respectively. The coordinates are usually obtained from the CAD model of the vehicle or by physical measurements of actual properties (e.g., exterior models or production vehicles) using computerized coordinate measurement machines. These coordinates can be converted into polar coordinates from the knowledge of the coordinates of the eye points.

Figure 8.20 shows the polar plot of the driver's view (from the vehicle data shown in Figure 8.19) obtained from the driver's eyes located at the centroids of the eyellipses. The polar view in Figure 8.20 extends horizontally from over -90 to +90 degrees and vertically from -45 to +45 degrees. The plot shows the angular size of each window opening from both the left and right eyes. The left eye's view is shown in dotted lines and the right eye's view is shown in solid lines.

The other advantage of the polar plot is that the polar coordinates of objects included in the plot provide angular locations which can be used to directly measure

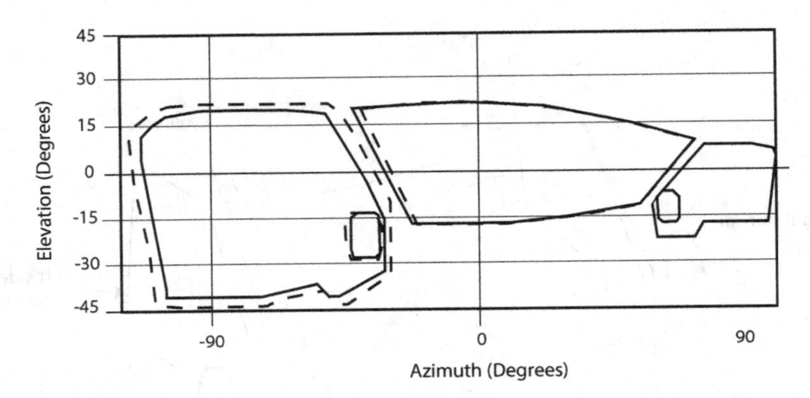

FIGURE 8.20 Polar plot of superimposed views of the window openings from left and right eyes.

driver eye movements from the straight ahead location (which is the origin of the polar plot and thus, it has the polar location of [0,0]). Similarly, the size of the object shown in the polar plot can be directly measured to determine sizes of monocular and binocular obstructions. Thus, the angular locations of the pillars, up and down angles and the binocular obscurations caused by each of the pillars can also be measured directly from the polar plot in Figure 8.20.

The roadway and many other external objects on the roadway can also be included in the polar plots. These external objects help in understanding many visibility issues in terms of what objects can be seen through the window openings and what objects are fully or partially obstructed by vehicle components. Through extensive photographic measurements of objects in the driver's view, Ford Motor Company (1973) developed targets that encompass the regions on the roadways where different objects appeared in the photographic data. These Ford targets represent different external objects such as overhead signs, side mounted signs, traffic signals, vehicles approaching from intersecting roadways, vehicles in adjacent lanes, vehicles ahead, vehicles behind the driver's car, and so forth. The targets can be placed in the polar plot to evaluate field of views from the vehicles. McIssac and Bhise (1995) describe the use of targets in polar plots. Their paper also describes the use of polar plots in determining indirect visual fields from plane and convex mirrors by plotting virtual images of objects seen in the mirrors and the outlines of the mirrors.

OTHER VISIBILITY ISSUES

Light Transmissivity

In the above-discussed field of view analyses, it was assumed that if any of the targets are located in the window openings, they will be visible to the driver. The visibility of the targets, however, depends upon the visual contrast of the targets against the luminance of the background against which each of the targets appear. The luminance of the targets and their backgrounds are affected by light transmission losses in the glass used for the windshield and other window openings. The light transmissivity depends upon the type of glass, glass thickness and installation (i.e., rake) angles. The light transmission losses increase with higher glass raked angle. The minimum light transmission (or tinting) requirements for different window openings are included in FMVSS 205 (NHTSA, 2024).

Other Visibility Degradation Causes

The visibility through different window openings also depends upon other light effects such as: a) unwanted reflections of interior lighted components in windshields, side glass, backlites (in pick-up trucks) and mirrors, and b) scattering of light from external and internal sources due to dirty or degraded windshields and veiling glare reflections. The visibility considerations and models to predict visibility are discussed in detail in Chapter 14.

Shade Bands

Shade bands are usually applied on the top parts of the windshield and backlite and also in some cases on the side glass panels to reduce unwanted sun glare. The lower edges of the shade bands are placed above the eye locations of the tall drivers using the 95th percentile eyellipse. SAE J100 Recommended Practice provides boundaries for shade bands on glazed surfaces in class "A" vehicles (SAE, 2009).

Plane and Convex Combination Mirrors

A number of truck products use a combination of both plane and convex mirrors on the driver's side. The plane mirror is useful to view directly behind and when the reflected sightline is close to the side of the vehicle. Convex mirror provides a wider field of view and reduces blind areas. The provision of two separate mirrors is easier to use from the viewpoint of visual strain, as the driver can view either the plane or convex mirror. This combination, thus, avoids difficulties similar to those the drivers experience while using aspherical mirrors.

Heavy Truck Driver Issues

The driver's SgRP in a heavy truck is located typically at higher distances (over about 2.5 m) from the ground. The high truck-driver eye location causes the following two unique problems: a) small cars are hidden in front of the long hood at intersections, and b) pedestrians, cyclists or small vehicles in the right adjacent lane can be hidden below the beltline (i.e., lower edge of the passenger side window). Some truck cab designs extend the beltline lower and also provide additional small window openings in the doors below the traditional side windows.

Many heavy trucks have tractor and trailer combinations. The tractor-trailer combination driver needs to view the rear tires of the tractor and the trailer during sharper turns. Therefore, they are equipped with unique plane and convex combination side view mirrors. The side view mirror is called the "west-coast mirror" and it typically includes a 152 mm wide x 406 mm tall (6 x 16 inch) plane mirror and a 152 x 152 mm (6 x 6 inch) convex mirror (mounted below the plane mirror). The tall plane mirror allows the driver to view the side of the entire vehicle and the rear wheels of the tractor and trailer. The convex mirror provides a wider rear view. Both the mirrors are especially useful while negotiating wide turns in urban areas and while parking in dock areas.

Cameras and Display Screens

A number of camera and sensor systems are available to provide the driver with additional information. Some examples of such systems are briefly described below.

1. Back-up cameras and sensors are currently available on a number of vehicles to provide the driver with a visual image and/or auditory signal to alert the driver about an object in the path of a reversing vehicle. The rear facing cameras within the vehicle body (e.g., on trunk lids, fenders, tail lamp housings) are used to provide a video image of the view directly behind the vehicle.
2. Side view cameras and sensors are also used on vehicles to provide the driver with a visual image and/or auditory signal to alert about the presence of a vehicle in an adjacent lane.
3. The display screens for camera field views in vehicles should be placed above 35 degrees down-angle zone. Some currently available systems have placed or integrated the display at locations such as: a) inside mirrors, b) center stack screens, c) instrument clusters, or d) side view mirrors.
4. Forward facing right side cameras (e.g., mounted on the right outside mirror) have been proposed as an aid to the driver in obtaining a better (angled) view of lead vehicles and objects ahead.

CONCLUDING REMARKS

Providing the driver with adequate views around the vehicle is very important and is an essential safety need. The fields of view available to the driver depend upon driver positioning in the interior package of the vehicle and the integration of window openings in the exterior design of the vehicle. Therefore, during the design process, the driver's field of view must be constantly evaluated to ensure that visibility issues and requirements covered in this chapter are met to accommodate the largest percentage of drivers.

After a customer uses his or her new vehicle, the visibility problems of the vehicle can be noticed very quickly. Examples of such problems are a) the inside or outside mirrors are too small, b) large obstructions caused by the steering wheel, pillars and headrests, and c) unwanted reflections occurring during nighttime or daytime. Unfortunately, most visibility problems cannot be easily fixed without major and expensive changes in the vehicle design – which generally means waiting for the next major model change. Thus, the importance of providing the right feedback on visibility issues during the very early stages of a vehicle design should not be underestimated.

REFERENCES

Ford Motor Company. 1973. Field of View from Automotive Vehicles. Report prepared under the direction of L. M. Forbes, Report no. SP-381. Presented at the Automobile Engineering Meeting held in Detroit, Michigan. Published by the Society of Automotive Engineers, Inc., Warrendale, PA.

McIssac, E. J. and V. D. Bhise. 1995. Automotive Field of View Analysis Using Polar Plots, SAE paper no. 950602, Society of Automotive Engineers, Inc., Warrendale, PA.

National Highway Transportation Safety Administration. 2024. *Federal Motor Vehicle Safety Standards*, Federal Register, CFR (Code of Federal Regulations), Title 49, Part 571, National Highway Transportation Safety Administration, U. S. Department of Transportation. www.ecfr.gov/current/title-49/subtitle-B/chapter-V/part-571/subpart-B (accessed March 11, 2024)

Society of Automotive Engineers, Inc. 2009. *The SAE Handbook.* Warrendale, PA: Society of Automotive Engineers, Inc.

9 Automotive Lighting

INTRODUCTION TO VEHICLE LIGHTING

Automotive Lighting Equipment

Vehicle lighting systems are primarily safety devices because they provide visibility under night driving and convey vehicle state information to other drivers under all driving conditions. The vehicle forward lighting system allows the driver to see the roadway, traffic control devices (e.g., roadway delineation markers [reflectors], traffic signs), route guidance signs and targets in the roadway. The signaling and marking lamps and devices provide vehicle visibility and information on the motion characteristics (braking, turning or backing; flashing hazard signals for slow moving or stopped vehicles) of the vehicle to other drivers.

Vehicle lighting is a broad area, and it includes a number of different lamps, lighting devices and reflex reflectors. Vehicle lighting equipment can be categorized as follows:

1. *Exterior Lamps and Lighting Devices*
   a) Roadway Illuminating Devices
      - Headlamps – Low and High Beams
      - Front Fog Lamps
      - Auxiliary Headlamps
      - Cornering Lamps
   b) Signaling and Marking Lamps
      - Parking (front), Tail, Stop, Turn and Hazard Warning Signal Lamps
      - High Mounted Stop Lamp
      - Marking: Side-marker Lamps, Identification Lamps for Trucks and Reflex Reflectors
      - Back-up (Reversing) Lamps
      - Daytime Running Lamps
      - Rear Fog Lamps

DOI: 10.1201/9781003485582-9

c) Security/Convenience Lighting
   - Under Mirror Flood Lamps
   - Cargo Lamps inside Truck Bed
   - Running Board Lamps for Trucks and SUVs

2. *Interior Lamps and Lighting Devices*

a) Illuminated Displays (Graphics/Labels) and Controls
   - Interior Displays
   - Lighted Labels on Controls for Identification and Setting Labels
   - Illumination Lamps (or LEDs) for Displays and Controls

b) Interior Illumination
   - Dome Lights
   - Map Lights
   - Courtesy/Convenience Lamps, for example, lamps mounted under instrument panels to illuminate floor, lamps on doors or sun visors

3. *Other Lamps/Lighting Devices*

a) Engine, Trunk, Cargo Area and Glove Box Lamps
b) Security/Peripheral Lighting
c) Emergency, Police and Service Vehicle Warning Lamps, for example, blue/yellow/amber/red/white flashing or rotating warning lamps

Designing or evaluating a vehicle lighting system is a systems problem because the performance of a driver using the lighting system depends upon the characteristics of a) the driver and other drivers on the roadway, b) the lighting system, and c) the driving environment (the roadway, traffic, lighting and weather conditions).

This chapter concentrates on headlighting and signal lighting in terms of human factors issues in their designs, photometric specifications, their effects on driver performance and methods used for their evaluations. The visibility and glare modeling and computational issues are covered in Volume 2, Chapter 14.

Objectives

The objectives of this chapter are to provide information for the following:

1. To understand ergonomic issues in night driving conditions and the considerations involved in target detection, disability and discomfort glare evaluation
2. To study the issues associated with headlamp beam pattern design
3. To study the methods to evaluate headlamp systems
4. To understand the ergonomic issues related to automotive signaling and marking devices
5. To study signal lighting evaluation methods
6. To review a few key ergonomic research studies in vehicle lighting
7. To review future technology trends and research issues to improve night visibility

HEADLAMPS AND SIGNAL LAMPS: PURPOSE AND BASIC ERGONOMIC ISSUES

HEADLAMPS

The purpose of the headlamps is to illuminate the roadway in front of the vehicle such that the driver can see the pavement, traffic control devices (e.g., lane lines, signs, reflectorized markers, work zone reflectorized cones and barrels) and other targets (e.g., objects in the roadway, other vehicles, pedestrians and animals) far enough ahead to safely drive the vehicle at night. The low beams are designed for driving in opposed traffic to minimize blinding or discomforting drivers in on-coming vehicles. The high beams are designed to provide higher visibility under unopposed driving situations (i.e., when on-coming vehicles are absent).

The basic ergonomic issues associated with the design of headlamps are as follows:

a) *Target detection and recognition.* Distance from the observer vehicle at which a target is detected (i.e., target visibility or detection distance) and recognized or identified (i.e., target recognition distance). Target detection distance depends upon the characteristics of the driver, the vehicle, the headlamps and the driving environment.
b) *Effects of glare from on-coming vehicle headlamps.* Glare causes two types of effects, namely, "Discomfort Glare" and "Disability Glare". The "Discomfort Glare" reduces visual comfort and is psychological in nature. It may affect the driver's behavior, for example, the driver may look away from the glare source or make a dimming request to the on-coming driver to switch his high beam to low beam. The "Disability Glare" is physical in nature as it affects functioning of the eye by scattering the light from the glare source into the observer's eyes, and reduces visibility of objects (targets) in the driver's visual field. The glare effects are also dependent on the characteristics of the driver (e.g., age), the vehicle, the glare source, its location with respect to the driver, and the driving environment.
c) *Trade-off between visibility and glare.* If headlamp output is increased to improve the visibility of the driver, the increased output can increase discomfort and disability glare experienced by other drivers in on-coming vehicles. Thus, low-beam patterns are designed by considering the trade-off between the subject driver's visibility and the glare effects experienced by the drivers in on-coming vehicles.

SIGNAL LAMPS

The purpose of signal lamps and marking devices is to provide information to other drivers on the presence (visibility), identification, location, orientation and motion characteristics of a vehicle and intent of its driver (e.g., turning, stopping, backing).

The basic ergonomic issues associated with the design of signaling devices are as follows:

a) *Visibility and recognition of vehicle state.* Distances at which a vehicle can be visible to other drivers and distances at which other drivers can correctly recognize the vehicle and determine its state or movements due to illuminated signal lamps (e.g., discriminating a tail lamp from a stop lamp, discriminating a stop lamp from a turn signal lamp, and determining its direction of movement and speed of the vehicle).
b) *Signal lamp conspicuity or effectiveness.* The visual strength of a signal lamp (its conspicuity or effectiveness) will depend upon its luminous intensity directed at the observer, signal lamp size, shape (e.g., length-to-width ratio), and luminance distribution on the lamp surface and luminance of its background.
c) *Effects of glare.* The effect of glare from an illuminated signal lamp on driver discomfort and driver disability (i.e., reduction in visibility distances) while viewing other objects will depend upon the signal lamp illumination directed at the other driver's eyes (drivers in other vehicles) and the angle of the glare source (signal lamps) from the driver's line of sight. For example, glare from a yellow turn signal or red stop lamps on a lead vehicle experienced by a driver in the following vehicle while waiting at an intersection at night.
d) *Daytime visibility of signal lamps.* For example, the distance from where a lighted stop lamp or turn signal lamp will be visible to the drivers in other vehicles on a bright sunny day with the sunlight falling directly on the signal lamp lens.
e) *Trade-off between visibility and glare.* For example, the visibility of the front turn signal should be increased to improve its visibility in the presence of the low or high beam headlamp illumination at night. However the increased intensity turn signal will increase discomfort from the glare to other drivers.

HEADLIGHTING DESIGN CONSIDERATIONS

In order to design or evaluate headlamps, one must have data on the characteristics and locations of the targets that the driver must see to safely drive the vehicle under dark nighttime conditions and in the presence of other drivers and glare sources. The headlighting design considerations are:

a) The driver must be able to see the roadway at least 2 s ahead to maintain lateral control (Bhise et al., 1977). At 100 km/h (62 mph), the vehicle will travel 56 m (180 feet) in 2 s. Therefore, the headlamp illumination must provide visibility of lane lines (i.e., reflectorized lane delineation markers) at least 56 m ahead to allow the driver to maintain the vehicle in the lane at 100 km/h.
b) The driver must be able to see stationery (e.g., potholes) and moving targets (e.g., pedestrians, animals and other vehicles) from far enough distances to avoid collisions. The visibility distance of a target should be larger than the distance the driver will travel during his response time to recognize the target and to complete a maneuver, for example, braking and/or steering, to avoid a collision with the target.

c) The driver must be able to see critical traffic control devices (i.e., delineation lines, roadside markers/reflectors, signs, traffic cones and barrels) from safe distances (i.e., distances within which the intended maneuver can be completed) on the highway. Some traffic control devices may be illuminated by other external light sources or reflectorized to be seen under headlamp illumination.
d) The headlamps on the observer vehicle should not discomfort other drivers (in on-coming vehicles, or in leading vehicles when viewing to the rear through their mirrors).
e) Visual information acquisition capabilities of the drivers vary due to factors such as their age (visual detection and glare thresholds), visual search behavior, state of alertness, and workload (presence of other tasks simultaneously performed) or inattention.
f) Vehicle characteristics affect the driver's visibility due to factors such as the driver's eye height above the pavement, headlamp location (headlamp mounting height and lateral separation distance between left and right headlamps), headlamp beam pattern, headlamp aim, vehicle loading, lamp voltage, headlamp lens cleanliness, light transmissivity of glazing materials (e.g., windshield) and their cleanliness.
g) Geometric, photometric and traffic characteristics of roadways affect driver visibility. Some variables related to the above three characteristics are: i) geometric characteristics: road topography, lane width, road curvatures, lateral separation distance between lanes, ii) photometric characteristics: retroreflectance and specular reflectance of pavements and road shoulders, ambient luminance, and street lighting, and iii) traffic characteristics: vehicle (traffic) density, distances between cars, ratio of cars to trucks.
h) Weather conditions (e.g., driving in rain, snow and fog; dry, wet or snow covered pavements) also affect visibility.
i) Target characteristics (e.g., target size, shape and target orientation, its motion, reflectance, luminance and color) also affect visibility.
j) Other vehicles, their relative locations, headings, velocities, headlamp beam patterns and other signaling and marking devices, driver eye locations and locations of headlamps also affect driver visibility.

Thus, headlamp design is a systems problem involving the consideration of many variables. A *c*omprehensive *h*eadlamp *e*valuation *s*ystems *s*imulation (CHESS) model developed by Bhise, et al. (1977) to account of above variables is presented later in this chapter.

TARGET VISIBILITY CONSIDERATIONS

A driver must be able to perform all driving tasks safely in night driving situations. Unfortunately, it is not possible to provide the driver the same level of visibility at night as in the daytime due to the following reasons:

a) Targets and areas required to be seen are spread in large fields around the vehicle (Refer to Chapter 8 on field of view from vehicles).
b) The level of illumination from a headlamp on a target placed in front of the headlamp falls off rapidly as the distance of the target from the headlamp is increased. The level of illumination falling on the target will depend upon the headlamp intensity directed at the target divided by the square of the distance between the headlamp and the target (Note: This is the inverse square law for illumination).
c) Target size (angle subtended by the target at the observer's eye point) decreases as the distance from the eye to the target increases. A small target requires larger visual contrast threshold for visibility as compared to a large target. The visual contrast is defined as the difference in the luminance between the target and its background divided by the luminance of the background (See Chapter 14 for more details on visibility prediction).
d) The visual contrast required to see a target increases as the luminance of the target's background (i.e., the driver's adaptation luminance) is decreased (Note that items (b) through (d) above are based on Blackwell's Visual Contrast Threshold Curves. See Figure 6.9 and Volume 2, Chapter 14).
e) Higher (more intense) headlamp output can benefit the driver in the observer vehicle, but on-coming drivers will be discomforted and disabled due to the increased glare from the observer vehicle's headlamps (assuming that the observer and the on-coming vehicles have the same headlamp system). The glare is affected by the aim of headlamps (i.e., headlamp misaim) and road geometric conditions, such as hilly terrain with curves and grades.

PROBLEMS WITH CURRENT HEADLIGHTING SYSTEMS

Some key problems with the current headlamp systems are:

a) Insufficient visibility, especially, with the low beam (e.g., insufficient seeing distances to relevant targets at highway speeds, inadequate light distribution – as different light distributions are ideally needed for different night driving situations).
b) Glare caused by headlamps due to factors such as headlamp location, beam pattern and headlamp misaim.
c) Variability in lamp output/beam patterns (e.g., manufacturing variations in light sources and lamp optics, fluctuating voltages, lens cleanliness, degradations in light output due to aging of materials used in headlamp lenses and reflectors).
d) Variability in headlamp aim retention (The headlamp aim can change in vehicle usage due to changes in load carried in the vehicle, changes in vehicle components and their alignments due to wear and accidents).
e) Non-uniformity (poor perception) of beam pattern on the roadway (e.g., non-uniform --splotchy or streaky appearance).

f) Lower visibility under adverse weather conditions due to attenuation and scattering of light directed above the lamp axis and changes in pavement reflectance characteristics such as wet versus dry. (Note: Retro-reflectance of wet pavements is much lower than that of dry pavements.)
g) Older drivers find night driving more difficult. The visual contrast thresholds and glare effects generally increase with increase in age.

NEW TECHNOLOGICAL ADVANCES IN HEADLIGHTING

A number of different improvements in the headlamps systems have been made due to advances in technologies related to optics, light sources, sensors, actuators and information technologies. Some key advances are briefly described below.

a) New LED light sources are more efficient, that is, they produce more light flux per unit energy input than the old tungsten-halogen lamps (Note: Light Emitting Diodes[LED] lamps produce about 75 lumens/watt as compared to about 24 lumens/watt produced by tungsten-halogen lamps. Lumen is a unit of light flux). LEDs are now used as standard headlighting equipment in luxury vehicles and available as optional equipment in most other vehicles.
b) The ability to produce smaller headlamps with reduced frontal area and reduce aerodynamic drag and require smaller packaging space.
c) New optical solutions (e.g., use of projector lamps, complex reflectors, axial filament light sources, LED sources, light engines with fiber optic cables, and multiple source lamps) improve: i) light distribution (e.g., providing sharper cut-offs in low beam patterns), ii) reduce glare experienced by on-coming drivers, and iii) provide smoother more uniform luminance distributions on the roadway.
d) Smart headlamps (adaptive headlamps) that can change the headlamp aim and/or beam pattern as functions of vehicle velocity, road geometry (e.g., approaching curves and intersections), and road lighting. Adaptive driving beam is now defined by NHTSA (2024) in the Federal Motor Vehicle Safety Standards (FMVSS 108) as a long-range light beam for forward visibility, which automatically modifies portions of the projected light to reduce glare to traffic participants on an ongoing dynamic basis.
e) Night vision systems that provide driving scene imagery (at farther distances than conventional headlamps) on a driver display are now available. A temperature sensitive infrared camera provides the driver a view of the road scene in a display screen or a head-up display.
f) Integral beam headlamp is defined by the NHTSA (2024) in FMVSS 108 as a headlamp comprising an integral and indivisible optical assembly including lens, reflector, and light source, except that a headlamp conforming to requirements on the vehicle headlamp aiming device (VHAD) may have a lens designed to be replaceable.

SIGNAL LIGHTING DESIGN CONSIDERATIONS

Important ergonomic considerations in designing and evaluating signal lamps are:

1. Signal lamps serve as displays indicating the presence of a vehicle and provide cues or information that other observers (e.g., drivers in other vehicles, pedestrians, and bicyclists) can use to determine how safely they can make their intended maneuvers.
2. Displays should provide needed information in the shortest possible time with minimal (or no) confusion or errors. Thus, the signal lamp should be able to provide clear information on a) vehicle presence (detection), b) vehicle location (position and distance), c) vehicle size (width, length, height) and d) vehicle movement (or state) of the vehicle (accelerating, decelerating, stopping, turning, and direction of travel).
3. Signal Effectiveness:
   a) Signal Visibility and Conspicuity: A signal lamp should be visible under all driving environments (daytime and with sunlight falling on the signal lens, dawn/ dusk, night, rain, and fog).
   b) The signal should be recognizable and interpretable with minimal confusion (e.g., a tail lamp should not be perceived as a stop lamp and vice versa).
   c) The driver's response time to detect, recognize and correctly interpret a given traffic situation indicated by vehicle signal lamps should be as short as possible (about 1 s).
   d) A signal lamp should not degrade a driver's ability to see and interpret other visual details or signals (e.g., masking of other signals and glare due to illumination from the signal lamp).
4. Signal coding methods:
   a) Color or change in signal color must conform to color specifications – red, amber, white, blue; defined by SAE International (Society of Automotive Engineers) J978, SAE (2009).
   b) Intensity and/or change in intensity (tail versus stop lamp, flashing lamps, low beam versus high beam) must conform to SAE and FMVSS requirements (SAE, 2009; NHTSA, 2024).
   c) Spatial cues on visibility and state of vehicle motion are obtained by the drivers from lamp locations on the vehicle. The lamp mounting location, position (height, distance from outer edges of the vehicle) and separation distances between lamps on the vehicle (i.e., distance between left and right lamps) must meet SAE and FMVSS requirements (SAE, 2009; NHTSA, 2024).

SIGNAL LIGHTING VISIBILITY ISSUES

More specific issues related to visibility of signal lamps are briefly described below:

a) Other drivers must be able to see the subject vehicle from distances larger than their perception–reaction and maneuvering distances to safely avoid collisions.
b) The visibility of a signal lamp depends upon a number of factors related to signal lamps, ambient lighting conditions, driver characteristics and the road environmental characteristics.
c) The signal lamp characteristics that affect its visibility are: luminous intensity directed at the observer (beam pattern), illuminated area and shape (e.g., length-to-width [aspect] ratio), distribution of luminance over the face on the lamp, spectral characteristics of the light emitted by the lamp, that is, its color, and the amount of dirt on the lamp, which reduces its output.
d) The ambient lighting conditions affect the visibility (or conspicuity) of the lamp. For example, a stop lamp should be visible under both nighttime and daytime conditions. Further, when direct sunlight falls on the signal lamp lens, its "on" and "off" states must be discriminable.
e) The primary driver characteristics that affect signal visibility are driver's visual thresholds (luminous intensity of point sources and visual contrast thresholds), driver age, confidence in detection and recognition, and driver alertness.
f) The driver's response time to recognize a change in signal (e.g., onset of a stop lamp or initiation of a turn signal) should be very short (close to a simple reaction time of 0.3 to 0.5 s).
g) Driver errors in recognizing a signal correctly must be minimized to avoid accidents resulting from not recognizing a signal or recognizing a signal too late.

PROBLEMS WITH CURRENT SIGNAL LIGHTING SYSTEMS

Since different vehicle models use different lamp designs and light sources with different technologies and manufacturing variations in lamps and light sources, there is potential to create confusion in correct recognition of signals. Issues related to such problems are presented below.

a) Confusion errors in identification of a red signal as a tail lamp, a stop lamp or a rear red turn signal
b) Recognition of status or change in vehicle motion (recognition that a vehicle is slowing, accelerating or has stopped)
c) Variations in lamp intensities between different brands and models of vehicles and within samples of the same vehicle make and model (i.e., manufacturing variations)
d) Equivalency of signal (visual strength) of lamps differing due to differences in lamp size, shape, light sources (e.g., tungsten, halogen, LED, neon, etc.), lamp color (e.g., equivalency of the red rear turn signal with the amber rear turn signal), flashing frequency, flash on versus off durations, and rise time (time to achieve 85% of the minimum required intensity).

NEW TECHNOLOGY ADVANCES AND RELATED ISSUES IN SIGNAL LIGHTING

Some issues related to the introduction of new technologies in producing future signal lamps are presented below.

a) Signal lamps can be created by using different lighting technologies. For example, the light sources within a signal lamp can be created by using technologies such as the tungsten filament bulb, neon tube, LEDs, electro-luminance film, and OLEDS (organic light emitting diodes). The light produced by different sources generally differ in their spectral distribution and rise time.
b) These technologies can also allow for the creation of lamps with different luminance distributions on the lamp lens. Lamps with a small single source generally causes a hot spot, that is, a brighter area centered around the lamp source, which improves its visibility as well as its effectiveness as a signal as compared to a lamp with a very uniform luminance over a larger illuminated area. Thus, a more uniform-looking LED lamp created by using many equally spaced LEDs is not perceived to be as effective as compared to a lamp with a single source with the same total luminous intensity (Bhise, 1981 and 1983).
c) Future lamps may also be used to create large displays on the back of the vehicle. For example, such displays can convey information to other drivers by flashing words or symbols to inform the vehicle state to other drivers.
d) The output of the lamp may be affected by its temperature. For example, the light output of the LEDs reduces with increase in temperature (SAE J 2650, SAE, 2009).
e) Since many of the new technology light sources do not require a metal filament (which can break with vibrations), their life is much longer than the traditional tungsten or halogen light sources. Further, such non-filament sources can potentially reduce variability in headlamp beam patterns that is typically associated in light sources with filaments.

PHOTOMETRIC MEASUREMENTS OF LAMP OUTPUTS

LIGHT MEASUREMENT UNITS

The light measurement variables and their units are given below.

a) Luminous (Light) Flux (Φ): It is the time rate of flow of radiant light energy and is measured in Lumen (lm).
b) Luminous Intensity (I): measured in Candela (cd) = light flux per unit solid angle (lumen/steradian). (Note: Steradian is a unit of solid angle).
c) Illumination or Illuminance (E): measured in foot-candle (fc) or lux.
   1 Foot-candle (fc) = 1 Lumen/ft^2 = 10.76 lux; Lux (lx) = Lumens/m^2.
d) Luminance (L) (Physical Brightness): measured in foot-Lambert (fL) or cd/m^2.

$$1\ \text{cd/m}^2 = 1\ \text{nit} = 0.29\ \text{fL};\ \text{cd}/[\text{ft}^2 \times \pi] = 1\ \text{fL}$$

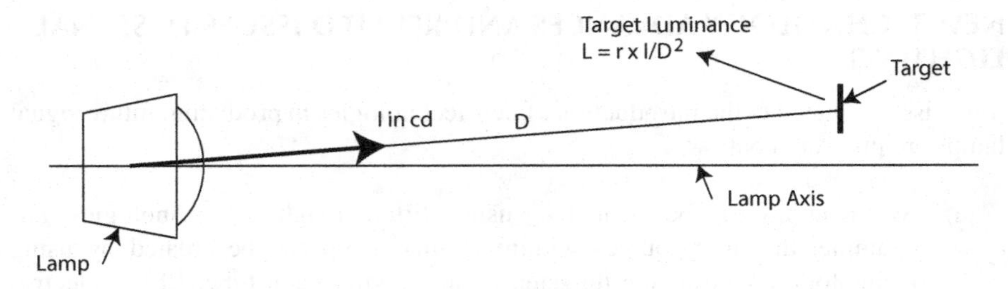

FIGURE 9.1 Luminous intensity of a lamp (I) producing luminance (L) of a target with reflectance (r).

e) Lamp Beam Pattern: Luminous intensity output (cd) over a range of angles with respect to the lamp axis. The angular locations are measured in degrees horizontal and degrees vertical with respect to the lamp axis.

   It should be noted that the Federal Motor Vehicle Safety Standard 108 (NHTSA, 2024) and SAE Standards (SAE, 2009) on vehicle lighting provide minimum and maximum luminous intensity (cd) requirements at different test point locations with respect to the lamp axis for different automotive lamps (e.g., tail lamp, stop lamp, turn signal lamp, headlamp, side marker lamp, and back-up lamp).

f) Total light source output flux is measured in Lumens. It is the integrated value of light flux emitted in all directions from the light source.

g) Illumination (E) at a distance (D) from a lamp with intensity (I) is computed as: $E = I/D^2$. E is measured in fc or lux, where D = Distance between light source and incident surface. The distance is measured in feet for illumination measured in fc and meters for illumination measured in lux.

h) Luminance (L) = r x E, measured in fL, where r =Reflectance of the surface (see Figure 9.1). Also, $L = (r \times E)/\pi$, where L measured in cd/m^2, and E is measured in lux.

HEADLAMP PHOTOMETRY TEST POINTS AND HEADLAMP BEAM PATTERNS

The photometric requirements on low- and high-beam patterns of replaceable bulb headlamps with optical aiming described in the FMVSS 108 (NHTSA, 2024) are presented in Tables 9.1 and 9.2, respectively. The requirements are specified in terms of minimum and maximum luminous intensity (specified in "cd") at different angular locations (called the test points, which are specified as: degrees left or right, and up or down with respect to the lamp axis which is parallel to the longitudinal axis of the vehicle for front and rear lamps). The test point locations and their output requirements, thus, shape the beam patterns of low and high beams (see Figures 9.2 and 9.3) drawn from FMVSS 108 requirements (NHTSA, 2024).

TABLE 9.1
Low Beam Photometry Test Points and Minimum and Maximum Intensity Requirements

Test Point Location: (azimuth, elevation) in degrees	Luminous Intensity (cd) Minimum	Maximum
10 up to 90 up	--	125
(-8,4) and (8,4)	64	--
(-4,2)	_135	--
(1,1.5) to (3,1.5)	200	
(1,1.5) to (90,1.5)	--	1400
(-1.5, 1) to (-90,1)	--	700
(-1.5, 0.5) to (-90,0.5)	--	1000
(1,0.5) to (3,0.5)	500	2700
(0,0)	--	5000
(-4,0)	135	--
(-8,0)	64	--
(1.3, -0.6)	10000	--
(0,-0.86)	4500	--
(-3.5,-0.86)	1800	12000
(2, -1.5)	15000	--
(-9,-2) and (9,-2)	1250	--
(-15,-2) and (15,-2)	1000	--
(0,-4)	--	7000
(4,-4)	--	12500
(-20,-4) and (20,-4)	300	--

Source: FMVSS 108, NHTSA 2023.

Figures 9.4 and 9.5 show the low and high beam patterns of typical U.S. headlamps. The candlepower values shown in these diagrams were obtained from measurements made at the University of Michigan Transportation Research Institute for a sample of 20 low- and 20 high-beam lamps from high-volume vehicles sold in the United States. (Schoettle et al., 2001).

The difference between the low and high beam patterns is primarily due to the location of the hot spot (or the highest intensity point). To limit the glare experienced by the on-coming driver, the low beam hot spot is directed lower by about (1.5 to 2.5 degrees) and to the right side by about 2 to 3 degrees (see Figure 9.4). The high beam has its hot spot directed straight ahead along the lamp axis at (0,0) (see Figure 9.5). Also, it has a much higher value of intensity as compared to the low beam.

PAVEMENT LUMINANCE AND GLARE ILLUMINATION FROM HEADLAMPS

Figure 9.6 shows the luminance of the pavement in the middle of the driving lane resulting from a pair of low and high beams mounted on a passenger vehicle on a straight level roadway. The figure shows that since the low beam hot spots are lower

TABLE 9.2
High Beam Photometry Test Points and Minimum and Maximum Intensity Requirements

Test Point Location: (azimuth, elevation) in degrees	Luminous Intensity (cd) Minimum	Luminous Intensity (cd) Maximum
(0,2)	1500	--
(-3,1) and (3,1)	5000	
(0,0)	40000	75000
(-3,0) and (3,0)	15000	--
(-5,0) and (5,0)	5000	--
(-9,0) and (9,0)	3000	--
(-12,0) and (12,0)	1500	--
(0, -1.5)	5000	--
(-9,-1.5) and (9,-1.5)	2000	
(0, -2.5)	2500	--
(-12,-2.5) and (12,-2.5)	1000	
(0,-4)	--	5000

Source: FMVSS 108, NHTSA 2024.

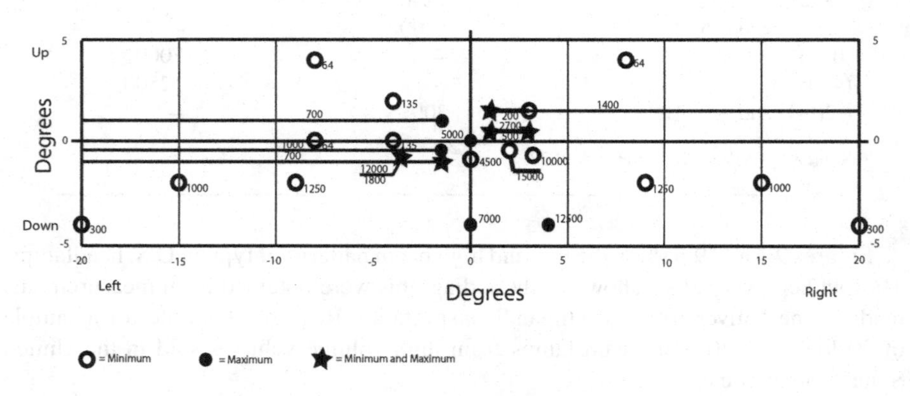

FIGURE 9.2 Low beam photometry test point locations with respect to the lamp axis.

Note: The numbers next to each test point symbol present the minimum, maximum or both intensity requirement values in cd at the test point. All test points falling on a given horizontal straight line have the same intensity requirement shown above the line.

than the hot spots of the high beam, the luminance of the pavement under the low beam is higher at closer distances on the pavement as compared to the high beam. The curves presented in Figure 9.6 were generated by using the beam patterns presented in Figures 9.4 and 9.5 and by assuming that the pavement retro-reflectance is 0.05 and the ambient luminance is 0.003 cd/m^2 (0.001 fL).

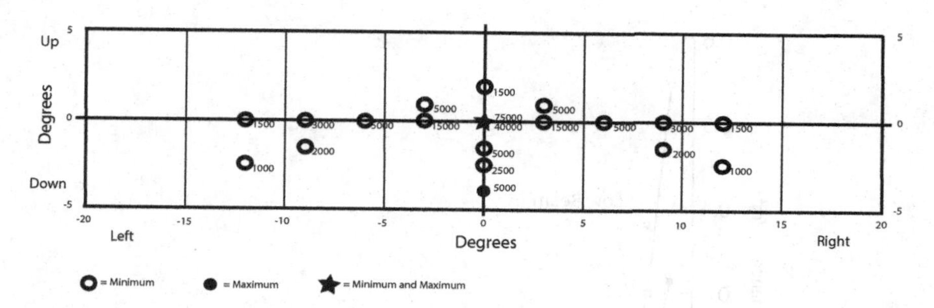

FIGURE 9.3 High beam photometry test point locations with respect to the lamp axis.

Note: The numbers next to each test point symbol present the minimum, maximum or both intensity requirement values in cd at the test point.

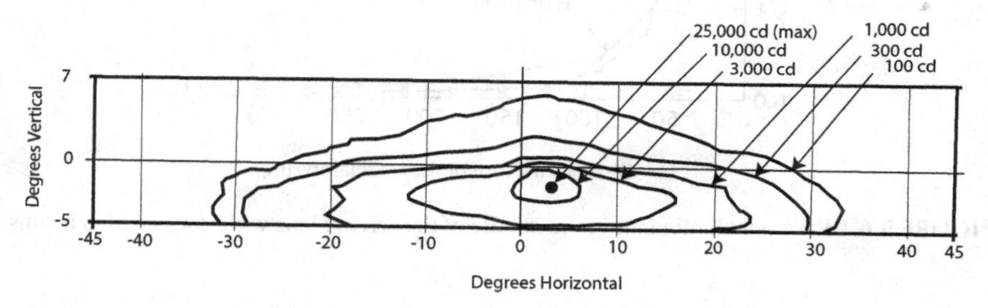

FIGURE 9.4 Distribution of a typical low beam pattern.

Source: Redrawn from: Schoettle et al., 2001.

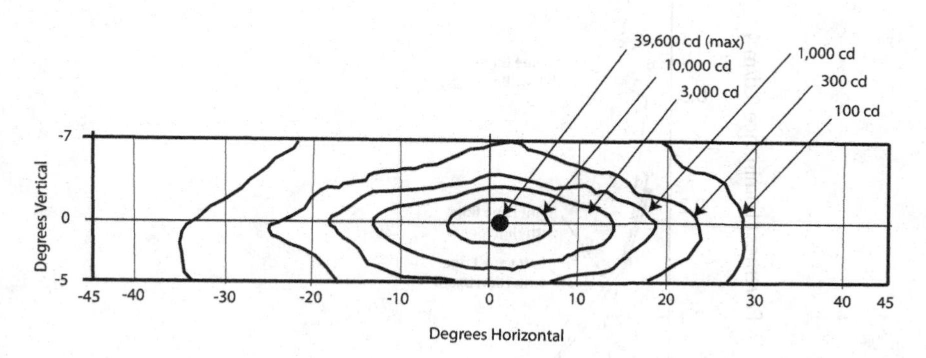

FIGURE 9.5 Distribution of a typical high beam pattern.

Source: Redrawn from Schoettle et al., 2001.

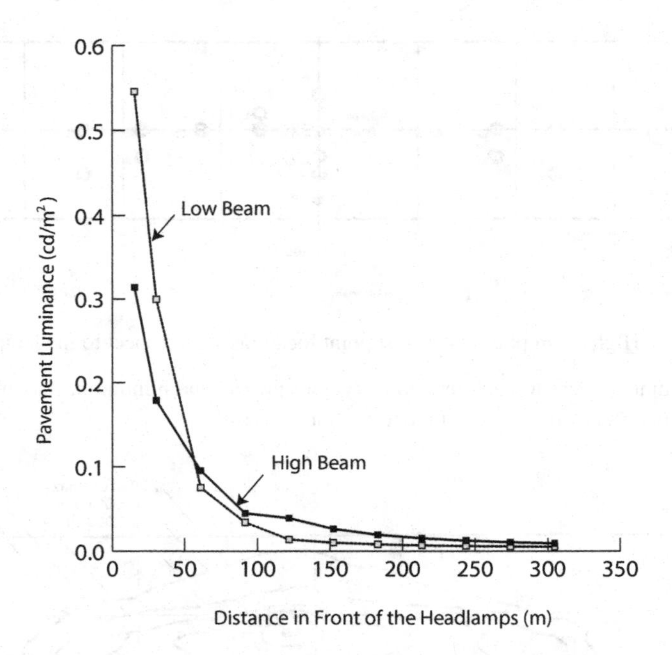

FIGURE 9.6 Pavement luminance in the middle of the driving lane with low and high beams.

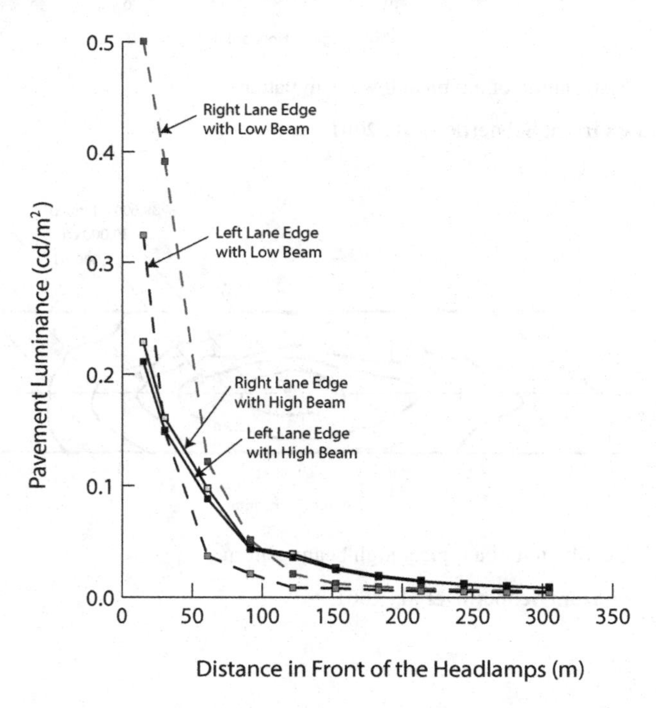

FIGURE 9.7 Pavement luminance at the lane edges of the driving lane with low and high beams.

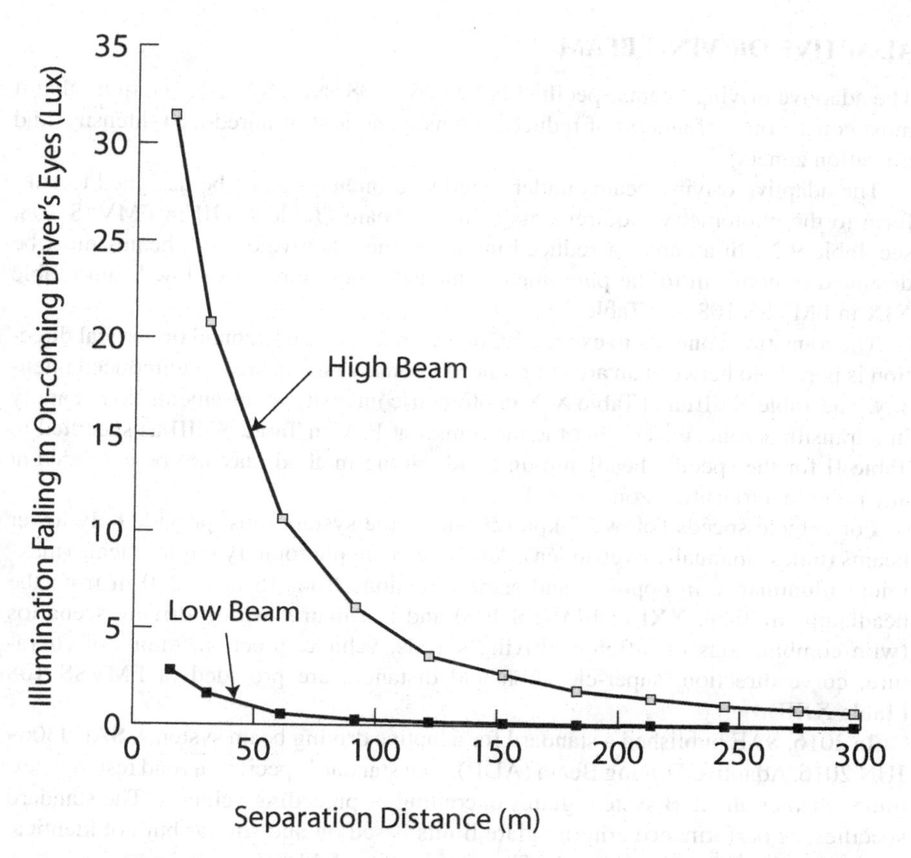

FIGURE 9.8 Illumination falling into the on-coming driver's eyes from properly aimed high and low beams on a 2-lane straight level roadway.

Figure 9.7 presents the luminance of the pavement on the left and right edges of a 3.66 m (12 ft) wide straight level driving lane for the low and high beams. The two curves for the low beam show that the pavement on the right hand side of the driving lane will be considerably brighter than the pavement at the corresponding distance on the left side of the driving lane. The figure also shows that the pavement will be brighter under a high beam beyond 100 m as compared to under the low beam. The above differences in brightness are due to less intense and lower and rightward location of the hot spot in the low beam pattern as compared to the high beam pattern (See Figures 9.4 and 9.5).

Figure 9.8 presents the amount of illumination falling into the on-coming driver's eyes from properly aimed high and low beams on a two-lane straight and level road as functions of separation distance between the observer and the on-coming vehicles. The figure shows that the on-coming driver will experience substantially less glare illumination from the low beam as compared to the high beam.

ADAPTIVE DRIVING BEAM

The adaptive driving beams specified in FMVSS 108 (NHTSA, 2024) require that it must consist only of area(s) of reduced intensity, area(s) of unreduced intensity, and transition zone(s).

The adaptive driving beams under unreduced intensity must be designed to conform to the photometry requirements of upper beam (Table XVIII in FMVSS 108; see Table 9.2). In an area of reduced intensity, the adaptive driving beams must be designed to conform to the photometric intensity requirements of low beam (Table XIX in FMVSS 108; see Table 9.1).

The transition zone not to exceed 1.0 degree in either a horizontal or vertical direction is permitted between an area of reduced intensity and an area of unreduced intensity. The Table XVIII and Table XIX photometric intensity requirements do not apply in a transition zone, except that the maximum at H-V in Table XVIII as specified in Table II for the specific headlamp unit and aiming method may not be exceeded at any point in a transition zone.

For vehicle speeds below 32 kph (20 mph), the system must provide only lower beams (unless manually overridden). Adaptive beam photometry requirements (maximum illuminance in opposite and same directions from 15 m to 220 m from the headlamps in Table XXI of FMVSS 108) and test matrix with 8 driving scenarios (with combinations of different driving speeds, vehicle direction, radius of curvature, curve direction, super-elevation and distance) are provided in FMVSS 108 (Table XXII).

In 2016, SAE published a standard for adaptive driving beam systems, SAE J3069 JUN 2016, Adaptive Driving Beam (ADB). The standard specifies a road test to determine whether an ADB system glares oncoming or preceding vehicles. The standard specifies, as performance criteria, glare limits based on and similar but not identical to the glare limits used in the ADB Test Report (See Table 3).

SAE J3069 specifies a straight test track with a single lane 155 m long. On either side of this test lane, the standard specifies the placement of test fixtures simulating an opposing or preceding vehicle. The test fixtures are fitted with lamps having a specified brightness, color, and size similar to the taillamps and headlamps on a typical car, truck, or motorcycle. The standard specifies four test fixtures: An opposing car/truck; an opposing motorcycle; a preceding car/truck; and a preceding motorcycle. In addition to simulated vehicle lighting, the test fixtures are fitted with photometers to measure the illumination from the ADB headlamps.

The standard specifies a total of 18 different test drive scenarios. The scenarios vary the test fixture used, the placement of the fixture (i.e., to the right or left of the lane in which the ADB-equipped vehicle is travelling), and whether the lamps on the test fixture are illuminated for the entire test drive, or are instead suddenly illuminated when the ADB vehicle is close to the test fixture. During each of these test runs, the illuminance recorded at 30 m, 60 m, 120 m, and 155 m must not exceed the specified glare limits. If there is no recorded illuminance value at any of these distances, interpolation is used to estimate the illuminance at that distance. For sudden appearance tests, the system is given a maximum of 2.5 seconds to react and adjust the beam. If any recorded (or interpolated) illuminance value exceeds the applicable glare limit, the standard provides for an allowance: The same test drive scenario is run, except now only the lower beam is activated. The ADB system can still be deemed to have passed the test as long as any of the ADB exceedances do not exceed 125 percent of the measured (or interpolated) illuminance value(s) for the lower beam. In addition to the dynamic track test, the standard contains a number of other system requirements, such as physical test requirements and requirements for the telltale. It also requires the system to comply with certain aspects of existing standards for lower and upper beam photometry as measured statically in a laboratory environment (for example, for the portion of the ADB beam that is directed at areas of the roadway unoccupied by other vehicles, the lower beam minimum values specified in the relevant SAE standard must be met).

PHOTOMETRIC REQUIREMENTS FOR SIGNAL LAMPS

The Federal Motor Vehicle Safety Standard 108 (NHTSA, 2024) and the SAE standards (SAE, 2009) provide photometric requirements in terms of minimum and maximum luminous intensity requirements for different lamps at different test points (angular locations with respect to the lamp axis). Table 9.3 provides the minimum and maximum intensity requirements for tail, stop, turn, parking and side marker lamps. The maximum intensity requirements apply to any test point within the beam patterns of any of the lamps, whereas, the minimum intensity requirements, shown in Table 9.3, apply to the test point located at (0,0) (i.e. on the lamp axis).

Figure 9.9 illustrates minimum luminous intensity requirements in terms of the percentage of minimum values shown in Table 9.3 for different test points for stop and rear turn signal lamps. The requirements for other signal lamps are specified similarly using different percentage values. In additional there are requirements on the sum of intensities for test points within different groups (zones). For more details the reader should refer to the FMVSS 108 and SAE standards (refer to SAE standards J222, J 585, J586, J588 and J592 in the SAE Handbook [SAE, 2009]). To reduce confusion between tail and stop signals incorporated in the same lamp, there are also requirements on the minimum ratio of stop-to-tail lamp intensities at different test points.

HEADLAMP EVALUATION METHODS

A number of objective and subjective methods are used in the industry to evaluate headlighting systems. Five of these methods are described below.

1. *Photometric Measurements and Compliance Evaluations*
   This headlamp evaluation method involves the measurement of headlamp intensity (candlepower) output at angular locations (test points) that define the photometric requirements (e.g., Table 9.1 for low beam and Table 9.2 for high beam). To conduct the measurements, a headlamp is mounted on a computer controlled goniometer and turned on using a regulated power supply at a specified voltage (12.0 or 12.8 V) and the lamp intensity (in candela) is measured

TABLE 9.3
Current Signal Lighting Requirements

Lamp	Lighted sections	Luminous Intensity(cd)	
		Minimum	Maximum
Stop(Rear Red)	1	80	300
	2	95	360
	3	110	420
Tall(Rear Red)	1	2	18
	2	3.5	20
	3	5	25
Red Rear Turn signal	1	80	300
	2	95	360
	3	110	420
Yellow Rear Turn signal	1	130	750
	2	150	900
	3	175	1050
Yellow Front Turn signal	1	200	--
	2	240	--
	3	275	--
Yellow front Turn signal (when signal lamp spaced less than 100 mm From lighted edge of headinglamp)	1	500	--
	2	600	--
	3	685	--
Front packing	1	4	--
Center High Mounted stop	1	25	160
Front Yellow sidemarker	1	0.62	--
Rear Red Sarker	1	0.25	--
Rear Red Sidemarker	95	3	

Source: FMVSS 108, NHTSA 2024.

by aiming the lamp in front of a photocell mounted at a distance at least 18.3 m (60 ft) from the lamp. The lamp axis is re-aimed after each measurement to cover all the required test points. This method is primarily used by the headlamp and the vehicle manufacturers to check if a given headlamp meets the lamp output requirement at each test point.

2. *Static Field Tests*
The static field test involves parking a vehicle equipped with a test headlight system on a roadway. A number of targets are placed at different longitudinal and lateral locations and selected drivers are asked to sit in the driver's seat. The field tests are generally conducted in dark areas at night away from any external light sources and populated areas to reduce the effect of ambient illumination caused by the scattered light from city lights (city glow). The driver is typically asked to identify the targets (e.g., full-size pedestrian silhouettes or rectangular targets of different dimensions and reflectance values) that the driver can see and is also asked to provide ratings on the glare produced by

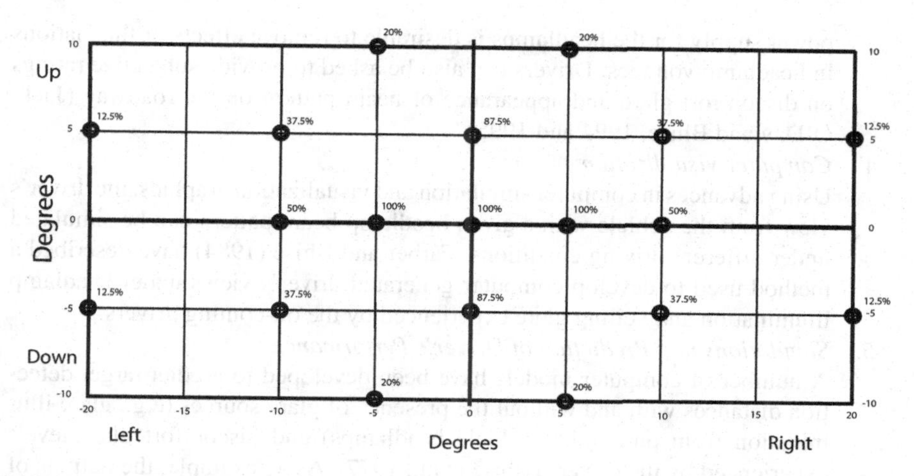

FIGURE 9.9 Rear signal test points grid showing percentages of minimum luminous intensity specified in Table 9.3 for stop and turn signal lamps.

the headlamps of a parked vehicle (if present) facing the evaluator. In this type of static test, the drivers get more time to make their responses (i.e., tell the experimenter if the driver could see each target) as compared to the time they would have in actual driving situations. Further, such static tests are generally conducted under somewhat unrealistic test conditions as they involve straight, level, and dry roadways, alerted and younger drivers, clean headlamps and windshields, and properly aimed headlamps. If such static tests involve the presence of a glare vehicle, the glare exposure that the subjects get is considerably different than the changing glare that the drivers experience from the motion of both the on-coming and subject vehicles in meeting situations. Thus, the static headlighting tests have limited usefulness but they are conducted because they are less time consuming, convenient to set up and they allow visualization of beam patterns on a roadway.

3. *Dynamic Field Tests* (Target detection, glare judgments and appearance of beam pattern). Dynamic tests are conducted to evaluate headlamps under more realistic driving conditions on public roads, proving grounds or airport runways where motion cues such as the pitching of the car, relative motion between the target and the driver, relative motion between the subject car and the glare car, and moving shadows of the stand-up targets are also available. For seeing distance tests, targets are usually located at different points on the roadway and the drivers are asked to indicate as soon as they can see a target and also identify the target. On-board recorders are generally used to record timings of events (e.g., subject response), vehicle velocity and distance travelled on the roadway. The driver's attentional demands can be controlled by providing needed instructions, but generally, such tests are conducted with "alerted" subjects (whereas in the real world the subjects generally are "unalert", that is, they are not aware of approaching targets). Use of regulated

power supply for the headlamps is desirable to remove effects of fluctuations in headlamp voltages. Drivers can also be asked to provide subjective ratings on discomfort glare and appearance of beam pattern on the roadway (Jack, O'Day and Bhise, 1994 and 1995).

4. *Computer Visualization*
   Using advances in computer simulation and visualization graphics, the driver's view from the vehicle with a given headlamp beam pattern can be simulated under different driving conditions. Farber and Bhise (1984) have described a method used to develop computer generated driver's views under headlamp illumination and veiling glare experienced by the on-coming drivers.
5. *Simulations and Prediction of Driver's Performance*
   A number of computer models have been developed to predict target detection distances with and without the presence of glare sources (e.g., glare illumination from on-coming vehicle headlamps) and discomfort glare levels experienced by the drivers (Bhise et al., 1977). As an example, the outputs of the seeing distance model developed by Bhise, Farber and McMahan (1977) are presented in Figure 9.10. The figure presents predicted visibility distances for 1.83 m (6 ft) high 7 percent reflectance pedestrian targets located on the right hand shoulder of a 2-lane roadway under low beam illumination (Bhise and Matle, 1989). The seeing distances were predicted for 5th, 50th and 95th percentile visual capabilities (based on distributions of contrast thresholds) for drivers ranging from 20 to 80 years of age. The figure shows that visibility distances of the pedestrian targets decreased as the driver's age increased. The

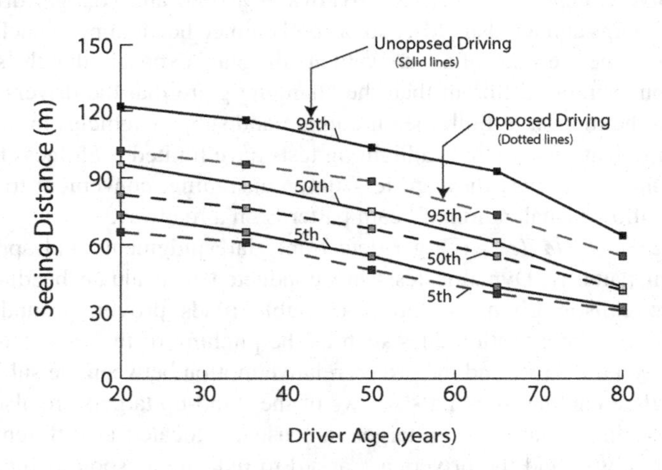

FIGURE 9.10 Comparison of seeing distances for 5th, 50th and 95th percentile visual capabilities of drivers as a function of driver age and unopposed and opposed driving with low beams.

Note: A glare vehicle with low beam was placed at 122 m (400 ft) separation distances in the opposing lane.

Source: Redrawn from Bhise and Matle, 1989.

visibility distances to pedestrian targets were lower under opposed situations (i.e., data shown in dashed lines; when a glare vehicle using the low beam was placed at 122 m (400ft) separation distance) as compared to under the unopposed situation (data shown in solid lines). Further it should be noted that the largest seeing distance with 95th percentile (best) visual capability of a 20 year old driver under unopposed situation was about 120 m (394 ft); whereas the shortest seeing distance of an 80 year driver with 5th percentile (worst) visual capability under opposed driving situation was about 30 m (98 ft). Thus, the ratio of best case to worse case seeing distances was 4 (120m/ 30m).

More details of the seeing distance prediction and discomfort glare models are presented in Volume 2, Chapter 14.

Bhise et al. (1977) developed a *C*omprehensive *H*eadlamp *E*nvironment *S*ystems *S*imulation (CHESS) model to evaluate headlighting systems. The model simulates night driving situations in a representative U.S. night driving environment (called the standardized test route with different types of roadways, traffic speeds, traffic density, road geometry and photometric variables to represent different reflectance [e.g., pavement reflectance], ambient lighting conditions and external light sources [e.g., street lights]) and creates thousands of night driving encounters involving drivers of different characteristics, pedestrian targets, lane marking targets and on-coming vehicles. The model evaluates input beam patterns by determining if the drivers can see lane lines far enough ahead to maintain lane position (at least 2 s ahead at the vehicle speed), detect pedestrian targets to avoid accidents (from perception and stopping distances), and not discomfort drivers (more than properly aimed U.S. low beams) in the on-coming vehicles. The output of the model is a figure of merit that represents the percentage of driving in which all the three visibility criteria simultaneously are met. Figure 9.11 shows a simple diagram illustrating the basic working of the model with inputs and outputs. Table 9.3 presents a list of variables simulated in the model. Detailed description of the model and its applications are available in Bhise et al. (1977, 1988).

The basic outputs of a low beam evaluation presented in Bhise et al. (1988) are summarized below:

a) Figure of merit of a properly aimed US low beam was 69.1. This means that, out of the over 6,400 random night driving encounters simulated by the CHESS model on different roadways, 69.1 percent of the encounters met all the three visibility criteria described in Figure 9.11.
b) Percentage of pedestrians detected in time to avoid accidents with the vehicles were 43.8 and 32.8 percent under unopposed (no on-coming vehicle present) and opposed situations (with on-coming vehicles), respectively.
c) 88.9 and 88.3 percentage of delineation lines (102 mm [4 inches] wide painted lane lines) were visible (over 2 s of driving distance) in unopposed and opposed driving encounters, respectively.
d) Percentage of drivers discomforted under opposed encounters were 2%.

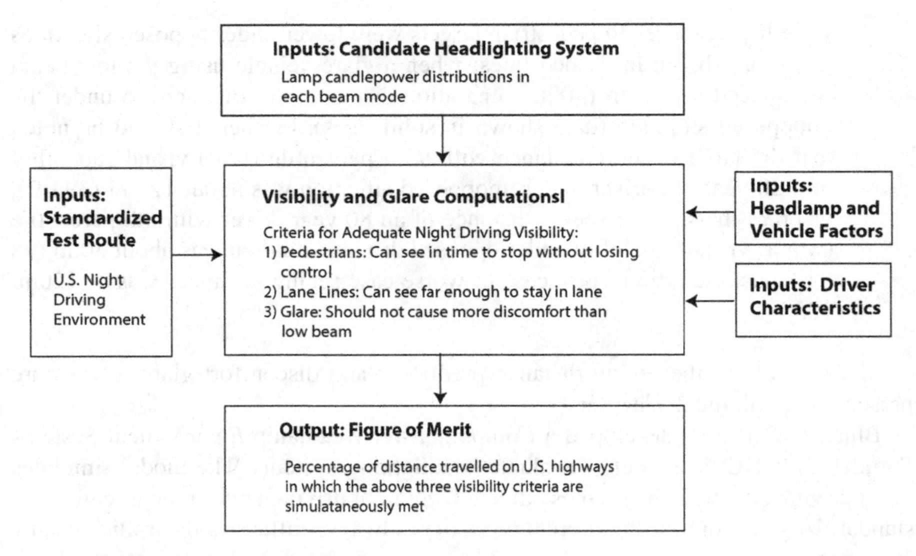

FIGURE 9.11 CHESS model inputs and outputs.

TABLE 9.4
List of Variables Simulated in the CHESS Model

Road-Environment characteristics
Area: Urban/Rural
Illuminated/Non-illuminated
Lane configuration
Lane Delineation
Road Topography/Curvatures
Ambient Luminances
Road Surface Type
Road Wetness
Pavement Friction
Traffic Density
Pedestrian Exposure
Pedestrian Characteristics
Pedestrian Action
Walking Speed
Clothing Reflectance: Top and Bottom
Dimensions: Height and width

TABLE 9.4 (Continued)
List of Variables Simulated in the CHESS Model

Vehicle Characteristic: Observer and On-coming
Velocity
Headlamp Height and Separation
Headlamp Aim
Beam Patterns
Intensity multiplies for Dirt and Voltage
Lamp manufacturing Variations
Driver Variables
Drive Age
Visual Contrast Thresholds
Percentile Visual Capability
Glare Sensitivity
Alertness
Reaction Time
Drive-Vehicle Variables
Drive Eye Location
Beam Mode Selection

The above results thus demonstrate that improving low beam performance still remains to be a challenge due to the basic trade-off between visibility and glare.

SIGNAL LIGHTING EVALUATION METHODS

1) *Photometric Measurements and Compliance Evaluation*: This signal lamp evaluation method involves the measurement of lamp candlepower output at different angles that define their photometric requirements (e.g., Figure 9.9). To conduct the measurements, a lamp is mounted on a computer controlled goniometer and powered on using a regulated power supply at a specified voltage (12.0 or 12.8 V) and the lamp intensity (in candela) is measured by aiming the lamp in front of a photocell mounted at a distance at least 3 m or at least 10 times the maximum linear extent of the effective projected luminous area of the lamp, whichever is greater. The lamp axis is re-aimed after each measurement to cover all the required test point locations. This method is primarily used by the lamp and vehicle manufacturers to check if a given signal lamp meets the lamp output requirement at each test point.
2) *Field Observations and Evaluations*: The most common method used for signal lighting evaluations involves asking a group of subjects (either individually or in

a group) to view test signal lamps of different signal characteristics (e.g., intensity, illuminated area, shape of lamp, and lamp source type/technology) in a series of trials from selected viewing locations and ambient lighting conditions and provide subjective ratings on visibility, conspicuity, or effectiveness of the signals when viewed from a given distance in a given situation. In some field observations, the subjects are presented with a pair of signals (a reference signal along with a test signal) and asked to compare the test signal to the reference signal. The method of paired comparisons is generally superior as humans are very good at discriminating between two signals presented at the same time, and thus, can provide more reliable ratings on equivalency or relative (less/more than the reference) visibility, conspicuity or signal effectiveness. (See Chapter 19 for more information evaluation methods using paired comparisons).

a) *Identification of Tail and Stop Lamps*: Figures 9.12 and 9.13 illustrate the results of the tail versus stop lamp discrimination studies based on lamp intensity conducted by researchers at the Ford Motor Company (Troell et al., 1978). Figure 9.12 shows the relationship between percent confusion in recognizing a lamp between two lamps (such a tail lamp and a stop) as a function of ratio of intensities of the brighter to dimmer lamp. The curve suggests that at least a 10:1 ratio of intensities between the two lamps should be maintained for less than 12% confusion error. It should be noted that current signal lighting standards for rear signal lamps with combined tail and stop signals require that the stop signal intensities to be at least 5 times more than the tail signal intensities in the middle photometric zone closer to the lamp axis (FMVSS 108, NHTSA, 2024; J586, SAE, 2009).

Figure 9.13 presents results of a field test in which a vehicle driven with a variable intensity red rear signal lamps was viewed by the subjects from over 300 m (Troell,

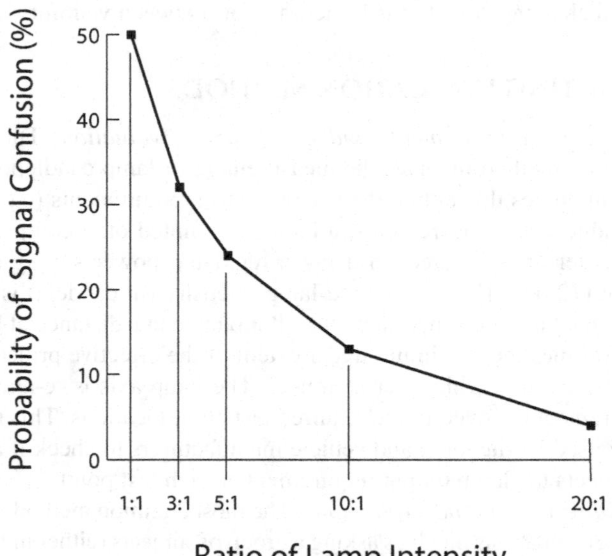

FIGURE 9.12 Probability of signal confusion as a function of stop-to-tail lamp intensity ratio.

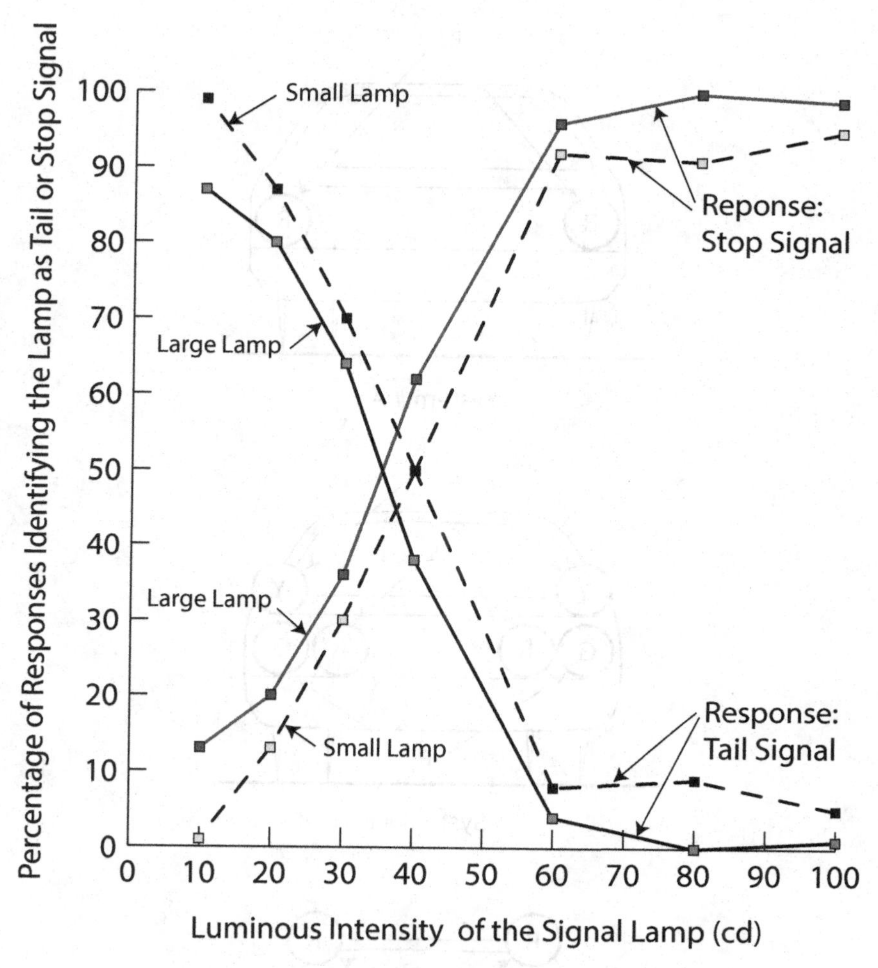

FIGURE 9.13 Recognition of tail or stop lamp as a function of lamp intensity (cd) and lamp area (small lamp of 25.8 cm² [4 in²] and large lamp of 51.6 cm² [8 in²]) at nighttime.

et al., 1978). In each trial the lamp intensity was varied, and the subjects were asked if the red signals were tail or stop lamps. The figure shows that lamps over 60 cd were judged as stop lamps in over 90% of the trials. Conversely, lamps under 20 cd were judged as tail lamps in about 80% or more trials. The recognition performance was also affected by the illuminated area of the lamp. The larger 51.6 cm² (8 in²) lamps were recognized as stop lamps in about 10 more percent of the trials as compared to the smaller 25.8 cm² (4 in²) lamps of the same intensity as the larger lamps.

b) *Measurement of Reaction Times and Signal Recognition Errors*: The following vehicle driver's reaction times to recognize various rear signals and their errors have been measured in actual driving as well as in studies using driving simulators. Figures 9.14, 9.15, 9.16 and 9.17 illustrate some

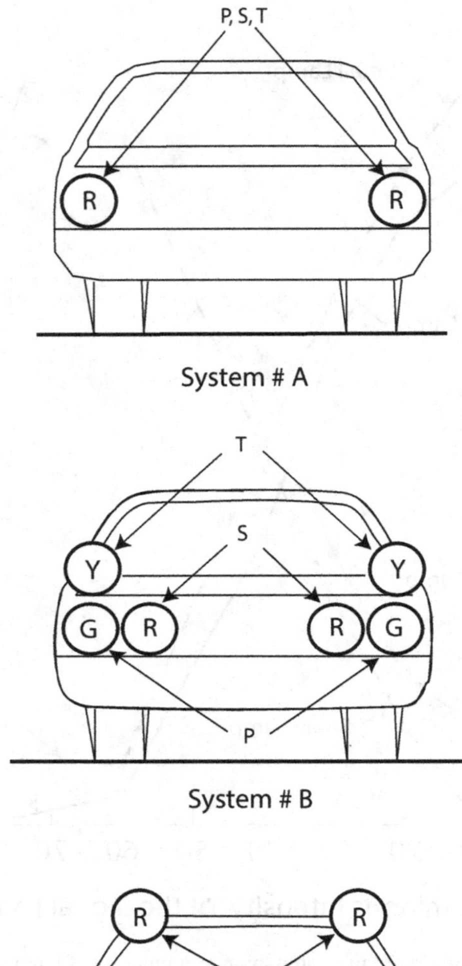

FIGURE 9.14 Three rear signal systems evaluated by Mortimer.

Note: P = parking(tail) lamps, S= stop lamps, T = turn signal lamps, R = red signal color, G = green signal color and Y= yellow signal color.

Source: Redrawn from Mortimer, 1970.

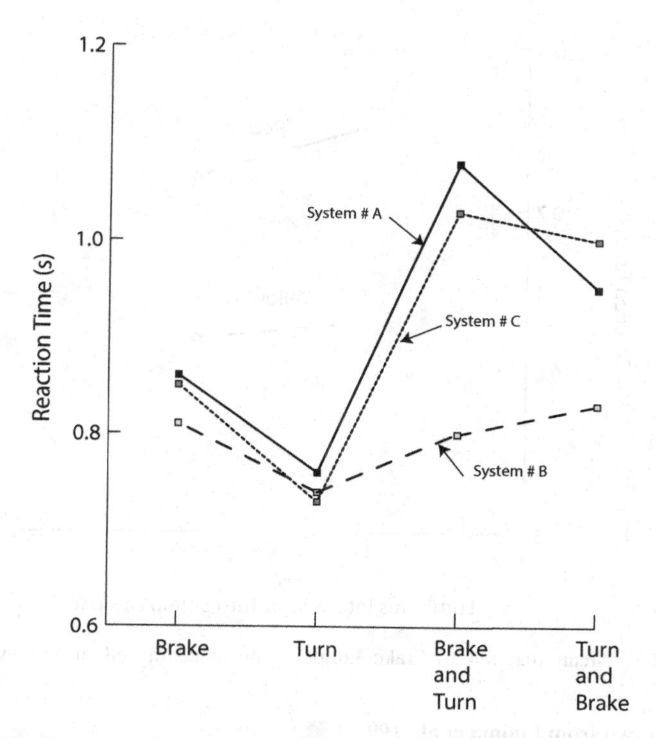

FIGURE 9.15 Response time of the following driver to different signals in rear signal systems.

Note: B= Brake (stop) signal only, T = Turn signal only, B+T= Brake signal changing to brake plus a turn signal, and T+B = Turn signal changing to turn plus brake signal.

Source: Redrawn from Mortimer, 1970.

examples from the studies conducted on driver reaction times to various rear signals (Mortimer, 1970; Luoma et al., 1995; Rockwell and Banasik, 1968). Figure 9.15 presents the response times of the following drivers to four different signal modes presented by evaluating three signal systems shown in Figure 9.14. The data in Figure 9.15 show that the response time of these drivers was reduced when the tail, stop and turn signals were coded by green, red and yellow colors respectively in system #B.

Figure 9.16 shows that the reaction time to a rear red brake (stop) signal in the presence of the turn signal is more affected by the color of the turn signal than its intensity. The brake signal reaction time was shortest in the presence of the 130 cd yellow turn signal. The turn signal color had a greater effect on the stop signal reaction time than the intensity because the yellow turn signal could be more quickly distinguished from the red stop lamps.

Figure 9.17 shows the reaction time of the following driver to stop signal and braking deceleration level while following a lead vehicle equipped with four different

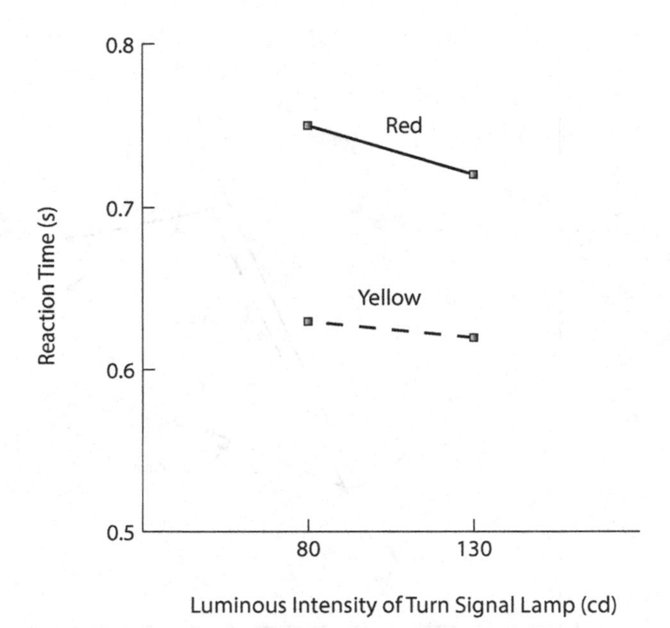

FIGURE 9.16 Mean reaction to brake lamps in presence of red or yellow turn signal activations.

Source: Redrawn from Luoma et al., 1995.

rear signaling system systems (Rockwell and Banasik, 1968). The tri-light system, which provided color coded information on the state of the lead vehicle to the following driver (green light for driver's foot on the gas pedal, red light for driver's foot on the brake pedal and yellow light when the driver's foot was not on brake or accelerator pedal), was found to be most effective in reducing the following driver's response time. The NIL systems (having no rear signals) had reaction times of 2 s or more. The AID system which displayed the level of acceleration of the lead vehicle by the number of green lamps and the level of deceleration by the number of red lamps did not reduce the following driver's reaction time as compared to the tri-light system.

It should be noted that in-spite of low reaction times for the Tri-light system, it could not be considered as a practical alternative because the tri-color equipped vehicles would create confusions in traffic situations involving presence of many lead vehicles. The problem was solved by incorporation of the center high mounted stop lamp, and it is covered in the next section.

c. *Analysis of Accident Data*: Rear-end accident data have been analyzed to determine if the rear end accident rates are affected by rear signal lamp characteristics such as the separation of tail lamps from stop lamps, incorporation of a center high mounted stop lamp, and rear turn signal lamp color (yellow versus red) (Malone et al., 1978). The fleet study is described in the next section.

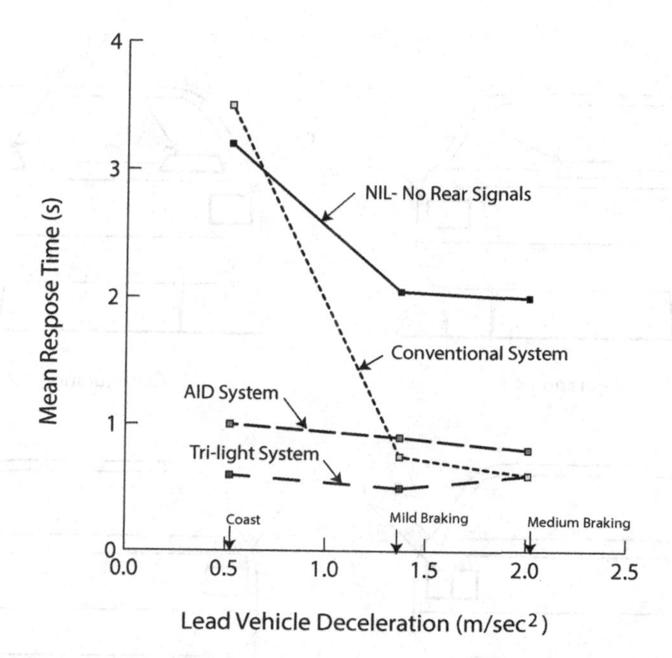

FIGURE 9.17 Following driver's response time to different rear signal systems as a function of lead vehicle deceleration under four rear signal systems.

Note: NIL=No rear signals presented; Conventional = Red tail lamp and red stop lamp system; Tri-light = Rear signal was green when the lead car was traveling at constant speed, the rear signal turned yellow when the lead car driver's foot released the gas pedal and the signal turned red when the brake pedal was pressed by the lead car driver; AID = The rear signal provided the level of acceleration and deceleration of the lead vehicle by lighting up a number of green or red lamps on the back of the lead vehicle.

Source: Redrawn from Rockwell and Banasik, 1968.

CENTER HIGH MOUNTED STOP LAMP (CHMSL) FLEET STUDY

The Essex Corporation conducted a 12-month fleet study of four rear signaling systems installed on 2101 taxicabs in the Washington DC area (Malone et al., 1978). The four signaling systems are shown in Figure 9.18. Configuration #1 involved an addition of a center high mounted stop lamp on the top of the trunk to taxicabs equipped with production red rear lamps which displayed the three signals, namely, tail(P), stop(S) and red turn signals(T) in red rear lamps mounted on each side of the car. Configuration #2 was similar to configuration #1 except that it had two high mounted stop lamps mounted near the outboard sides of the trunk. Configuration #3 did not have any high mounted stop lamps. But its tail lamps (P) were in separate outboard compartments and the stop and turn signals were combined in inboard compartments as shown in Figure 9.18. Configuration #4 had the conventional rear signal system

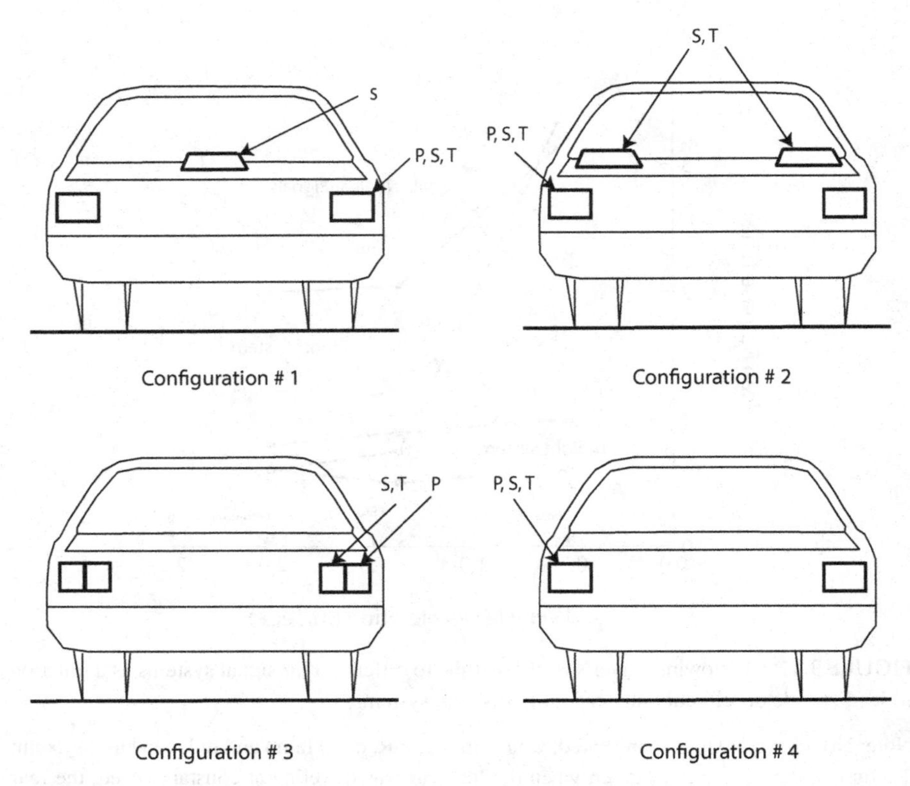

FIGURE 9.18 Four rear signal lamp configurations evaluated in the fleet study.

Note: P= Red tail lamp, S= Red stop lamp and T = Red turn signal lamp.

Source: Redrawn from Malone et al., 1978.

with all the three signals, that is, the tail (P), stop (S) and turn (T) combined into the red rear lamps mounted on each side of the car. The four signal systems installed in 2101 taxicabs with about equal installation of each configuration (about 525 taxicabs with each configuration). The four signal systems were also distributed evenly among the taxicabs available from eight different taxicab companies that provided their vehicles for the study.

The research team monitored each taxicab in terms of the mileage and accidents experienced over the 12-month study period. The taxi cabs accumulated a total of 60 million vehicle miles and 1470 accidents over the 12-month period. Out of the 1470 accidents, 217 accidents involved taxicabs being struck in the rear. The data on the rate of the rear-end accidents (rear-end accidents per million vehicle miles traveled) showed that the vehicles equipped with the center high mounted stop lamps had about a 50 percent reduction in their rear-end accident rate as compared to the rates for the other three rear-end configurations (see Figure 9.19).

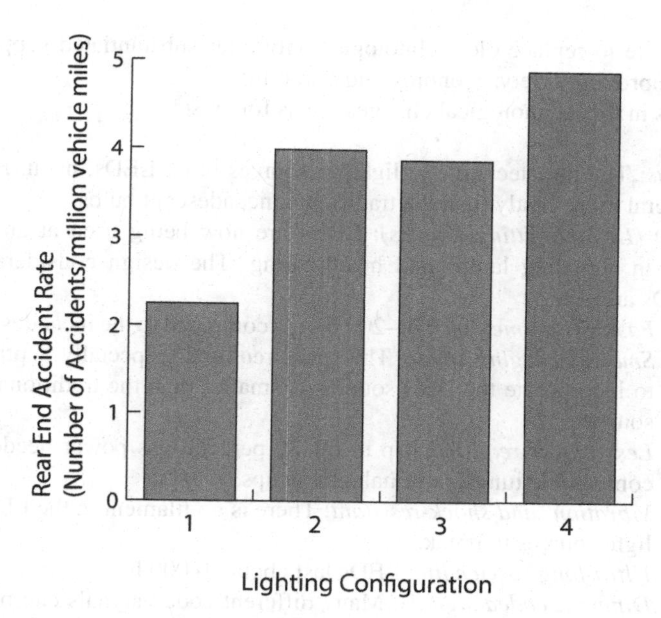

FIGURE 9.19 Rear end accident rates by configuration of rear signal system.

Source: Redrawn from Malone et al., 1978.

OTHER SIGNAL LIGHTING STUDIES

The SAE Vehicle Lighting Committee and MVMA (Motor Vehicle Manufacturers Association until about mid 1990s) Lighting Task Force have conducted a number of studies (generally conducted as lighting demonstrations to show proposed lighting configurations, intensities, and so forth to familiarize the committee members prior to ballots for approval or disapproval of proposed SAE practices) (Mortimer, 1970; Mortimer et al., 1973; Bhise, 1981 and 1982; Moore and Rumar, 1999) . The above-mentioned and other research studies have included evaluations of signals from different viewing locations on issues such as a) Signal effectiveness through measurements of driver ratings and responses, for example, response time, correct identification of signal modes, b) Minimum perceptible difference in signal intensities, c) Equivalency of signal value, d) Effect of lamp size, shape, length-to-width ratio, e) International Harmonization between SAE, ECE and Japanese lighting requirements, and f) effectiveness of the Center High Mounted Stop Lamp (CHMSL).

CONCLUDING REMARKS

With the advances in new lighting technologies, the lighting systems in vehicles have been improving continuously. The progress in the implementation of new developments has been somewhat limited because of the cost of the new equipment and the lengthy approval process of government regulations. Since the lamps are directly related to safety and are regulated by government agencies, new technologies

have been able to replace old technologies only after substantiated supporting evidence for improving safety, economy and durability.
Some issues in the technological changes are as follows:

1. *Costs*: The new technology lighting sources (e.g., LEDs, neon, HID) are in general more costly than the traditional incandescent bulbs.
2. *LED (Light Emitting Diodes)*: LEDs are now being used at an increasing rate in signaling lamps and headlighting. The design considerations with LEDs are:
    a) *Faster rise time*: (by 170–200 ms) as compared to the incandescent lamps.
    b) *Small packaging space*: The space required (especially depth of lamps) to incorporate the LED sources is smaller than the traditional bulb type sources.
    c) *Less power required*: Up to 60–80 percent less power needed than the comparable tungsten or halogen lamps.
    d) *Vibration- and shock-resistant*: There is no filament in the LED and HID light sources to break.
    e) *Ultra-long service life*: LEDs last about 10,000 h.
    f) *Different coded signals*: Many different coded signals can be displayed (based on combinations of lighted patterns, colors and words/messages/symbols).
    g) *No complete burnout*: The LED lamp will not completely burn out as the percentage of inoperative LEDs during the useful vehicle life will be very small.
    h) *Lamp output*: The LED lamp is affected by temperature.
3. *Neon:* Neon lamps are used on a limited basis, primarily in rear signal lamps (e.g., high mounted stop lamps). The design considerations with neon technology are as follows:
    a) Faster rise time.
    b) Longer life 2500 hours with 1 million on/off cycles.
    c) Design flexibility, continuous smooth uniform appearance and seamless lamp.
4. *Fiber Optics and Light Piping*: Fiber optics (and other light piping materials) allow for the piping of light flux from a larger, more efficient light source placed at a less proximate location. The design considerations with this technology are as follows:
    a) Packaging flexibility.
    b) One energy efficient light source can distribute light flux to different locations (called the light engine concept).
    c) Total electrical isolation – Immune to electrical interferences (e.g., electro-magnetic and radio frequency interferences).
5. *New signal functions*: Besides the presence, parking, stop and turn information, signal lamps based on new technologies have the potential to provide additional information to other drivers. Thus, some future application issues are these:

a) A new "stopped vehicle signal" may have the potential to reduce rear-end collisions.
b) Improving effectiveness of stop lamps by incorporating a "stopping" vehicle signal that provides information on the rate of closure during stopping (i.e., providing information on the level of lead vehicle deceleration).
c) Displaying state of other decelerating devices (e.g., retarders on heavy trucks, vehicle stability control systems).
d) Displaying other information (e.g., disabled vehicle signal).
e) Car following aids (e.g., safe headway/following distance signal).
f) Variable intensity signal lamps (that adjust according to the changing ambient lighting conditions, for example, bright sun sunlight (100,000 lux of incident illumination) to dusk to night).
g) If future lighting system concepts do not comply with the existing FMVSS 108 requirements (NHTSA, 2024), then the developer must make a petition to the NHTSA to allow the use of such systems.

REFERENCES

Bhise, V. D. 1981. Effects of Area and Intensity on the Performance of Red Signal Lamps. A report on the SAE lighting Committee demonstrations conducted at the Ford Dearborn Proving Ground on September 9–10, 1981.

Bhise, V. D 1983. Effects of Luminance Distribution, Luminance Intensity and Lens Area on the Conspicuity of Red Signal Lamps. A report on the SAE Lighting Committee Demonstrations conducted at the Rockcliffe AFB, Ottawa on October 3, 1982.

Bhise, V. D. and C. C. Matle. 1989. Effects of Headlamp Aim and Aiming Variability on Visual Performance in Night Driving. *Transportation Research Record*, No. 1247, TRB, Washington, D.C.

Bhise, V. D., E. I. Farber and P. B. McMahan. 1977. Predicting Target Detection Distance with Headlights. *Transportation Research Record*, No. 611, TRB, Washington, D.C.

Bhise, V. D., E. I. Farber, C. S. Saunby, J. B. Walnus and G. M. Troell. 1977. Modeling Vision with Headlights in a Systems Context. SAE Paper No. 770238, Presented at the 1977 SAE International Automotive Engineering Congress, Detroit, Michigan, 54 pp.

Bhise, V. D., C. C. Matle and D. H. Hoffmeister. 1984. CHESS Model Applications in Headlamp Systems Evaluations. SAE Paper No. 840046, Presented at the 1984 SAE International Congress, Detroit, Michigan.

Bhise, V. D., C. C. Matle and E. I. Farber. 1988. Predicting Effects of Driver Age on Visual Performance in Night Driving. SAE paper no. 881755 (also no. 890873), Presented at the 1988 SAE Passenger Car Meeting, Dearborn, Michigan.

Blackwell, H. R. 1952. Brightness Discrimination Data for the Specification of Quantity of Illumination. *Illuminating Engineering*, 47(11) : 602–609.

Farber, E. I. and V. D. Bhise. 1984. Using Computer-Generated Pictures to Evaluate Headlamp Beam Patterns. *Transportation Research Record*, No. 996, TRB, Washington, D.C., pp. 47–53.

Jack, D. D., S. M. O'Day and V. D. Bhise. 1994. Headlight Beam Pattern Evaluation – Customer to Engineer to Customer. SAE paper no. 940639. Presented at the 1994 SAE International Congress, Detroit, Michigan.

Jack, D. D., S. M. O'Day and V. D. Bhise. 1995. Headlight Beam Pattern Evaluation – Customer to Engineer to Customer – A Continuation. SAE paper no. 950592. Presented at the 1995 SAE International Congress, Detroit, Michigan.

Luoma, J., M. J. Flannagan, M. Sivak, M. Aoki and E. Traube. 1995. Effects of Turn-Signal Color on Reaction Time to Brake Signals. Report no. UMTRI 95-5.The University of Michigan Transportation Research Institute, Ann Arbor, Michigan.

Malone, T. B., M. Kirkpatrick, J. S. Kohl and C. Baler. 1978. "Field Test Evaluation of Rear Lighting Systems", NHTSA, DOT Report no. DOT-HS-5-01228, February 1978.

Moore, D. W. and K. Rumar. 1999. Historical Development and Current Effectiveness of Rear Lighting Systems. Report no. UMTRI-99-31. The University of Michigan Transportation Research Institute, Ann Arbor, Michigan.

Mortimer, R. G. 1970. Automotive Rear Lighting and Signaling Research. Final Report, Contract no. FH-11-6936, U.S. Department of Transportation, Highway Safety Research Institute, University of Michigan, Report HuF-5.

Mortimer, R. G., Moore, Jr., C. D., Jorgeson, C. M. and J. K. Thomas. 1973. Passenger Car and Truck Signaling and Marking Research: I. Regulations, Intensity Requirements and Color Filter Characteristics. Report no. UM-HSRI-HF-73-18, Highway Safety Research Institute, The University of Michigan, Ann Arbor, Michigan.

National Highway Transportation Safety Administration. 2024. Federal Motor Vehicle Safety Standards. *Federal Register*, CFR (Code of Federal Regulations), Title 49, Part 571, U. S. Department of Transportation. www.ecfr.gov/current/title-49/subtitle-B/chapter-V/part-571#571.108 (accessed March 11,, 2024).

Rockwell, T. H. and R. C. Banasik. 1968. Experimental Highway Testing of Alternate Vehicle Rear Lighting Systems. RF Project 2475. Systems Research Group, Department of Industrial and Systems Engineering, The Ohio State University, Columbus, Ohio.

Schoettle, B., M. Sivak and M. Flannagan. 2001. High-Beam and Low-Beam Headlighting Patterns in the U.S. and Europe at the Turn of the Millennium. Report no. UMTRI-2001-19, The University of Michigan Transportation Research Institute, Ann Arbor, Michigan.

Society of Automotive Engineer, Inc. 2009. *SAE Handbook*. Warrendale, PA: SAE.

Troell, G., D. H. Hoffmeister and V. D. Bhise. 1978. "Identification of Steady Burning Red Lamps as Tail or Stop Signal in Night Driving", Presented at the Annual Meeting of the Human Factors Society, Detroit, Michigan, October 1978.

10 Entry and Exit from Automotive Vehicles

INTRODUCTION

The driver and occupants should be able to enter and exit from the vehicle quickly and comfortably, without any awkward postures or highly physical efforts that may involve excessive bending, turning, twisting, stretching, leaning, and hitting of body parts on the vehicle components. In this chapter we cover many problems that drivers and passengers experience during entry and exit from vehicles and relate these to different vehicle package dimensions.

Drivers experience different problems while entering and exiting vehicles with different body styles, from low sports cars to sedans to SUVs to pickups and heavy trucks. Assuming that a vehicle or a physical buck is available, the best method to uncover these problems would be to ask a number of male and female drivers with different anthropometric characteristics (e.g., tall, short, slim, and obese) to get in and out of the vehicles, preferably after they have adjusted the seat and steering wheel to their preferred driving positions, and observe (or video record and replay in slow motion) and ask them to also describe the difficulties that they encountered. Such exercises are usually performed by package engineers during the evaluation of the physical buck of a new vehicle. Comparisons are also made with different benchmarked vehicles to understand the differences in vehicle dimensional characteristics and assist features (e.g., door handles, grab handles, steps, and rocker panels) used during entry/exit performance and difficulty/ease ratings to determine if a given vehicle package will be acceptable or will need improvements.

PROBLEMS DURING ENTRY AND EXIT

The problems drivers experience while entering or exiting from a passenger car depend upon their gender and anthropometric characteristics.

1. *Drivers with short legs* (predominantly women) will complain, saying this:
    a. The seat and step-up (top of the rocker panel) are too high. The rocker panel is the lower part on the side of the vehicle body under the doors

DOI: 10.1201/9781003485582-10

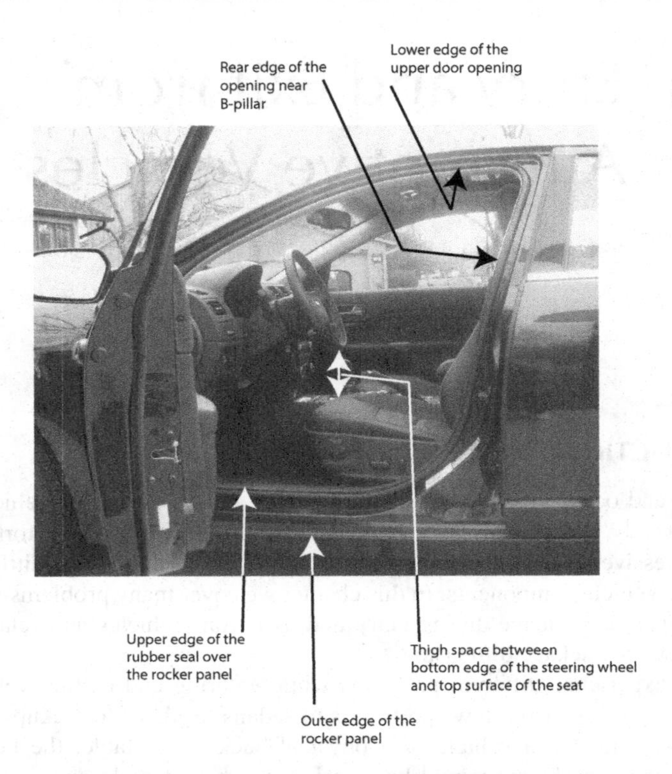

FIGURE 10.1 Space available for the driver to slide into the seat is limited by the height and outer edge of the rocker panel, the thigh space between the steering wheel and the top of seat, head and torso space between the top of seat to the lower edge of the upper body opening, and the B-pillar location.

(which in effect creates the lower part of the of the door frame) over which the occupant's feet move during entry and exit (see Figure 10.1 for a picture of the rocker panel and Figure 10.2 for its lateral cross section at the seating reference points SgRP).

b. The step-over is too wide (rocker panel section is too wide). The lateral distance between the outer edge of the seat cushion and the driver centerline and/or the outer edge of the vehicle (i.e., the rocker panel) from the driver centerline is too far out to move the lagging leg during entry from the ground to inside the vehicle (see dimensions "S" and "W" in Figure 10.2).

c. The clearance between the driver's knees and the instrument panel and/or the steering column (due to seat moved forward to accommodate the driver's short legs to reach the pedals) is insufficient.

2. *Older, obese, mobility challenged drivers* will complain, saying this:
    a. The seat is either too high or too low (see dimension H5 in Figure 10.2). This indicates that the driver had difficulty climbing up into the seat (e.g., strain in the knees) or sitting down into the seat due to larger muscular forces needed in the leg and back muscles to move the driver's body onto the seat during entry.
    b. The upper part of the body door opening is too low (entrance height defined as H11; see Figure 5.9). The person will experience difficulty in moving his/her head under the lower edge of the upper body opening (see Figure 10.1).
    c. The step-over is too wide (Dimension "W" in Figure 10.2).
    d. The thigh clearance (between the bottom edge of the steering wheel and top surface of the seat) is insufficient (SAE dimension H74, SAE J1100, SAE, 2009).
    e. The steering wheel-to-stomach clearance (between the bottom edge of the steering wheel and the driver's stomach) is insufficient.
    f. The door does not open wide enough (i.e., the space between the opened door [i.e. the inner door trim panel] to vehicle body-side is insufficient).
3. *Drivers with a tall torso* will complain, saying this:
    a. The upper body opening (entrance height, H11) is too low.
    b. The A-pillar is too close to their head while bending the torso forward (point H in Figure 10.6).
    c. The seat bolsters (i.e., the raised sides of the seat cushion) are too high.
    d. The head clearance is insufficient.
4. *Drivers with long legs* will complain, saying this:
    a. The seat track does not extend sufficiently rearward (seat track is too short and placed more forward in the vehicle).
    b. The front edge of the B-pillar (roof pillar between the front side window and the rear side window) is too far forward. The seatback in this case is moved rearward of the front edge of the B-pillar requiring the driver to brush past the B-pillar to get into the seat (See Figure 10.1).
    c. The lower rear edge of the cowl side is too far rearward (points F in Figure 10.6). (Note: Cowl panel is the body panel behind the rear edge of the engine hood and lower edge of the windshield). The legroom for the driver is not sufficient to move his legs from the ground to inside the vehicle. This problem usually results in the drivers' shoes hitting the rear edge of the cowl side (look for shoe scuffmarks on the trim parts on the top of the cowl side and rocker panels).
    d. The door does not open wide enough (i.e., the space between the opened door to vehicle body is insufficient).

Thus, the design of the door opening size and shape created by rocker panel, B-pillar, roof rail (vehicle body component above the doors and mounted on the sides of the vehicle roof), A-pillar (front roof pillar), cowl side and the knee bolster, the door opening angles, positioning of the seat, the steering wheel, and door grab and opening handles – all affect the driver's ease of entry and exit.

VEHICLE FEATURES AND DIMENSIONS RELATED TO ENTRY AND EXIT

Vehicle features and vehicle dimensions that the designers and engineers should pay attention to that facilitate ease during entry/egress are as follows:

Door Handles

1. *Height of the Outside Door Handle*: The short 5th percentile woman should be able to grasp the door handle without raising her hand over her standing shoulder height and the tall 95th percentile male should be able to grasp the handle without bending down (i.e., not below his standing wrist height).
2. *Longitudinal Location of the Outside Door Handle*: The handle should be placed as close to the rear edge of the door as possible to avoid the lower right corner of the driver's door hitting the driver's shin during door opening.
3. *Insider Door Handle Location*: While closing the door (after the driver has entered the vehicle and is seated in the driver's seat), the inside door grasp (or grab) handle should not require the driver to get into the "chicken winging" type wrist posture. This means that the inside door handle should be placed: a) forward of the minimum reach zone (see Chapter 5); b) rearward of the maximum reach zone (see Chapter 5); c) placed not below the door armrest height; and d) placed not above the seated shoulder height. While the driver is exiting, the inside door-opening handle location should also meet the above location requirements.
4. *Handle Grasps*: The grasp area clearances should be checked to ensure that the outside door handle and the inside grasp handle (or pull cup) allow for the insertion of four fingers of the 95th percentile male's palm (by considering palm width, finger widths and finger thickness; see Table 4.1). Further, to facilitate gloved-hand operation, additional clearances would be needed depending on the winter usage and type of gloves worn by the population where the vehicle is to be marketed. Further, additional clearances should be considered to avoid scratches on nearby surfaces due to finger rings and long fingernails.

Lateral Sections at the SgRP and Foot Movement Areas

The following vehicle dimensions shown in Figure 10.2 are important for ease during entry and exit.

1. Vertical height of the Seating Reference Point (SgRP) from the ground (H5)
2. Lateral distance of the SgRP from outside edge of the rocker (W)

3. Lateral distance of the outside of seat cushion to outside of rocker (S)
4. Lateral overlap thickness of lower door (T)
5. Vertical top of rocker to the ground (G)
6. Vertical top of the floor to the top of the rocker (D)
7. Curb clearance of doors at design weight (C)

To improve the ease of the driver's entry and exit, the magnitudes of the above dimensions (separately and in combinations) should be evaluated during the early stages of the vehicle design.

The H5 dimension should allow drivers to easily slide in and out of their seats without climbing up onto the seat or sitting down into the seat. Thus, H5 should be about 50 mm below the standing buttock height of most of the users (by considering the up/down adjustment of the seat cushion). The top of the seat to ground distance of about 500–650 mm is generally considered to facilitate easy ingress and egress for the U.S. population.

The dimension "W" should be as short as possible (See Figure 10.2). This means that the dimension "S", lateral distance from the outer edge of the seat and the outer edge of the rocker, should be short enough to allow the driver's foot to be placed easily on the ground during entry/exit. The lateral distance of the outer edge of the rocker panel to the SgRP (i.e., distance "W") should be about 420–480 mm to accommodate most drivers. Figure 10.3 shows a picture of a vehicle with a wide rocker and a large "W" dimension. A smaller width rocker section or a hidden rocker design (where the door extends down and overlaps the rocker and the rocker is tucked in [i.e., its width is reduced] toward the vehicle centerline; see Figure 10.4) will help in shortening dimension "W". The height of the lower door edge "C" (at maximum

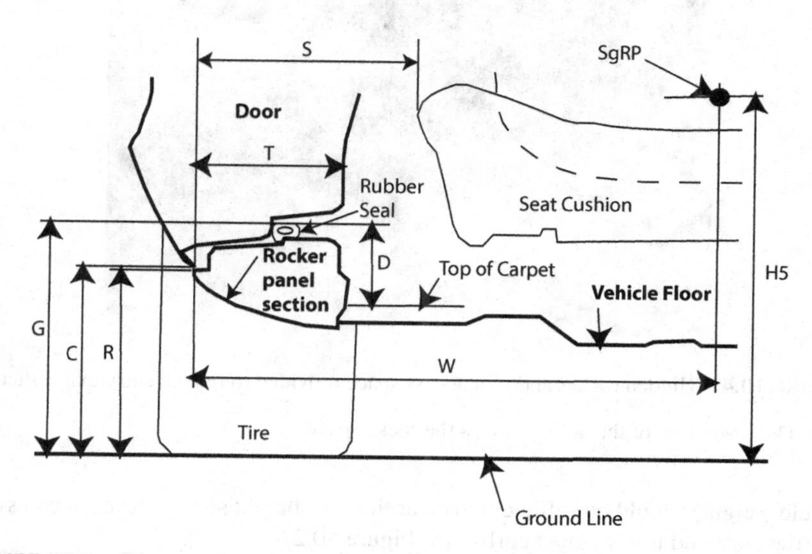

FIGURE 10.2 Cross section of the vehicle at SgRP (in the rear view) showing dimensions relevant for entry and exit.

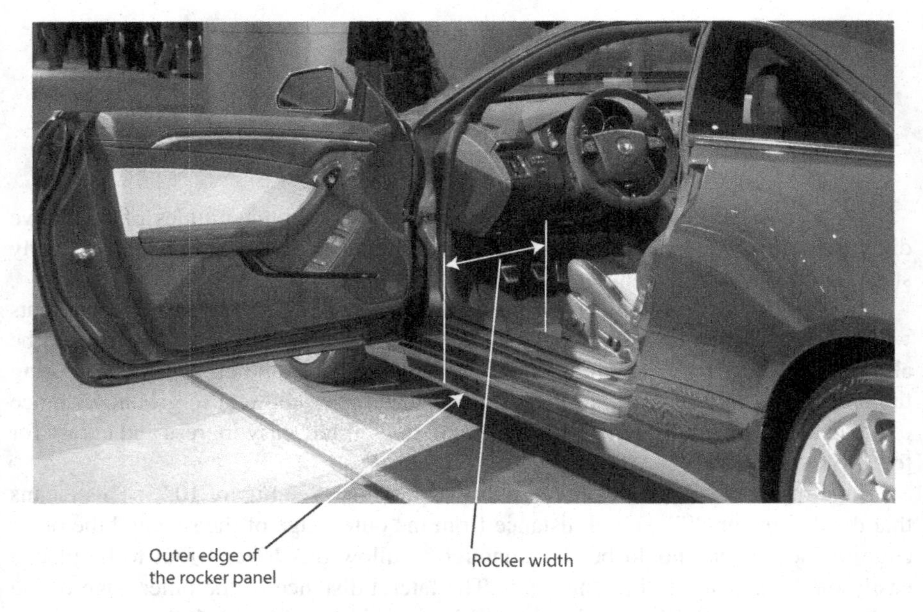

FIGURE 10.3 Wide rocker panel makes the entry/exit more difficult as the driver needs to take a wider lateral step from the outer edge of the rocker to move into the seat.

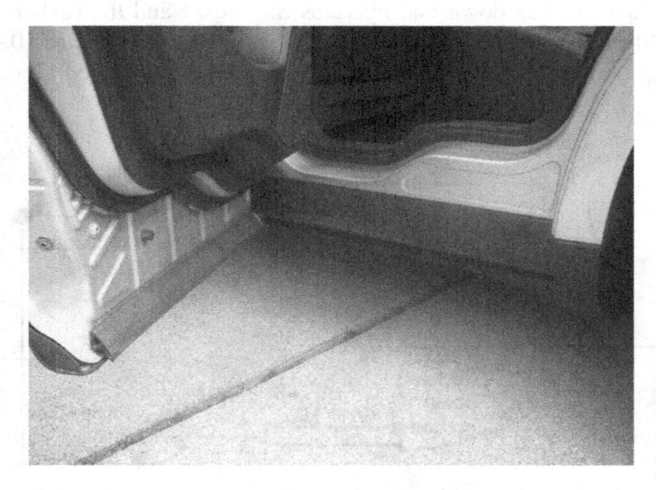

FIGURE 10.4 Hidden rocker and foot access space provided to reduce entry/exit difficulty.

Note: The lower part of the door overlaps the rocker panel.

vehicle weight) should be sufficient to clear the curb height so that the door can swing over the curb and not hit most curbs (see Figure 10.2).

The top of the rocker from the ground (dimension "G") and rocker top from the vehicle floor (dimension "D") should be as small as possible to reduce foot lifting during entry and exit. These dimensions are dependent upon ground and curb clearances (see Figure 10.2).

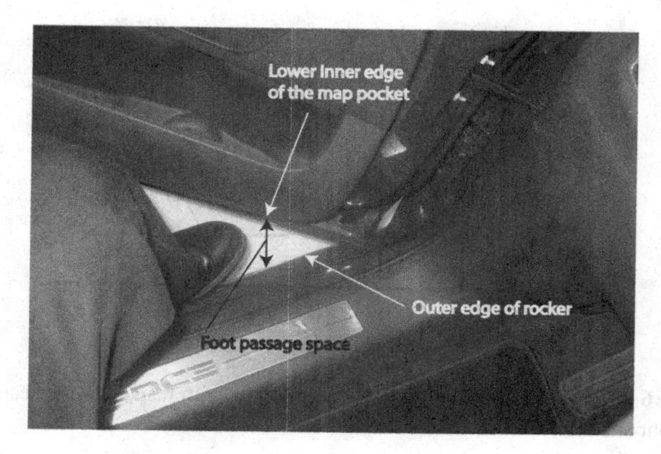

FIGURE 10.5 Foot movement space between the outer edge of the rocker and the inner edge of the map pocket.

The width dimension "T", which is the lateral dimension from the outer edge of the rocker to the lower edge of the inner door trim panel (generally the lower inward protruding edge of the map pocket), should be as small as possible. The smaller the dimension, the more foot passage space will be available during entry and exit (see Figure 10.5). This is especially important when the door cannot be opened wide due to restricted space on the side, such as in garages and when parking close to another vehicle (side-by-side).

Body Opening Clearances from SgRP locations

Figure 10.6 shows a number of points on the door openings and the instrument panel measured from the seating reference points (SgRPs) of the front and rear occupant positions. Larger distances of these points from the respective SgRPs will allow more room during entry and exit. For the driver's door opening, the point "T" defines the entrance height for the head clearance, point "H" defines the head swing clearance during leaning and torso bend, point "K" defines the knee clearance to the lower portion of the instrument panel, point "F" defines the foot clearance, point "B" defines the top of the rocker and point "R" defines the forward edge of the B-pillar. Similarly points T', F', B' and R' define clearances for the rear passenger entry and egress. The package engineers generally compare the dimensions to the above points and the SgRP locations of the vehicle being designed to other benchmarked vehicles.

Doors and Hinge Angles

1. Door fully opened angles should be designed to provide the following:
   a) Normally about 65–70 degrees for front doors and about 70–80 degrees for rear doors.

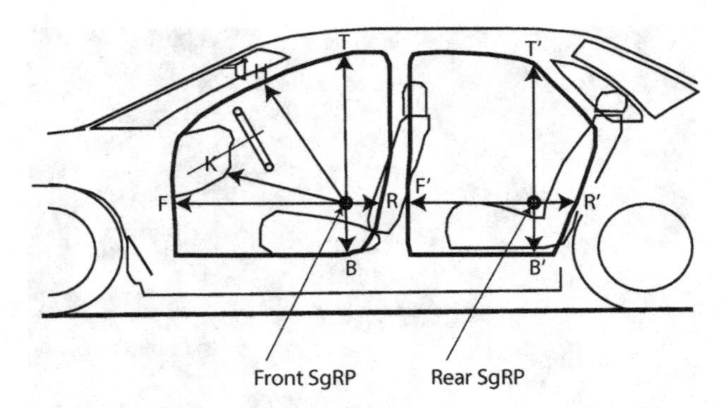

FIGURE 10.6 Side view showing body opening points with respect to the seating reference points of front and rear seats.

   b) Too large of an opened angle will increase the difficulty of reaching the door grab handle/area to close it.

2. Door Hinge Centerline
   a) The amount of space available between the open door and the vehicle body is affected by the position and orientation of the hinge centerline of the door. The hinge centerline is the axis passing through the upper and lower hinges.
   b) The top of the door hinge centerline should be slightly inward and forward to improve head and shoulder clearance to upper front corner of opened door. This will also reduce the door closing effort.

Seats

1. Very high (raised) side bolsters on the seat cushion and the seat back make entry/exit task more difficult as the occupant needs to slide over the raised bolsters.
2. The outer edge of the seat cushion should be located as outboard as possible for easier entry/exit by reducing the lateral foot movement distance between the occupant centerline and the outer edge of the rocker.
3. The surface friction characteristics of the seat material and clothing worn by the occupants affect how easily the occupants can slide in and out of the seat. Higher friction upholstery fabrics make entry/exit task more difficult. Leather seating surfaces, thus, aid in sliding and repositioning during entry/exit.

Seat Hardware

1. The recliner handles, side seat shields (trim parts on the side of the seat cushion and lower part of the seat), and other seat controls should be located below the point where the seat compresses during entry/exit.

2. The seat track should be located out of way of the rear occupant's foot passage area to reduce intrusions during entry/exit.
3. The automatic movement features,f or example, rearward movement of the driver's seat and upward and forward movement of the steering wheel after the engine is turned-off and the driver's door is opened, can reduce entry/exit difficulties.

TIRES AND ROCKER PANELS

The current design trend is to increase the tread-width of the front and rear wheels as much as possible to improve vehicle stability and external appearance. Further, locating the outer side surfaces of the wheels flush with the outer sides of the fenders will also help in reducing aerodynamic drag associated with the fenders and wheels. However, more outward positioning of the wheels causes another interesting problem – called "stone-pecking". The term "stone-pecking" refers to the action of the spinning tires to pick-up road dirt and small stones and spray them against the lower sides of the vehicle. Stone-pecking can cause erosion (or sanding due to spraying of stones and dirt) of the paint on the lower body side of the vehicle and also make the rocker panel dirty. One solution to the paint erosion problem is to increase the width of the rocker panels. However, the increase in the width of the rockers will increase the step-over width and make the entry/exit tasks more difficult. Further, leg contact with the wider and slushy/muddy/wet rockers can stain pant legs and rough rocker surfaces or debris on the rocker can also snag stockings. Thus, this design trend creates conflicts in improving the occupant's ease and reducing customer complaints related to entry and egress. It is thus a trade-off issue that ergonomics engineers have to resolve with the exterior styling and body designers.

RUNNING BOARDS

The entry/exit ease from taller vehicles such as SUVs and pickup trucks can be improved by providing running boards. Shorter drivers can benefit more from the running boards as compared to some tall drivers who may find the running boards to be an intrusion in the foot placement space on the ground next to the vehicle.

Figure 10.7 shows a lateral cross-sectional through a running board with important dimensions related to entry and exit. The dimension "L" should wide enough (at least 50–55 mm) to allow foot placement. The dimension "T" should be large enough to accommodate the toe height of most shoes (about 50 mm or more) and dimension "W" should allow for a shoe support width of at least 125–150 mm. The height of the step (H) will depend upon the state of suspension (i.e. loading) and tires. However, from the viewpoint of ease in climbing the step, the dimension "H" should not be greater than about 450 mm from the ground. The dimension "X" (not shown in Figure 10.7) represents the longitudinal length of the running board. It should be designed to ensure that sufficient longitudinal foot support space is provided forward of each occupant's SgRP.

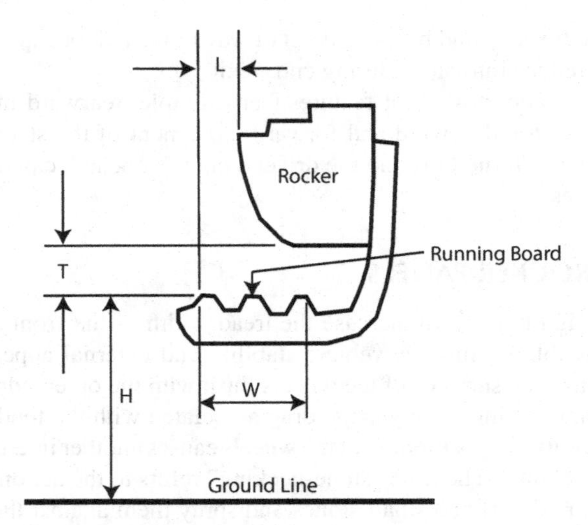

FIGURE 10.7 Running board dimensions.

THIRD ROW AND REAR SEAT ENTRY FROM 2-DOOR VEHICLES

1. Entry/exit to the 3rd row of minivans (or large SUVs) or the back seat of 2-door coupes is challenging for many customers because of a) lack of direct door access, b) the need to climb up over and through a small opening to reach the rear seat, and c) the need to reposition one's body before sitting on the seat.
2. Provision of inside grab handles and their locations are important in improving ease during entry and exit. The grab handles can be located as follows: a) locate high on the B-pillar for 2nd row, and the C-pillar for 3rd row entry/exit assistance, b) particularly helpful for 3rd row exit to pull an occupant up and out of the seat, and c) also a grab handle will aid during repositioning and transferring weight between sitting, standing, and climbing.

HEAVY TRUCK CAB ENTRY AND EXIT

The heavy truck cab floor is generally high above the ground so accessing the cab usually requires climbing up about three steps (see Figure 10.8). The basic rule in designing steps and grab handles is that the driver or the passenger must be able to always be in "three points of contact" with the truck during the entire entry or exit process. The contact points are defined as the user's foot contacts and hand contacts with the truck. Thus while entering the cab from the ground, two hands on the grab bars (driver's left hand on the door grab bar or the steering wheel, and the right hand on the inside or outside rear grab bar) and one foot on a step, or two feet on one or more steps and one hand on a grab bar must be minimally accommodated. The cabs should have two long vertical grab bars on the rear side of the door opening (one

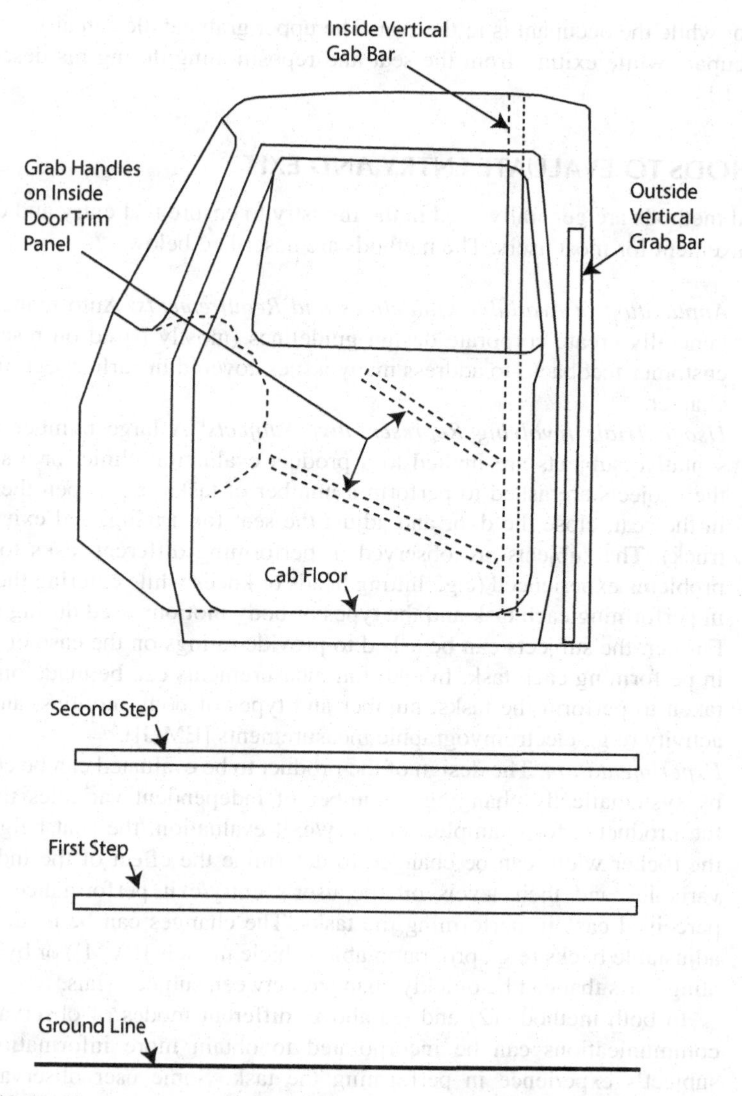

FIGURE 10.8 Heavy truck cab showing two steps and four grab bars used to facilitate entry and exit.

mounted outside and one inside the cab; see Figure 10.8). Further, at least two grab handles or grasp areas must be incorporated in the interior door trim panel. One grab handle should be located near the lower rear corner of the interior door trim panel so that the user can grab the door handle while on the ground or the first step. The second grab handle should be high on the door trim panel which can be easily reached (must be placed rearward of the maximum reach envelope. See Chapter 5) to close the door while the occupant is in the seat. The upper grab handle can also be used by the occupant while exiting from the seat and repositioning during his descend (see Figure 10.8).

METHODS TO EVALUATE ENTRY AND EXIT

Several methods are generally used in the industry to ensure that entry and exit tasks are convenient for most users. The methods are described below.

1. *Application of Available Guidelines and Requirements*: Auto manufacturers generally create corporate design guidelines (mostly based on research and customer feedback) to address many issues covered in earlier sections of this chapter.
2. *Usage Trials Involving Representative Subjects*: A large number of representative subjects are invited to a product evaluation clinic, or a study and the subjects are asked to perform a number of tasks (e.g., open the door, sit in the seat, close the door and adjust the seat for driving, and exit from the truck). The subjects are observed in performing different tasks to uncover problems experienced (e.g., hitting heads or knees while entering the vehicle) in performing each task and the types of body motions used during the tasks. Further, the subjects can be asked to provide ratings on the ease or difficulty in performing each task. In addition measurements can be made on the time taken to perform the tasks, number and types of body motions, and muscle activity (e.g., electromyographic measurements [EMG]).
3. *Experimentation*: The design of the product to be evaluated can be configured by systematically changing a number of independent variables that define the product – for example, for entry/exit evaluation, the seat height and/or the rocker width can be changed to determine the effect of the independent variables and their levels on the user's entry/exit performance and their perceived ease in performing the tasks. The changes can be made by using adjustable bucks (e.g., programmable vehicle models [PVM]) or by incorporating parts that can be quickly changed between subject trials.

   In both methods (2) and (3) above, different modes of observations and communications can be incorporated to obtain more information on the subject's experience in performing the task. Some user observations can include a) observe what subjects do, b) observe how they do it, c) record hand/foot contact points and body postures, d) track errors committed, e) measure time taken to complete the task. Some user communication methods can include a) ask the subjects about problems encountered/complaints, b) ask them to provide feedback on dimensions of certain details, also, ask them to rate the ease or difficulty in performing the task (e.g., Is the grab bar diameter too large, too small or just right to hold? Is the step too wide, too narrow or just right?); and c) ask the subjects about product features or items that they liked or disliked very much (see Volume 2, Chapter 19 for more information on methods of vehicle evaluation).

4. *Use of Mathematical and Manikin Models*: Use of motion analysis, biomechanical and manikin models to predict entry and exit performance have had limited reliability because of complex changes in body postures performed by subjects with different anthropometric characteristics and the complex three dimensional space in the vehicle body. Some design guidelines based on such applications of biomechanical principles are generally more useful (see item (1) above) than the use of complex modeling required to predict and analyze the body motions.
5. *Task Analysis*: Task analysis is a simple but powerful method to determine user problems in product designs. The next section provides additional information on this technique.

TASK ANALYSIS

Task analysis is one of the basic tools used by ergonomists in investigating and designing tasks. It provides a formal comparison between the demands each task places on the human operator and the capabilities the human operator possesses to deal with the demands. Task analysis can be conducted with or without a real product or a process. But it is easier if the real product or equipment is available, and the task can be performed by actual (representative) users under real usage situations to understand the subtasks.

The analysis involves breaking the task, or an operation into smaller units (called the subtasks), and analyzing subtask demands with respect to the user capabilities. The subtasks are the smallest units of behavior that need to be differentiated to solve the problem at hand, for example, grasp a handle, read a display, and set speed. The user capabilities that are considered here are generally as follows: sensing, use of memory, information processing, response execution (movements, reaches, accuracy, postures, forces, and time constraints).

Task analysis can be conducted by using different formats. Table 10.1 presents a format presented by Drury (1983) that the author found to be useful in analyzing products during their design process. The left hand side of Table 10.1 describes the subtasks involved in the task along with the purpose of each subtask and the event that triggers each subtask. Thus, the description of each subtask makes the analyst think about the need for each subtask and, perhaps, to even suggest a better way to do the task. The right hand columns of the table force the analyst to consider different human functional capabilities such as searching and scanning, retrieving information from the memory, interpolating (information processing) and manipulating (e.g., hand finger movements) required in performing each subtask. The last column requires the analyst to think about possible errors that can occur in each subtask. This last column is the most important output that can be used to improve the task and/or improve the product to reduce errors, problems and difficulties involved in performing the task

If the task analysis is performed on a product that already exists (or on its mock-up, prototype or a simulation), then a number of user trials can be performed, and information can be gathered on how different users perform each subtask and the problems, difficulties and errors experienced by the users. The information then can

TABLE 10.1
Tabular Format of Task Analysis

Task Description					Task Analysis				
Task	Subtask No.	Subtask/Step Description	Subtask Purpose	Trigger	Scanning and Seeing	Memory	Interpolating	Manipulating	Possible Errors
Egress from Driver's Seat of a Heavy Truck	1	Pull door latch handle	Disengage the door latch mechanism	Need to open the door	Locate the inside door opening handle	Remember the location of the door handle	Determine if the door latch is already disengaged	Pull handle until the latch disengages	Door latch handle not pulled far enough. Latch opening force too high
	2	Push door open	Position door to facilitate exit from the cab	Need to exit the cab	See when the door is open enough to egress from the cab	Remember door opening angle	Determine the position of the opened door	Push door until opened a sufficient angle/ distance to exit the cab	Door is pushed hard and damaged. Door is not opened far enough.
	3	Rotate torso to face inward	To leave the driver's seat and be properly positioned for descending down the steps	Need to leave the cab	See that the cab floor and steps are clear	Remember to face the proper direction and hold grab handle or steering wheel	Determine if you are ready to egress	Look to see if you are clear of obstructions	Operator faces wrong direction and slips while exiting

4	Grab handles	To anchor body while climbing down and avoid fall	Need to prevent injury from falling	Look for the grab handles and foot placement areas on the steps	Remember the grab handles and steps	Determine the positions of the grab handle and steps	Grab handle and position feet	Operator slips due to not able to grasp a handle
5	Climb down the cab step	To reach ground	Need to leave the vehicle	Look for the cab steps	Remember to climb down by holding on both hands. Remember to use 3-points contact with the cab all the time during egress.	Determine the positions of the grab handles and steps with respect to your body	Transfer weight to cab step	Operator misses cab step and falls off
6	Climb down to ground level	To reach ground	Need to leave the vehicle	Look for the ground	Remember to get feedback from the leading foot when it touches the ground	Determine the position and status of the ground	Transfer weight to ground	Operator steps in incorrect place and injury occurs

be used in creating the task analysis table (Table 10.1). Even if a product is not available, the task analysis can be performed on an early product concept by predicting the possible sequence of subtasks needed to use the product in performing each task (or product usage).

The task analyzed in Table 10.1 is for egressing from the driver's seat of a heavy truck. The last column of possible errors can provide insights that will assist in designing and locating components such as door opening handles, seats, grab handles, and steps.

EFFECT OF VEHICLE BODY STYLE ON VEHICLE ENTRY AND EXIT

Bodenmiller et al. (2002) conducted a study to determine differences in older (over age 55) and younger (under age 35), male and female drivers while entering and exiting vehicles with three different body styles – namely, a large sedan, a minivan, and a full-size pickup truck. The test vehicles were: a) 2001 Pontiac Bonneville 4-Door Sedan: Rocker Height: 381 mm (15"), Seat Height: 521 mm (20.5"); b) 2001 Oldsmobile Silhouette Minivan: Rocker Height: 419 mm (16.5"), Seat Height: 686 mm (27"); and c) 2000 Chevrolet Full-Size Silverado, 4WD Extended Cab, Long bed, Pickup (1500 Series), Rocker Height: 533 mm (21"), Seat Height: 889 mm (35").

Thirty-six drivers (male and female, ages 25 to 89 years) who participated in this study were first measured for their anthropometric, strength and body flexibility measures relevant to the entry/exit tasks. They were asked to first get into each vehicle and adjust the seat to their preferred seating positions. Then, they were asked to get in the vehicle, and their entry time was measured. Their entry maneuver was also recorded on video, and they were asked to rate the level of ease/difficulty (using a 5-point scale) in entering. A similar procedure was conducted during their exit from each vehicle.

In addition to the expected effects of differences due to gender (males stronger than females, and decrements in strength and flexibility with increase in age), the following results were observed on the entry/exit tasks: 1) Overall, the minivan was rated as the easiest vehicle to get in and out of, and the full-size pickup was the most difficult vehicle to get in and out; 2) The shortest average entry time was 6.4 s for the large sedan and the shortest average exit time was 5.9 s for the minivan; 3) Both entry and exit performance (as measured by total time to enter or exit) and ratings of older women were significantly worse than males and younger females.

The minivan was found to have the highest ratings for both entry and exit. The 5-point rating system was based on ease in entry or exit with a rating of 1 representing "very difficult" and a rating of 5 signifying "very easy". The mean entry ratings were as follows: 4.2 for the minivan, 4.0 for the sedan, and 3.1 for the pickup truck. The corresponding mean exit ratings were: 4.2 for the minivan, 4.0 for the sedan, and 3.3 for the pickup truck.

The mean times of entry and exit, however, did not exactly mirror the mean ratings. The mean entry times were as follows: 6.6 s for the minivan, 6.4 s for the sedan and 6.5 s for the pickup truck. The corresponding mean exit times were: 5.9 s for the minivan, 6.2 s for the sedan and 5.9 s for the pickup truck.

On average, the subjects took slightly longer (a couple tenths of a second) to get into the minivan than the sedan. The researchers attributed this minor difference to the fact that many of the test subjects were observed to simply drop right into the lower rocker and seat heights of the sedan. The exit from the minivan was the quickest. The average time differences between the three vehicles were not very large. The sedan took the longest average exit time. The researchers attributed this to the low rocker and the low seat height of the sedan. The sedan package appeared to require the test subjects to utilize their legs to push themselves up and out of the vehicle as compared to the other two vehicles with higher seat heights.

Analyzing the mean entry ratings and times for each of the three test vehicles by four subject groups (younger males, older males, younger females and older females), the following conclusions were made:

a) The minivan received the highest ratings from three of the four test subject groups, with the exception being the older females. The older females rated entry into the sedan easier than that of the minivan. However, it actually took longer for the older females, on average, to enter the sedan than the minivan. The researchers theorized that the senior female subjects were more familiar with sedans as they all own and drive them. None of the senior female subjects in this field study owned a minivan, so it appears that they were reluctant to give it a higher rating than the sedan, despite the quicker entry times associated with the minivan.
b) The older female test subjects gave the lowest ratings to the pickup truck, which also had the highest entry times for them. This was consistent with the observations made throughout the field testing on the pickup truck that many of the female seniors had extreme difficulty getting both into and out of the truck due to higher sill and seat heights of the pickup.

For additional data, plots and analyses, interested readers should refer to the paper by Bodenmiller et al. (2002).

CONCLUDING COMMENTS

The entry and exit performance and preferences of drivers and occupants are affected by combinations of many vehicle dimensions as well as combinations of the characteristics of subjects. Therefore, ergonomics engineers should understand the basic issues and guidelines presented in this chapter and develop a physical mock-up of the proposed vehicle for further verification. The only reliable method to verify entry and exit performance is to conduct a study where the representative subjects (i.e., potential customers/users) would be asked to rate overall ease in entry and exit as well as collect additional data by measurements such as time taken, EMG and observational measures such as number of problems observed, and problems reported by the subjects. In addition, the inclusion in the test of other existing vehicles (prior models of the vehicle being designed and its competitors) would provide useful information for reference and comparisons.

REFERENCES

Bodenmiller, F., J. Hart and V. Bhise. 2002. Effect of Vehicle Body Style on Vehicle Entry/Exit Performance and Preferences of Older and Younger Drivers. SAE Paper no. 2002-01-00911. Paper presented at the SAE International Congress in Detroit, Michigan.

Drury, C. 1983. Task Analysis Methods in Industry. *Applied Ergonomics*, 14: 19–28.

Society of Automotive Engineers, Inc. 2009. *The SAE Handbook*. Warrendale, PA: Society of Automotive Engineers, Inc.

11 Automotive Exterior Interfaces

Service and Loading/ Unloading

INTRODUCTION TO EXTERIOR INTERFACES

This chapter covers the issues that vehicle users and service personnel experience with exterior items during situations or operations, such as opening the hood, servicing the engine, loading and unloading items in the trunk and cargo areas, refueling the vehicle, changing a flat tire, and cleaning the vehicle exterior. The vehicle should be designed to ensure that customers can do all such tasks easily and without problems.

The objectives of this chapter are (a) to ensure that the exterior design allows the users to perform all interfacing tasks easily, quickly, and without error, and (b) to suggest improvements to the vehicle during early design stages.

EXTERIOR INTERFACING ISSUES

Some of the tasks and subtasks associated with servicing items located in different vehicle areas are listed below:

1. *Engine Compartment Service*
   a) Opening the hood involves the following steps:
      i. Finding and operating the "Inside Hood Release Control" (located at or near the instrument panel)
      ii. Finding and operating the "Secondary Hood Opening Lever" (located at the hood opening; see Figure 11.1 for the visibility of the secondary hood release lever). (Note: Some vehicles have an integrated hood release mechanism that operates when the inside hood release control is pulled twice.)
      iii. Raising the hood (lifting the hood weight and reaching to its raised location) (Note: Some vehicles have a different spring loaded hood holding-up hinge mechanism.)

DOI: 10.1201/9781003485582-11

FIGURE 11.1 Illustration of difficulty in finding the secondary hood release lever. The top picture shows the partially open hood which does not allow good visibility of the hood release handle. The bottom figure shows the inside hood release lever that is located over 100 mm deep from the edge of the hood and thus requires a palm insertion of about 125 mm to activate the lever.

  iv. Finding, positioning and inserting the hood prop-up rod (see Figure 11.2 for the use of a rubberized and color-coded sleeve to assist in finding, grasping and inserting the prop rod)
  v. Turning on a hood lamp for roadside engine service at night.

b) Checking engine oil involves the following steps:
  i. Finding the engine oil dipstick (see Figure 11.3 for color-coded grasp handle of the dipstick)
  ii. Reaching, grasping and pulling out the dipstick
  iii. Cleaning the oil on the dipstick

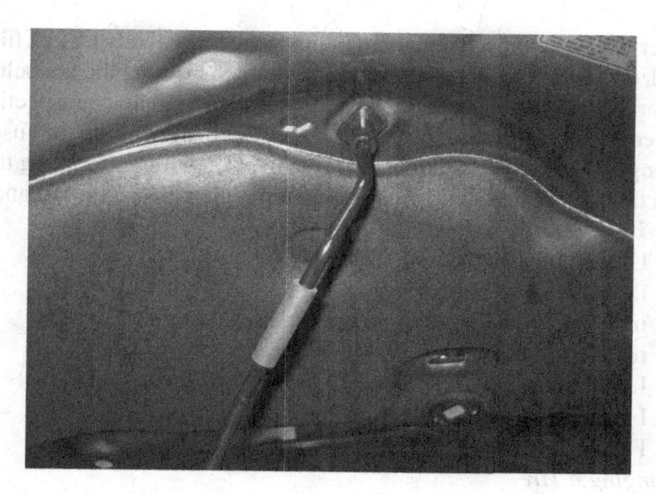

FIGURE 11.2 Illustration of rubberized and color-coded grasp sleeve around the prop-up rod and the rectangular insertion bracket to improve the usability of the prop-up rod.

FIGURE 11.3 Illustration of easy-to-find tall yellow dipstick handle and color-coded washer fluid and radiator fluid caps.

iv. Inserting the dipstick in the engine
v. Pulling out the dipstick
vi. Reading the oil level. (Note: easy-to-find holes or notches on the dipstick will facilitate reading the oil level.)
vii. Inserting the dipstick in the engine
viii. Removing the engine oil filler cap
ix. Filling engine oil

Other engine compartment tasks include checking washer fluid, filling washer fluid, checking coolant level, filling coolant, inspecting the fan belt, inspecting the brake fluid level, inspecting power steering fluid level, inspecting air filter, inspecting battery terminals, finding the fuse box, opening the fuse box, identifying problem fuse location and replacing a new fuse, checking the transmission fluid level, replacing a headlamp bulb, changing oil filter, and so forth.

2. *Trunk Compartment Related Tasks*
   a) Unlocking the trunk
      i. Using the remote key fob
      ii. Using the inside trunk release lever/button
      iii. Using the key
   b) Loading or unloading items
   c) Finding the spare tire
   d) Finding the jack
3. *Changing a Tire*
   a) Removing the spare tire, the jack and the wheel nut wrench
   b) Positioning the jack and jacking up the vehicle
   c) Locating special wrench used for wheel nuts removal and tightening
   d) Loosening the wheel nuts
   e) Removing the flat tire
   f) Installing the spare tire
   g) Replacing the jack, wrench and the flat tire in the trunk compartment
4. *Front Luggage (Frunk) Compartment Related Tasks.* Many battery electric vehicles have no need for the traditional front IC engine space under the hood because their front electric motors occupy relatively smaller space as compared to the IC engine. The electric motor also does not need additional accessories such as an alternator, water pump, vacuum pump and radiator. The front space under the hood is thus used as additional storage space. It also houses sockets and switches for 120/240V AC power supply for operating external equipment such as lights, power tools and entertainment gear (e.g., 2023 Ford F-150 Lightning pickup truck). Using the *frunk* compartment thus involves:
   a) Unlocking the frunk
      i. Using the remote key fob
      ii. Using the inside frunk release lever/button
      iii Using the key
   b) Using the inside secondary release handle
   c) Raising the hood
   d) Loading or unloading items
   e) Finding and using electrical outlets
   f) Flat working surfaces for placing items for purposes such as picnics, entertainment and small repair jobs.
   g) Lowering and locking the hood
5. *Refueling the Vehicle*
   a) Finding and operating the fuel door release lever/button

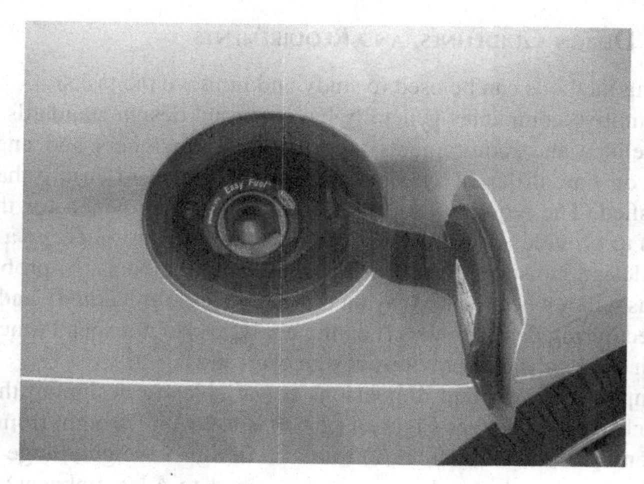

FIGURE 11.4 Cap-less fuel filler.

   b) Opening the fuel door
   c) Removing the cap (if not cap-less filler)
   d) Inserting the fuel pump nozzle (height of the filler above the ground and its orientation angle are important variables influencing comfort while filling gas)
   e) Fueling
   f) Removing the nozzle
   g) Replacing the cap and closing the fuel door

   With the incorporation of the cap-less fuel fillers in newer vehicles, many cap turning motions, cap removal and installation times and hand injuries (from sharp edges on the surrounding sheet metal) are eliminated (see Figure 11.4).

5. *Other Interactions with the Vehicle*
   a) Cleaning the windshield, backlite, headlamps, and tail lamps
   b) Aiming the headlamps
   c) Checking tire pressure and filling air
   d) Removing the snow
   e) Installing wiper blades

The above list of tasks and the steps within each task were provided purposely for the reader to realize that many of the steps can be simplified or even eliminated by improving the design of the interfacing equipment.

METHODS AND ISSUES TO STUDY

To ensure that All the tasks such as those listed above can be performed easily by the users, the ergonomics engineer must analyze each of the tasks, conduct necessary evaluations and provide necessary guidelines, and design suggestions to other designers and engineers associated with the design of the components and systems involved with the tasks.

STANDARDS, DESIGN GUIDELINES, AND REQUIREMENTS

The following methods can be used to study and improve the tasks.

The automotive companies generally have internal design standards that include design guidelines and requirements to provide the designers and engineers with information on how different systems are to be designed to ensure that customers will be satisfied. The requirements in the standards are provided for the following purposes: a) to provide uniformity in design so that the customers of a given brand can expect design consistency across different vehicle models, b) problems in previous designs will not be repeated (by incorporating lessons learned), and c) to reduce time involved during design stages (i.e., the engineers do not spend time researching issues which were solved in previous design exercises).

For example, the important dimensions or variables in designing the trunk of a passenger vehicle include (see Figure 11.5): a) liftover (sill) height from the ground, b) trunk opening width and height, c) trunk lid design – weight, hinge design (e.g., gooseneck hinges occupy trunk space as compared to 4-bar linkages), reach/grasp height to close the opened lid, head injury protection against sharp protruding edges, corners, latches, and so forth, d) sill design (e.g., sill width to rest and reposition lifted items during moves, durability of the sill and seal material, interference and protection of protruding seals, etc.), e) sill-to-load floor depth (which may require excessive leaning and torso bending to reach inside the trunk), f) trunk depth (i.e., longitudinal load length and ability of person to reach all corners of the trunk space and to slide and move loads), g) spare wheel location and access, h) stowage space (space for jack and wheel-nut wrench, tool boxes, first-aid kit, tie down latches, etc.), i) release and locking controls and mechanisms for: spare wheel cover, storage compartments,

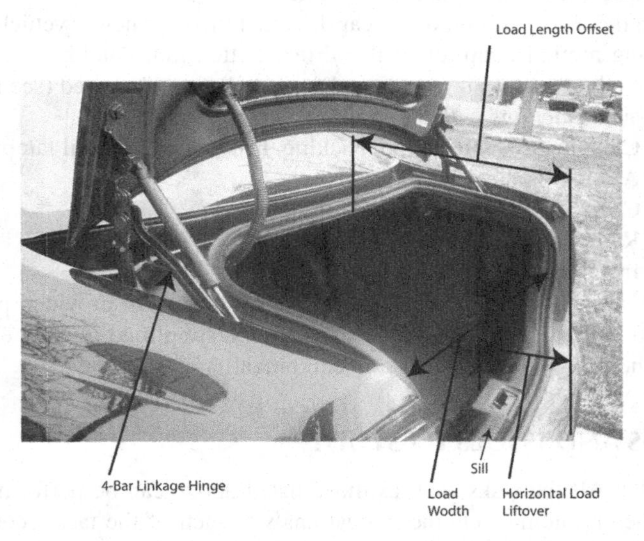

FIGURE 11.5 Illustration of trunk opening space for loading and unloading. The space has a wider opening from front and top, lower sill height, smaller horizontal liftover height and space-saving 4-bar linkage hinges.

rear seat folding, fuel door, child lock-out trunk opening release, and so forth, and j) access and clearance space to change rear lamp bulbs.
Most electric vehicles have a frunk, that is, the front storage compartment under the hood area. The space is available to store items because the front mounted electric motor does not require tall packaging space as compared to the IC engine. Thus the loading and unloading issues are very similar to the design of the trunk opening, sill height and floor depth covered in the previous paragraph.

Checklists

The checklists included in Chapter 7 for controls and displays (see Tables 7.1 and 7.2) will be useful to evaluate various service-related items because to service, that is, to use an item, the following sequence of ergonomic issues need to be considered: a) findability – visibility, identification and expected locations; b) reading and understanding labels to determine what needs to be done, c) accessing and reaching a needed item (e.g., 5th percentile female should able to reach and 95th male should be able to fit within the space available under the trunk lid or liftgate to provide clearances for his hands, head, torso, etc.), d) grasping and operating (i.e., turning, pulling, pushing, etc.), and e) obtaining feedback on completed movement or operation.

Biomechanical Guidelines for Loading and Unloading Tasks

For analyzing the tasks involving loading and unloading items into/from the trunk and cargo compartments, biomechanical guidelines related to lifting and carrying will be useful. Some guidelines available in this area (Konz and Johnson, 2004; Chaffin and Andersson, 1999) are provided below:

a) *Keep load close to body*: This guideline will reduce the moment created by the load being lifted and stored in the vehicle. Leaning forward to reach or place a load deeper (further forward in a trunk or cargo area) will increase the moment and load in the back muscles and the spinal column (disc between the L5/S1 joint. L5 = 5th lumbar vertebrae and S1 = 1st sacral vertebrae). The guideline also suggests that reducing longitudinal distance from the bumper edge to load area or engine compartment will reduce bending involved in placing loads and lifting hoods.
b) *Bend at knees rather than bend at back*: This guideline will help reduce stresses in the lower back (L5/S1 region) and place more load in the leg muscles during lifting.
c) *Work at knuckle height*: Lifting a load from a lower height (lower than standing knuckle height) will increase load in the L5/S1 disc. Lifting a load above the standing chest and shoulder levels will increase load in the upper extremity muscles (shoulder and neck regions). Therefore, working at about the knuckle height (or about 50 mm below the elbow height) is generally less stressful on the human body. (Note: The load floor heights of minivans, SUVs and pick-up are higher and more convenient for loading/unloading than lower trunk floor height in passenger cars.)

d) *Do not twist body during the moves*: Turning and twisting of the body will place more stresses in one side of the body (i.e., it will create asymmetrical loading which will be greater at some joints than if the load is held in a symmetric position with respect to the body). Provision of wider load-access openings in trunks and cargo areas will reduce turning and twisting during loading and unloading.
e) *Don't slip or jerk*: Any slipping or jerking motion while carrying or lifting a load will increase dynamic loading on the body (remember a "jerk" involves acceleration, and the stress in the body will be proportional to the mass moved times its acceleration).
f) *Get a good grip*: A good grip of the load and gripping on the vehicle (body or grab handles) during loading will reduce the load in the spinal column. A good grip of the load will also reduce sliding and slipping of the load which can create larger dynamic loading in the body. Supporting the body during bending (by a hand gripping on a firm support on the vehicle body) will also reduce stress in the spinal column.

Applications of Manual Lifting Models

Many biomechanical models are available (e.g., University of Michigan 2D and 3D static strength models (Chaffin and Andersson, 1999) and the NIOSH Lifting formula (Konz and Johnson, 2004) which can be used to analyze lifting situations involved in lifting/opening vehicle body closures (hoods, trunk lids and liftgates in hatchbacks, wagons, minivans and SUVs) and loading/unloading items (e.g., boxes, suitcases, golf bags, and grocery bags) in the vehicles. The vehicle dimensional parameters related to load floor heights, forward offset (leaning) distances (e.g., rear bumper to trunk storage area, front bumper to a service point in the engine compartment) and weights of hood, trunk lid, and so forth, along with the lift support provided by the struts can be evaluated by use of the biomechanical lifting models.

Task Analysis

Task analysis is a simple but powerful method to determine user problems in product designs. It involves an ergonomics engineer to break down a tasks into a series of subtasks or steps and analyze the demands placed on the user in performing each step and compare against the capabilities and limitations of the users. The analysis reveals a number of possible user problems and errors that the users can make in using the product. Table 11.1 provides an example of a task analysis for checking engine oil level. The description of the Task Analysis technique is provided in Chapter 10 (Drury, 1983).

Methods of Observation, Communication, and Experimentation

In order to understand the problems encountered by people in interfacing with a vehicle from the exterior, the basic methods of data collection involving observation

TABLE 11.1
Illustration of a Task Analysis for Checking Engine Oil Level

Task Description					Task Analysis				
Task	Subtask No.	Subtask/Step Description	Subtask Purpose	Trigger	Scanning and Seeing	Memory	Interpolating	Manipulating	Possible Errors
Opening the hood and checking engine oil	1	Pull hood release lever from inside the vehicle	To open the hood	To check fluid levels	Look for the inside release lever	Remember the location	Determine how to operate	Use fingers to pull the inside lever	Released the parking brake instead
	2	Lift the hood release lever under the hood	To open the hood	To check fluid levels	Look for the hood release lever	Remember the location of the hood release	To move fingers in the gap between the hood and the grill to find the lever	Use fingers to lift the hood release lever	Difficulty in seeing and finding the lever
	3	Prop up the hood	To keep hood open	To check fluid levels	Look for the prop rod and the hole in the hood	Remember the locations of the prop rod and hole in the hood	Guess which hole in the hood is closet to the top end of the prop rod	Keep holding the hood with the left hand and insert the prop rod in the hole with the right hand	Placed prop rod in the wrong hole

(*continued*)

TABLE 11.1 (Continued)
Illustration of a Task Analysis for Checking Engine Oil Level

Task Description					Task Analysis				
Task	**Subtask No.**	**Subtask/Step Description**	**Subtask Purpose**	**Trigger**	**Scanning and Seeing**	**Memory**	**Interpolating**	**Manipulating**	**Possible Errors**
	4	Locate oil dipstick	To check oil level	Need to check oil	Look for the dip stick	Remember how the dipstick looks and its location	Guess which side of the engine to look for	Use fingers to grasp the dipstick handle and pull it out	Selected the transmission oil dipstick instead
	5	Pull out, clean and reinsert the dip stick and pull again and check the oil level	To check oil level	Need to check oil	Look for min and max marks on the dipstick and oil line	Remember to clean the dip stick and reinsert	Reinsert clean dip stick, hold for 10 s and pull it out	Reorient the dipstick to get a good look at the oil level	Oil dripped on the floor
	6	Reinsert the dipstick	To place the dipstick back	Need to put back the dipstick	Look for the hole to insert the dip stick	Remember the hold location	Orient the dip stick over the hole	Slide the dipstick into the hole	Missed the hole two times before correctly inserting the dip stick in the engine

and communication would be very useful. In the methods of observation, the user is observed by a trained human observer and the observer records problems experienced by different users under selected usage situations. The user actions can be recorded using different methods such as video recordings or other measurements (e.g., outputs of posture angle sensors) and the records can be subsequently studied by trained analysts. The information gathered during the observations will help in identifying problems of different types (e.g., took long time to find an item, could not find an item, difficulty in reaching to a location, awkward postures, and could not comprehend operation of a device) and determining the frequency and severity of occurrences of the problems. In the methods of communication, an interviewer can ask the user to describe the problems encountered during usage of the product (e.g., opening the trunk and replacing a burnt-out tail lamp bulb) and/or ask to provide ratings (e.g., using a 10-point scale, where 10 = Very Easy and 1 = Very Difficult) on one or more important issues.

Further, experiments can be set up to evaluate important parameters of the vehicle design along with other comparators (existing or competitor's vehicles); and a number of representative subjects will be asked to perform a number of tasks and data can be collected from the methods of observation and communication as well as from the measurements on each subject's performance. The analyses of collected data would help determine if improvements can be justified and design superiority can be established.

Additional information on the methods of observation, communications and experimentation is provided in Volume 2, Chapter 19 on Vehicle Evaluation Methods.

CONCLUDING REMARKS

The vehicle should be designed to make all its user interfaces easy to use and service. A designer once came to the author and suggested the ergonomics engineers should do a study to help design wheel covers that would be easier to clean. The designer, who liked to keep his own vehicle spotless, complained that he was tired of spending time cleaning each spoke on his beautiful wire wheels with a set of toothbrushes. The point is that no one in the design studio had thought about the problem until other designers realized the need of his colleague. "Cleanability" of wheels is an important consideration to the wheel and wheel cover designers. The ergonomics engineers must therefore interact with the users and find out all their needs, prioritize these needs, and come up with the solutions by applying methods and information provided in this book to eliminate the customer complaints and make the users' experience more delightful.

REFERENCES

Chaffin, D. B., G. B. J. Andersson and B. J. Martin. 1999. *Occupational Biomechanics*. New York, NY: John Wiley & Sons, Inc.

Drury, C. 1983. Task Analysis Methods in Industry. *Applied Ergonomics*, 14: 19–28.

Konz, S. and S. Johnson. 2004. *Work Design-Industrial Ergonomics*. Sixth edition. Scottsdale, AZ: Holcomb Hathaway, Publishers.

12 Automotive Craftsmanship

CRAFTSMANSHIP IN VEHICLE DESIGN

OBJECTIVES

The objective of this chapter is to provide the reader with an understanding of the concept of craftsmanship, its importance, and product characteristics that affect craftsmanship.

CRAFTSMANSHIP: WHAT IS IT?

Craftsmanship is a relatively new technical area of increasing importance to ergonomics engineers. The whole idea behind craftsmanship is that the vehicle should be designed and built such that the customers will perceive the vehicle to be built by expert craftsmen who apply their skills to enhance the perceptual characteristics of the product such as its looks (appearance: i.e., shape, fit, finish, color and texture harmony between adjacent parts), tactile feel (feel in movements of controls, tactile feel of various interior materials, e.g., hardness/softness, compressibility, slipperiness/roughness), sound (how the vehicle sounds when you operate its different features: e.g., sound of the engine, sound of a door closing, and parking brake engagement's clicking sound), smell (the odors: e.g., smell of the leather used in the interiors, and "new car smell"), ease during use (all ergonomic considerations), and other features that customers associate with craftsmanship. The vehicle should also be perceived by customers to "belong" to the family of the brand it represents. For example, the customers will expect an expensive vehicle to be extra well made with all the features and overall perception of luxury and quality.

There is no agreement among automotive experts on what is exactly meant by "craftsmanship". Different customers also expect different product features and characteristics but, when asked, based on their internal conceptualization of what constitutes a well-crafted vehicle, they can always tell if a given product looks and feels like it is made by craftsmen.

DOI: 10.1201/9781003485582-12

The attribute of craftsmanship, thus, is based on the following:

1. Customer perceptions of the product after the customer experiences (i.e., sees and uses) the product. It is assumed that only customers can tell if a product has superior craftsmanship performance, that is, it has more up-scale characteristics as compared to other similar products in the market.
2. It is how well the product is executed (i.e., made, or perceived by the customers) to possess visual quality (appearance), sound quality, touch and feel quality, smell quality, usability/ergonomics, and features, especially those that delight the customer or create impressions of "Wows".
3. It is how the product affects its perception of quality (i.e., image, brand; some examples of semantic differential scales – see Chapter 19, Figure 19.1, scales (d) and (f) – that can be used to measure the product perception by using adjective pairs such as Solid/Flimsy, Cheap/Expensive, Fake/Genuineness of materials, Quality/Shoddy, Comfortable/Uncomfortable, Pleasing/Non-pleasing, and Like/Dislike).

Importance of Craftsmanship

The importance of craftsmanship can be better understood by studying the following two models: a) The Ring Model of Product Desirability and b) The Kano Model of Quality.

The Ring Model of Product Desirability

The concept presented by Peters (1987) as the Levitt Rings (Levitt, 1980) to describe product service characteristics was applied here to describe product desirability to explain the concept of craftsmanship. In that concept, Levitt described the basic service as "generic", which is enhanced by adding "expected" features and further "augmented" by some unique service features. The model was described by showing the "generic" portion of the service as the core, which was surrounded by two rings – the "expected" ring around the "generic" core, and the "augmented" ring around the expected ring. The overall desirability of the service (or product) is indicated by the size (diameter) of the outer ring.

Figure 12.1 shows a product concept. It is hypothesized that at the core of the product there exists a "functional" design (a mechanism that does some basic product function, e.g., transports four adults). This is shown by the inner core of the product. The functional product design is enhanced by modifying or adding certain features that improve the users' comfort and convenience (i.e., ergonomic characteristics by adding features such as: 8-way power seats with 4-way power lumbar support, and smart headlamps). The "comfort and convenience", in essence, adds a layer (or a ring) on the core of the product. The product's desirability can be further improved by improving its craftsmanship qualities or "pleasing perceptions" by providing a) materials that have better look and feel characteristics, b) better fitting parts with smaller gaps between parts, c) colors and textures between mating components that look like they were made by the same supplier, and so forth. The "pleasing perceptions" are also shown in the figure as adding another, or second, layer (or outer

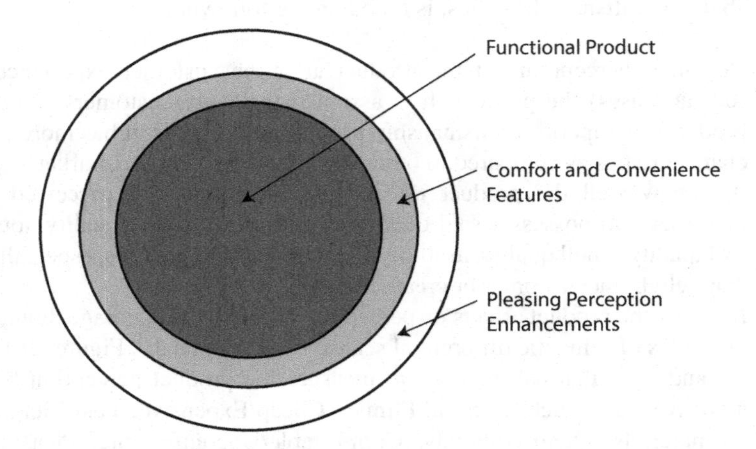

FIGURE 12.1 The ring model of desirability.

Note: Overall Desirability or value of the Product = Size of the Outer Circle = Core + Comfort Convenience Ring + Pleasing Perception Ring.

ring) to the product. The overall size of the outer ring is assumed to be proportional to the overall desirability (or value) of the product to the customer. The ergonomics engineers can help in increasing the sizes of both the outer rings.

Figure 12.2 shows two product concepts, A and B. The size of the inner core of product concept B is shown larger than the size of the inner core of product A. This indicates that product concept B had a better functional design. However, the overall size (with both the rings) of product A is larger than the overall size of product B, because the product B design did not offer more comfort and convenience (shown by thinner layer of comfort convenience ring as compared to product A) and had less pleasing perception-related enhancements (also shown by the thinner outer ring of pleasing perception than product A).

The figure, thus, illustrates that to improve the overall product appeal it is important that all the three parts (i.e., its core and the two rings) must be carefully designed with attention to all details that are perceived by the customers to make the product more desirable.

It is important to realize that, as the basic manufacturing technologies that affect the inner core of the product are known to all vehicle manufacturers, the task of improving the product appeal by improving its two outer layers, i.e., comfort and convenience and pleasing perceptions, is now more important in discriminating between products made by different manufacturers.

It is important to realize that human factors engineers are in a unique situation to help in improving the sizes of both outer rings. These outer rings are also largely responsible for increasing product appeal and for brand differentiation.

Kano Model of Quality

The model of quality proposed by Kano conceptualizes that the customer satisfaction is affected by three types of product features called (a) the removal of "dissatisfiers",

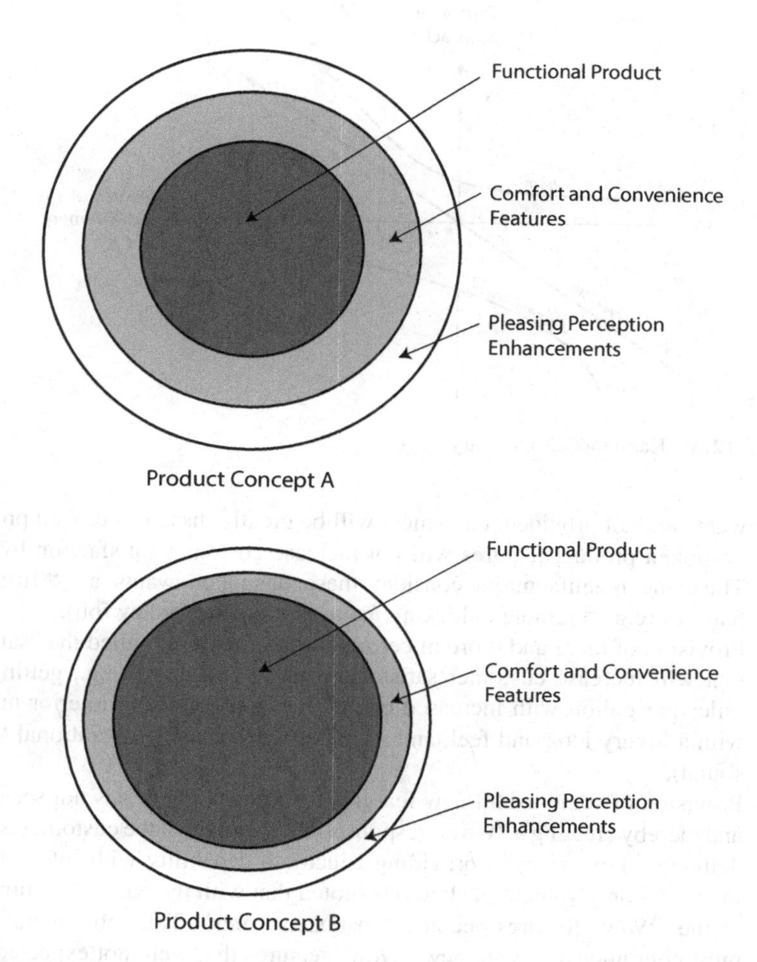

FIGURE 12.2 Illustration of two product concepts with different levels of functional, comfort and convenience, and pleasing perception characteristics.

(i.e., providing the "unspoken wants", otherwise dissatisfaction arises if the product does not provide what the customer expects); (b) "satisfiers" (giving more of these features increases satisfaction with the product); and (c) "delighters" (which create "Wows" when present but do not cause dissatisfaction if not there) (Yang and El-Haik, 2003). The Kano model presented in Figure 12.3 shows the effects of these three types of features (represented by the x-axis as the degree of critical to satisfaction [CTS]) on the customer satisfaction (shown on the y-axis).

The model shows that customer satisfaction will increase by doing the following:

a) Providing product features (or characteristics) that satisfy the "unspoken wants" (i.e., these are wants the customer expects in the product, and therefore, the customer does not even mention these needs). But if these unspoken

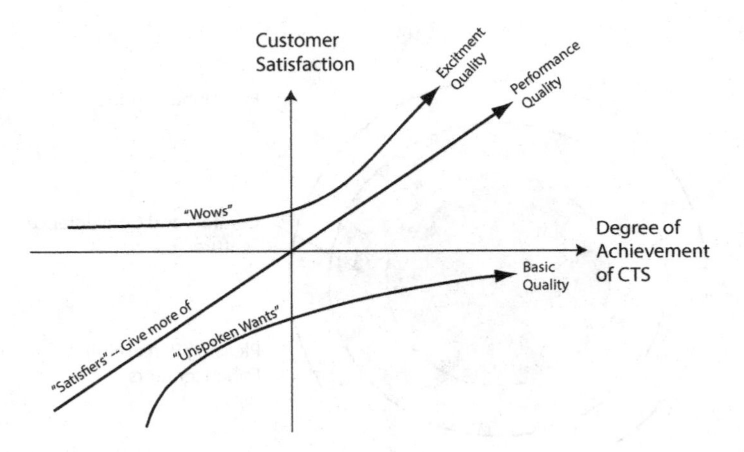

FIGURE 12.3 Kano model of quality.

wants are not provided, customers will be greatly dissatisfied. And providing unspoken product features will not increase customer satisfaction by much. Thus, the manufacturers consider these unspoken wants as "Must have" features (e.g., a remote unlocking feature on the key or key fob).

b) Provision of more and more of certain types of features called the "satisfiers" that will increase customer satisfaction more and more (e.g., getting more miles per gallon with increased engine horsepower, using interior materials with a luxury look and feel, and switches with pleasing operational feel and sound).

c) Provision of some useful new features that the customer has not seen before and thereby creating a "Wow" response – indicating that the customer is clearly delighted. For example, providing Bluetooth capability with Internet access in an economy vehicle. It should be noted that with the passage of time, most of the "Wow" features become "unspoken wants". Thus, the manufacturers must continuously create new "Wow" features that were not expected by the customers at the time of their introduction.

A careful study of all the above three types of product features (either available in the market currently or new features that could be developed by studying design and technological trends) and their incorporation into the product is essential to enhance craftsmanship and product quality.

Attributes of Craftsmanship

The following characteristics of products have been considered in the automotive industry to be related to perception of quality. The following list was prepared by the author through interviews with a number of vehicle owners/users and via discussions with engineers and designers working for different vehicle manufacturers and suppliers.

Visual Quality

1. Excellent fit between any two components in visible regions (characterized by smooth parting lines and small, thin gaps and non-perceptible mismatches (misalignment) associated with an even gap width along all joints, flush mating surfaces, parallel edges, smooth but rounded (non-sharp) edges, low variability/unevenness (e.g., warpage or distortions), and no see-through areas between joints).
2. Visual Harmony (e.g., similarity of look between adjacent components/materials based on color, brightness, texture/grain, gloss/reflectivity, and finish quality).
3. No evidence of degradation on visible surfaces (e.g., rust, fading, hazing, cracking/fracturing, peeling, wear, and scratches).
4. No exposed fasteners (e.g., visible screws, clips, and wires). The underlying concept is that product surfaces with invisible (or unexposed) fasteners should look clean and uninterrupted. And, thus, the designer should provide the feeling that the product is well made. On the other hand, some screws with a "machined look" can give the impression of well-crafted and solid joints.
5. No annoying visual distractions (e.g., glare from light sources or reflections of other brighter surfaces into glazing or reflective surfaces, waviness or distortions in reflected or transmitted images)

Touch Feel Quality

1. Vehicle interior surfaces that are touched (e.g., grasp areas of controls such as buttons, knobs, and handles, and seats, instrument panels, door trim panels, arm rests, and consoles) should have a pleasing feel by considering touch characteristics that can be described and scaled by using adjective pairs such as softness/hardness, smoothness/roughness, textured/non-textured and slippery/sticky).
2. Pleasing operational feel of switches (e.g., feel or feedback received during switch movements characterized by length of movements (e.g., right amount of switch travel or deflections, low pressures on the fingertips), low slop, sags or looseness at joints, crisp detent feel (measured by force-deflection curves), presence of sound or tactile feedback, and vibrations.

Sound Quality

1. Pleasing sounds (solid, smooth, non-harsh or not tinny) emanating from functional equipment (e.g., sound characteristics of engine, door slam, warning signals, and auditory feedback on driver-executed controls).
2. Absence of unwanted/annoying sounds (e.g., squeaks, rattles, and harshness).

Harmony

1. Harmony across systems, subsystems and components (e.g., similarities in appearance and operational feel of controls between radio and climate

controls. Note: Some differences are needed to discriminate between functions and reduce driver errors during equipment use. However, various operational features should provide the impression design consistency, that is, all systems should look and feel as if they were designed by the same designer, and they should exhibit certain theme or brand characteristics.)
2. Harmony between materials and their finishes within a vehicle of a given brand (to create brand image).
3. Smaller number of dissimilar materials in close proximity (e.g., avoid many components with dissimilar materials with many parting lines placed within a small region as they create perception of clutter, mismatches, and unevenness).

Smell Quality

1. Non-smelling materials (avoid materials that produce unpleasant and unsafe odors).
2. Use materials that produce pleasing odors (e.g., smell of genuine leather, flowery/fruity, and spicy).

Comfort and Convenience

Ergonomic considerations involved in designing all vehicle systems, subsystems and components (essentially the entire contents of this book) to ensure that the vehicle is comfortable, safe and easy to use.

Measurement Methods

A number of different methods are used in the automotive industry for measuring attributes of craftsmanship listed in the above section. The methods can be categorized as follows:

1. Checklists.
2. Objective measurements (e.g., gloss levels, compressibility, roughness, friction).
3. Subjective evaluations using rating scales or paired comparisons with acceptable reference samples.
4. Customer clinics or studies to understand customer preferences, ability to perceive, discriminate and categorize craftsmanship characteristics.

Various applications of the above methods are covered in Volume 2, Chapter 19.

SOME EXAMPLES OF CRAFTSMANSHIP EVALUATION STUDIES

Craftsmanship of Steering Wheels

A study was conducted to evaluate craftsmanship characteristics of steering wheels with a focus on rim and spoke feel. The study involved driving a vehicle in a driving simulator with eight different steering wheels in random orders, and the subjects were asked to provide ratings on the following details on each of the eight production steering wheels:

1. *Rim*: grasp comfort, thickness/thinness of the rim, rim surface softness/hardness, rim surface slipperiness/stickiness.
2. *Top side of upper spoke*: surface feel like/dislike, softness/hardness, smoothness/roughness.
3. *Stitches* (in the wrapped leather): visual appearance like/dislike, grasp feel comfortable/uncomfortable, protrusion/recessing.
4. *Top spoke size near rim*: small/large, easy or difficult to hold.
5. *Rim grasp area between top and bottom spokes*: small/large.
6. *Seams*: grasp feel over the seams comfortable/uncomfortable, visual appearance like/dislike.
7. *Characteristics* especially liked or disliked.

Figure 12.4 presents sketches of the eight steering wheels used in the study. The steering wheels were obtained from production vehicles made by different automobile manufacturers. The hub areas of the steering wheels were removed and masked with matte black cardboard to hide the identity of the manufacturers from biasing the subjects.

Figures 12.5 and 12.6 present examples of the outputs of the ratings data obtained from the study involving over fifty drivers of passenger cars.

Figure 12.5 presents the mean and 95 percent confidence intervals of ratings on how the steering rim grasp comfort was perceived by the subjects while using the eight steering wheels. The data from Figure 12.5 indicated that steering wheels G and N were perceived to be most comfortable.

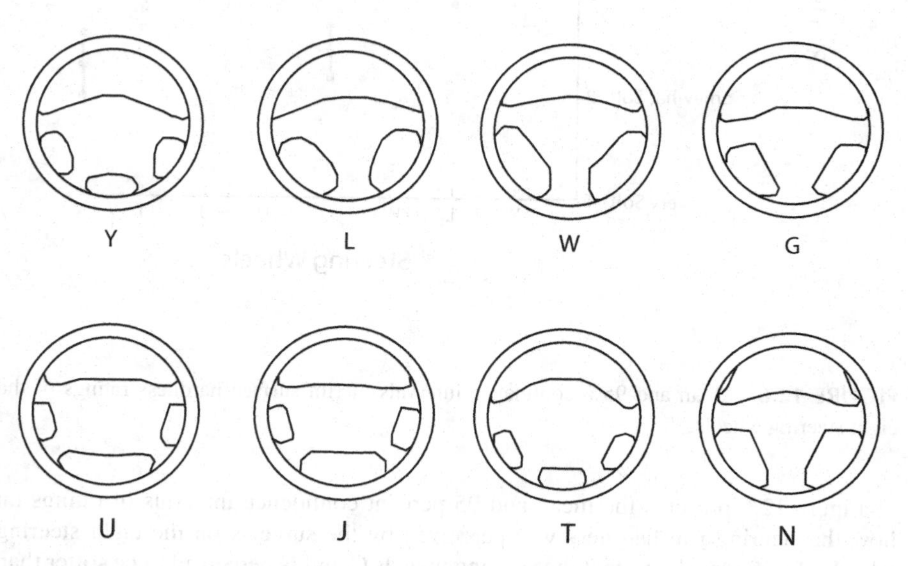

FIGURE 12.4 Eight steering wheels used in the study.

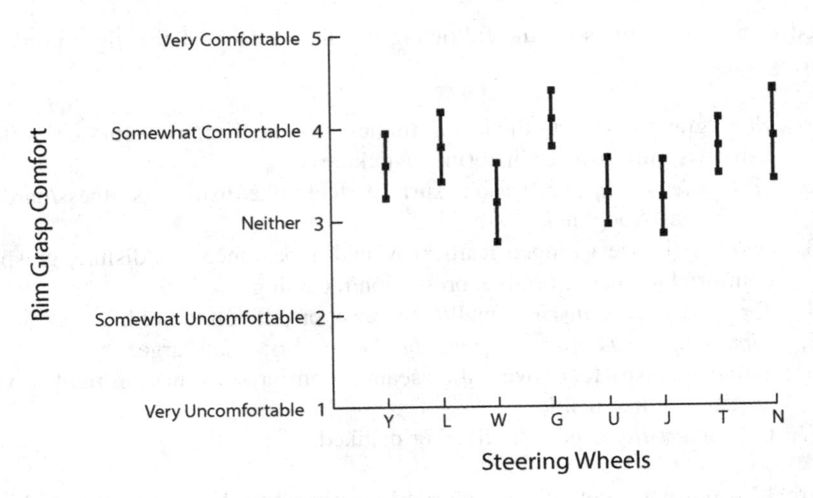

FIGURE 12.5 Mean and 95% confidence intervals of rim grasp comfort ratings of the eight steering wheels.

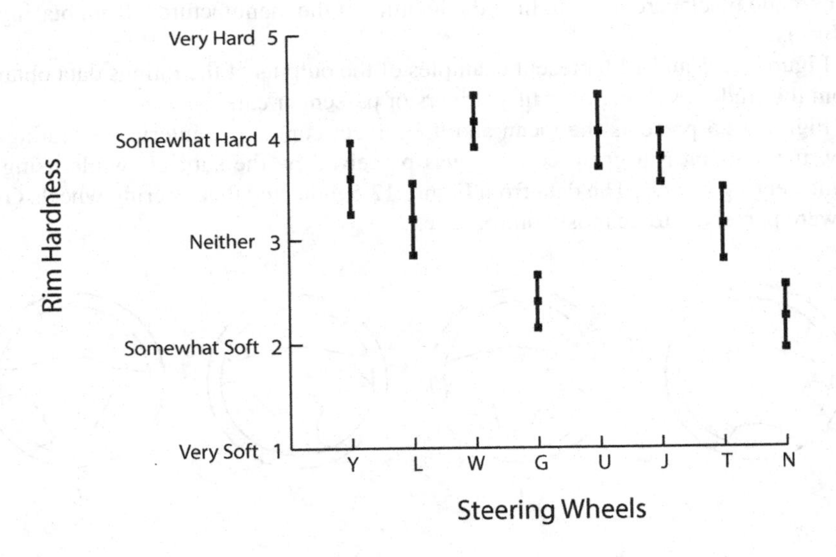

FIGURE 12.6 Mean and 95% confidence intervals of rim surface hardness ratings of the eight steering wheels.

Figure 12.6 presents the mean and 95 percent confidence intervals of ratings on how the steering rim hardness was perceived by the subjects on the eight steering wheels. The figure shows that the steering wheels G and N perceived to be softer than all other steering wheels.

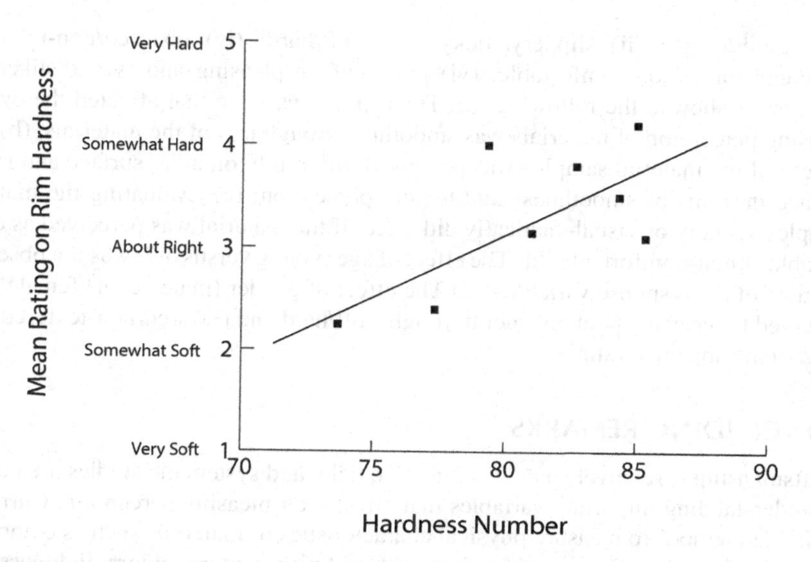

FIGURE 12.7 Relationship between mean ratings on rim surface hardness and hardness.

The relationship plot of mean hardness ratings presented in Figure 12.6 to the objective measure of hardness (called the hardness number) is presented in Figure 12.7. The plot suggested that if the steering wheel rim surface is made with hardness number below 77.5, the drivers will feel that the rim surface is softer. Similar correlation analyses allowed the manufacturer to determine the most preferred characteristics (e.g., rim diameter, rim slipperiness, type of leather, and type of stitches) of the steering wheel.

The above example thereby illustrates how studies can be conducted to obtain preferred characteristics of touch and grasp to improve customer satisfaction.

Other Studies

Similar studies conducted to determine customer-pleasing characteristics of interior materials used in armrests, seats, instrument panels, door trim panels, and other interior areas conducted by the author are available in the literature (Bhise, Hammoudeh, Nagaraj, Dowd and Hayes, 2006; Bhise, Onkar, Hayes, Dalpizzol and J. Dowd, 2008; Onkar, Hayes, Dalpizzol, Dowd, and Bhise, 2008).

A study conducted by Bhise, Sarma and Mallick (2009) evaluated effects of seven variables on leather and cloth materials used in automotive interior materials. The variables included: 1) compressibility (padded versus non-padded), 2) smoothness (or roughness), 3) shape (flat versus 30–35 mm diameter round grasp handle), 4) material type (leather versus cloth), 5) gender of subjects, 6) age of subjects, and 7) evaluation (visual only versus visual and tactile). The response measures used for the evaluations involved obtaining ratings using seven different 5-point semantic differential rating scales. The seven response variables were defined by the following adjective pairs

(i) smooth/rough; (ii) slippery/sticky; (iii) soft/hard; (iv) textured/non-textured; (v) comfortable/non-comfortable; (vi) pleasing/non-pleasing and (vii) dislike/like. The results showed the following: (a) The primary variable that affected the overall pleasing perception of materials was smoothness/roughness of the materials; (b) The texture of the material samples was perceived differently on a flat surface and round surface in terms of smoothness and texture perception; (c) Evaluating the material samples visually or visual–tactically did affect if the material was perceived as comfortable or non-comfortable; (d) The effect of age (young versus old) was not observed on most of the response variables. (e) The effect of gender (male versus female) was observed to have an effect on smooth/rough, soft/hard and textured/non-textured, and soft/harsh subjective ratings.

CONCLUDING REMARKS

Craftsmanship is relatively a new automotive field, and systematic studies are needed on understanding important variables that affect their pleasing perception. Currently available methods to measure physical characteristics of materials such as compressibility, hardness, coefficient of friction, surface finish, texture, colors, lightness, and gloss values are of limited usefulness in predicting pleasing perception qualities of interior materials. Additional research studies are needed on the effects of many objective and subjective characteristics of interior components and their materials and their relationships. Studies are also needed to determine how these effects are affected by characteristics of customers and users from different market segments.

REFERENCES

Bhise, V., R. Hammoudeh, R. Nagaraj, J. Dowd and M. Hayes. 2006. Towards Development of a Methodology to Measure Perception of Quality of Interior Materials. *Journal of Materials and Manufacturing*, SAE (Also published as SAE Paper no. 2005-01-0973).

Bhise, V. D., S. Onkar, M. Hayes, J. Dalpizzol and J. Dowd. 2008. Touch Feel and Appearance Characteristics of Automotive Door Armrest Material. *Journal of Passenger Cars – Mechanical Systems*, SAE 2007 Transactions (Also published as SAE Paper no. 2007-01-1217).

Bhise, V. D., V. Sarma and P. K. Mallick. 2009. Determining Perceptual Characteristics of Automotive Interior Materials. SAE Paper no. 2009-01-0017. Warrendale, PA: Society of Automotive Engineers, Inc.

Levitt, T. 1980. Marketing Success Through Differentiation of Anything. *Harvard Business Review*, Jan–Feb. Brighton, MA: Harvard Business Publishing.

Onkar, S., M. Hayes, J. Dalpizzol, J. Dowd and V. Bhise. 2008. A Value Analysis Tool for Automotive Interior Door Trim Panel Material and Process Selection. *Journal of Materials and Manufacturing*, SAE 2007 Transactions. (Also published as SAE Paper no. 2007-01-0453).

Peters, T. 1987. *Thriving on Chaos*. ISBN: 0-06-097184-3. New York, NY: Harper R Row.

Yang, K. and B. El-Haik. 2003. *Design for Six Sigma*. ISBN: 0071412085. New York, NY: McGraw-Hill.

13 Role of Ergonomics Engineers in the Automotive Design Process

INTRODUCTION

Designing a new vehicle is generally accomplished by using the Systems Engineering approach, which involves creating an organization of design teams involving different disciplines. The teams are also co-located at one site or in one building so that they can meet formally and informally to communicate on many issues and trade-offs related to attributes and interfaces between various systems, their lower-level systems and components in the vehicle. The ergonomics engineers assigned to the vehicle program follow the development process from its earliest stages of defining the concept until the vehicle is produced and used by the customers. During each stage of development, the ergonomics engineers conduct many evaluations to ensure that the vehicle being designed will be perceived by its customers to be ergonomically superior. In this chapter, we will review the following: (a) the Systems Engineering V model; (b) the goal of the ergonomics engineers; (b) methods used to achieve the goal; (c) evaluation measures used; (d) the responsibilities of the ergonomics engineers; and (e) their problems and challenges.

SYSTEMS ENGINEERING MODEL DESCRIBING THE VEHICLE DEVELOPMENT PROCESS

The Systems Engineering "V" model introduced in Chapter 2 is presented here again with more details to get a better understanding of involvement of ergonomics engineers in the automotive design process (see Figure 13.1; Bhise, 2017). The model shows basic phases of the entire vehicle development program on a horizontal axis, which represents time in months before "Job #1"; Job #1 in the auto industry refers to the event when the first production vehicle rolls out of the assembly plant. The project (or the vehicle program) generally begins many months prior to Job #1 (typically 12 to 48 months, depending on the complexity of the program).

In the early stages prior to the official start of a vehicle program, the advanced vehicle planning activity determines the vehicle type (i.e., body style, powertrain type, performance characteristics, and so forth.), the size of vehicle (e.g., size class, number of occupants and loads, weight and cargo volume), the intended market (i.e., countries), and a list of reference vehicles (or competitors) that the new vehicle may

DOI: 10.1201/9781003485582-13

replace or compete with. A small group of engineers and designers from the advanced design group are selected and asked to generate a few early vehicle concepts to understand design and engineering challenges. Ergonomics engineers are asked to meet with the team to evaluate the vehicle concept and provide recommendations to improve its ergonomic characteristics. A business plan including projected sales volumes, the planned life of the vehicle, program timing plan, facilities and tooling plan, manpower plan, and financial plan (including estimates of costs, capital needed, revenue stream and projected profits) are developed and presented to the senior management along with all other vehicle programs (to show how the proposed program fits with the overall corporate plan) planned by the company. The vehicle program, in most automotive companies, begins officially after the approval of the business plan by company management. This program approval event is considered to occur x-months prior to Job #1 in Figure 13.1.

At minus x-months, the chief vehicle program manager is selected, and each functional group – such as design, body engineering, chassis engineering, powertrain engineering, electrical engineering, climate control engineering, vehicle packaging and ergonomics/human factors engineering, and manufacturing and assembly engineering – within the product development and corporate activities is asked to provide personnel to support the vehicle development work. The personnel are grouped into teams, and the teams are organized to design and engineer various systems and subsystems of the vehicle. Systems engineering personnel develop SEMP to ensure that the tasks of all the teams are coordinated so that the SE process is followed to meet the vehicle program timings. Ergonomics engineers assigned to the vehicle program work with the SE to ensure that they understand the process and attend meetings of different teams to identify ergonomics issues and provide necessary support in resolving ergonomic issues during the vehicle development process.

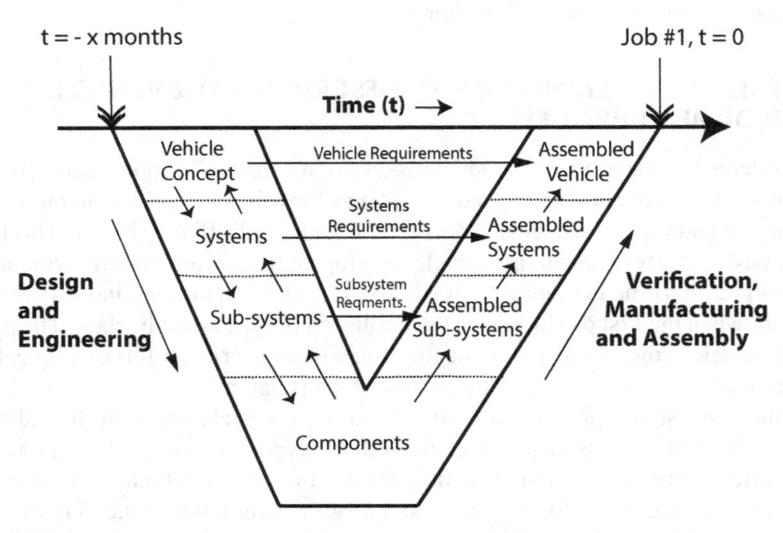

FIGURE 13.1 Systems engineering "V" model of the vehicle design process.

The left top part of Figure 13.1 shows that the first major task after the team formation is to create an overall product concept. During this phase, the designers (industrial designers) and the vehicle package engineers work with different teams to create several alternate vehicle concepts. The concepts are illustrated by creating a) drawings or CAD models of the proposed vehicle, b) computer generated 3-D life-like images or videos of the vehicle (fully rendered with color, shading, reflections and textural effects), and c) physical mock-ups (clay models; foam-core, wooden or fiberglass bucks to represent the exterior and interior surfaces). Ergonomics engineers are present during many of the design reviews to identify ergonomic problems with each product concept and discuss alternatives and potential solutions.

Market research clinics also are held for customers to come in and review different vehicle concepts and provide feedback on what they like and dislike about each concept. Images of the vehicle concepts and/or models are shown to prospective customers in these market-research clinics. Their feedback is used to select a leading concept for development of the proposed vehicle. The selected concept is further refined to ensure that real production vehicles can be produced to meet all customer, company and government requirements, The ergonomics engineers along with other team members conduct many analyses and evaluations in areas covered in earlier and succeeding chapters of this book, such as occupant positioning, entry/egress, instrument panel, door trim panels and console design, and so forth.

As the vehicle concept is being refined, each engineering team decides on how each of the vehicle systems can be configured to fit within the vehicle space and how the various systems can be interfaced to work together to meet all the functional and ergonomic requirements. This phase is shown as "Systems" in Figure 13.1. As the systems are being designed, the next phases involve a more detailed design, i.e., design of subsystems of each system and components within each of the sub-systems (see Volume 2, Chapter 21). These subsequent phases, straddled in time, are shown as "Sub-systems" and "Components". The above phases form the left half of the "V", and thus, represent the time and activities involved in Design and Engineering.

The right half of the "V", moving from the bottom to the top, involves testing, assembly and verification where the components are individually built and tested to assure that they meet their functional characteristics. The components are assembled to form sub-systems that are tested to verify they meet their functional specifications. Similarly, the sub-systems are assembled into systems and finally the systems are assembled to create a vehicle. At each of the phases, the corresponding assemblies are tested to ensure that they meet the requirements considered during their respective design phases (i.e., the assemblies are verified). These requirements are shown as the horizontal arrows between the left and the right sides of the "V" in Figure 13.1. Ergonomics engineers do not generally get into evaluations of component and lower level systems. However, they are asked to judge the operation of systems such as climate control, entertainment, and navigation that have human interfaces.

The ergonomics engineers assigned to the vehicle program work throughout all the above phases and continuously evaluate the vehicle design to ensure that the vehicle users can be accommodated, and they will be able to use the vehicle under all foreseeable usage situations.

VEHICLE EVALUATION

Goal of Ergonomics Engineers

The primary goal of ergonomics engineers is to work with the vehicle design teams to produce ergonomically superior vehicles. Some criteria that can be used to establish ergonomic superiority are:

1. Best-in-class or best in the industry – that is, the product will be perceived by the users to be the best within the selected class, or market segment, of vehicles or in the industry as determined by the company management, for example, countries or group of countries.
2. x percent better than products in the "class" (e.g., all driver interface components of the new vehicle should be at least 10 percent better than corresponding items in all other selected reference (or benchmarked) vehicles. The evaluation measures are generally determined by the program planning organization and/or attribute managers.
3. All important and pre-selected ergonomic requirements and all applicable standards must be met. Ergonomic requirements are selected by ergonomics attribute manager in concurrence with the program management.
4. User accommodation goals set in the early program definition phase must be met. The goal for accommodation can be set as 90–95% of the population, 5 percentile female and 99th percentile male. Accommodation goals are set by package, and ergonomics attribute managers in concurrence with the program management.
5. Minimum acceptable levels of ratings based on attributes such as easy to learn, easy to find and use, non-distracting and safe, comfortable and convenient must be met. The acceptable rating levels are set by attribute managers in concurrence with the program management.

Evaluation Measures

The following are examples of measures that can be used to evaluate the vehicle:

1. Percentage of ergonomic guidelines and requirements met in each evaluation category (e.g., driver accommodation, field of view, entry exit, and exterior use items (Note: All requirements mandated by the government, e.g., Federal Motor Vehicle Safety Standards, must be met.
2. The weighted sum of ergonomic guidelines met in each category (function or part of the vehicle) The categories can be weighted by frequency of use and importance of items or vehicle features.
3. Objective measures: Percentage of users accommodated by various task performance measures such as task completion times, error rates, eye involvement times, number of glances, standard deviation of lane position, mean velocity, and standard deviation of velocity.

4. Subjective measures: For example, percentage of users satisfied, percentage of users liked the vehicle (or preferred the vehicle over other benchmarked vehicles. The percentage of features rated as expected, pleasing and delightful to use as compared to features in other benchmarked vehicles, and a percentage of users liked usability of each feature (e.g., control, display). Other quantitative measures based on ratings such as averages of ratings, percentage of ratings above 8 on a 10-point scale, percentages of ratings below 5 on a 10-point scale, can also be used.

The exact definitions and acceptance levels of the above mentioned measures are generally set by the ergonomics engineers in concurrence of the attribute managers and the program management.

Tools, Methods, and Techniques

A number of tools, methods and techniques are used for evaluation of vehicles and systems during the vehicle development process (see Volume 2, Chapter 19 for more information). These tools and methods include:

1. Conduct benchmarking of selected existing vehicles to understand different designs and ergonomic issues with the designs [Bhise (2017, 2023].
2. Understand customer needs and translate them into functional specifications by using techniques such as quality function deployment (QFD) (Besterfield, Besterfield-Michna, Besterfield, and Besterfield-Scare, 2003).
3. Checklists and scorecards based on ergonomics requirements (in standards) and design guidelines.
4. Use of models: For example, physical models (bucks) and mathematical/computer models: Bhise, Mamoola, Devraj, Pillai and Shulze (2004), 200b, Bhise and Pillai (2006), Kang, Bhat and Bhise (2006), visualization/CAD (packaging studies – 3D models with occupants/manikins and other vehicle systems and components, field of view simulations) and CAE applications, for example, vehicle lighting evaluations, legibility models.
5. Task analysis, failure mode and effects analysis (FMEA) and cost-benefits analysis.
6. Laboratory and/or field studies using methods of observation, communication, and experimentation, for example, evaluations using prototypes and driving simulators, use of instrumented vehicles for driver performance measurements.

Ergonomics Engineer's Responsibilities

1. Provide the program teams the needed ergonomic design guidelines, requirements, data/information and results from analyses and experimental research, scorecards and recommendations for product decisions at right

points (timing, gateways) in front of right level of decision makers (teams, program managers, chief engineers, and senior management).
2. Apply methods/models/procedures available in the Society of Automotive Engineers (SAE), corporate and regulatory standards.
3. Conduct quick-react studies/experiments to resolve issues where sufficient information from prior research or vehicle program is not available.
4. Evaluate product/program assumptions, concepts, sketches/drawings, physical models/mock-ups/bucks, CAD models, mules (other production models equipped with new vehicle components or systems; sometimes called mechanical prototypes), prototypes, production vehicles and competitors.
5. Participate in the development of static and/or drive tests/market research clinics with existing products (leading competitors or other reference vehicles) as comparators, or controls.
6. Obtain, review and act on the customer feedback data – from complaints, warranty data, customer satisfaction surveys, J. D. Power's ratings/data, inspection surveys with owners, product review articles from automotive magazines, press – to improve the product.
7. Prepare ergonomics scorecards on alternative design concepts, benchmarked vehicles and the vehicle being designed.
8. Provide ergonomics consultations within product development and engineering activities.
9. Long term: Conduct ergonomic research, translate research results into design guidelines, develop design tools, evaluation procedures and databases for application to future vehicle programs.

STEPS IN ERGONOMICS SUPPORT PROCESS DURING VEHICLE DEVELOPMENT

Most automotive companies have a well-developed ergonomics support process that is synchronized with the overall vehicle development timing plan. Therefore, the ergonomics engineers supporting the program must understand the vehicle design process in terms of its phases, the work performed in each phase, functional areas involved in each phase, team structure, people and methods involved in the performing of different tasks, team communication methods, management review and the approval process.

The overall flow of a vehicle development program begins with the customers and ends with the customers. In the early stages even before the program plan is developed, the needs of the customers are compiled and understood by the product planning and the design team. And after Job #1, the customer feedback is continuously sought and reviewed to improve the product by removing production defects and planning for future design related product changes. Thus, basic vehicle development process flow involves the following major steps, many of which are usually conducted concurrently, using what is known as simultaneous engineering or concurrent engineering to reduce time and to avoid redesign/rework:

1) Customers (Understanding customers and their needs)
2) Product Planning
3) Design (Styling and Engineering)
4) Detailed Engineering
5) Prototyping, Testing, and Validation
6) Tooling Design
7) Plant Design, and Construction,
8) Production of Vehicles
9) Customers (Obtaining Feedback after Product Usages)

SOME DETAILS IN THE ERGONOMICS ENGINEER'S TASKS

Providing ergonomics support during product development involves performing a number of tasks. This section provides additional details on a few major tasks.

1. Understand Customer Needs and Translate into Vehicle Design Requirements.
   a) Review product letter that describes the proposed product specifications and features. Study new vehicle program assumptions. Predict ergonomic issues. Study customer feedback data from current vehicles.
   b) Study/Understand the market segment.
   c) Benchmark competitive and current products that will be replaced.
   d) Identify the customers and their characteristics, capabilities and limitations.
   e) Determine customer wants.
   f) Translate customer wants into product attributes and target specifications.
   g) Cascade vehicle level specifications to system, subsystem and component levels.
   h) Evaluate trade-offs between different product attributes
   i) Conduct market research clinics.
2. Understand Methods to Communicate with Different Teams
   a) Learn the team structure (co-located, dedicated, multi-disciplinary), team levels, meeting schedules and procedures for resolutions of issues.
   b) Implement Simultaneous/Concurrent Engineering.
   c) Deliver vehicle attribute related information to teams to satisfy customers.
   d) Understand vehicle systems, sub-systems, components and their interfaces and trade-offs with other attributes.
   e) Implement the Systems Engineering "V" process in program management.
3. Reviewing Early Product Concepts
   a) Review early exterior design sketches and renderings (e.g., see Figures 13.2, 13.3 and 13.4) and mock-ups to identify and resolve issues related to entry/exit, head clearance, field of view (pillar locations, pillar obstructions, windshield and backlite rake angles), locations of exterior lights, body cut lines (defining body openings and exterior body panels), fuel filler location, trunk opening and loading/unloading, and so forth.

FIGURE 13.2 Side view sketches of alternate SUV designs.

FIGURE 13.3 Exterior view of a selected SUV concept details.

FIGURE 13.4 Fully surfaced vehicle with shading and reflections details.

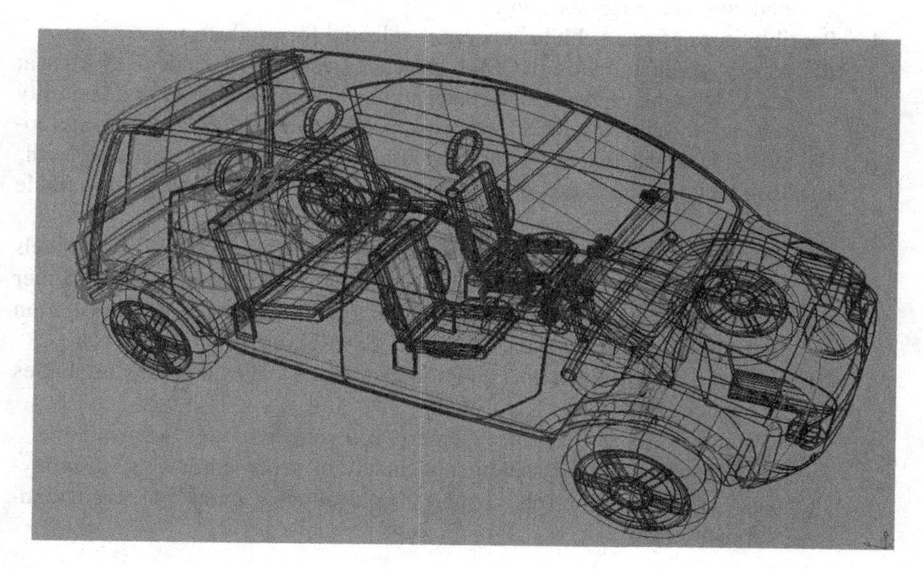

FIGURE 13.5 Wireframe diagram showing interior details.

Review ergonomic recommendations and suggested changes to the design team and other affected vehicle attribute teams.

b) Review interior sketches or 3-D models (e.g., see Figure 13.5) for driver accommodation issues (e.g. seat positioning, eye location, instrument panel layout, etc.) (Bhise, Shulze, Mamoola and J. Bonner, 2004a; Bhise, Kridli, Mamoola et al., 2004b). Review findings with the package engineering and design teams.

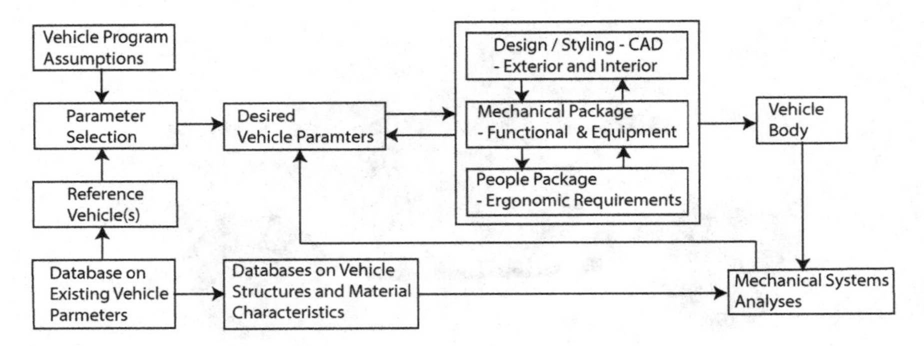

FIGURE 13.6 Flow diagram of a parametric model developed for vehicle packaging.

   c) Conduct package analyses, for example, use parametric model (Bhise et. al., 2004b) to evaluate occupant accommodation. (see Figure 13.6) and conduct customer clinics.

4. Possible steps after a vehicle concept is selected for production.
   a) Verify application of the SAE practices related to placement of seat track, location of eyellipses, reach and head clearance contours, visibility through the steering wheel, controls locations, direction of motion stereotypes, labeling, exterior lighting photometric requirements, and so forth.
   b) Conduct task analyses on items and issues related to different vehicle usages.
   c) Conduct studies to evaluate special issues, for example, legibility of labels and graphics, evaluation of unwanted interior reflections, new center stack units, entry/exit, obstructions due to pillars, operation of navigation units, entertainment systems, and exterior lighting photometric issues.
   d) Follow design refinements and conduct additional evaluations or studies as needed
   e) Evaluate hardware as prototype parts and vehicle models are available
   f) Provide ergonomics assessments/ maintain score cards (e.g. "Smiley-face" charts) on controls, displays, field of view, entry/exit, craftsmanship, and so forth.

PROBLEMS AND CHALLENGES

a) Insufficient data (incomplete and constantly changing/evolving data) available during early design phases.
b) Insufficient time and resources available to thoroughly research problems.
c) Need to consider many users and markets, usages, needs and expectations; it is difficult to study many issues, prioritize or weigh, and trade-off different considerations.
d) Fierce competition in fast changing and chaotic environment.
e) High cost of research studies.
f) Difficult to predict future design and technology trends.

g) Difficult to predict future plans of your competitors (i.e., "futuring" – what would your competitors do?)
h) Need to know limits of applicability of your tools, procedures and data.
i) Insufficient understanding of trade-offs between different vehicle attributes.

Many of the above challenges can be overcome through activities such as

a) in-house and continual training of ergonomists;
b) continual communication within members of the ergonomics departments on various projects under study and completed by various design and research teams;
c) information exchanges with other engineering departments on methods and completed projects;
d) creating databases and libraries on reference information and past studies;
e) planning activities of future research projects (see Chapter 25);
f) attending meetings and conferences where ergonomic research methods and studies are presented (e.g., SAE Technical Committee Meetings, SAE International Congress, Transportation Research Board Annual Meetings, Human Factors and Ergonomics Society Meetings).

CONCLUDING REMARKS

The product development process is rarely smooth because of many unknown issues involving hundreds of interfaces between many systems within the vehicle that need resolution within the tight program timing plan. In many major automotive companies, more experienced ergonomics experts are generally available to guide the ergonomics engineers working in the vehicle development teams. In addition, the ergonomics process and tools to be used in product development are also well documented and set-up for quick implementation. An extensive library of previous research studies, literature, reference books and databases is maintained within most ergonomics departments.

REFERENCES

Besterfield, D. H., C. Besterfield-Michna, G. H. Besterfield and M. Besterfield-Scare. 2003. *Total Quality Management*. ISBN: 0-13-099306-9. Upper Saddle River, NJ: Prentice Hall.

Bhise, V., R. Shulze, H. Mamoola and J. Bonner. 2004a. Interior Design Process for UM-D's Low Mass Vehicle. SAE Paper no. 2004-01-1709. Also in publication # SP-1848, SAE International, Warrendale, PA.

Bhise, V., G. Kridli, H. Mamoola, S. Devraj, A. Pillai and R. Shulze. 2004b. Development of a Parametric Model for Advanced Vehicle Design. SAE Paper no. 2004-01-0381. Also in publication # SP-1858, SAE International, Warrendale, PA.

Bhise, V. and A. Pillai. 2006. A Parametric Model for Automotive Packaging and Ergonomics Design. Proceedings of the International Conference on Computer Graphics and Virtual Reality, Las Vegas, Nevada.

Bhise, V., R. Boufelliga, T. Roney, J. Dowd and M. Hayes. 2006. Development of Innovative Design Concepts for Automotive Center Consoles. SAE Paper no. 2006-01-1474. Presented at the SAE 2006 World Congress, Detroit, Michigan.

Bhise, V. D. 2017. *Automotive Product Development*. Second Edition. ISBN: 978-1-4987-0681-0. Boca Raton, FL: CRC Press.

Bhise, V. D. 2023. *Designing Complex Products with Systems Engineering Processes and Techniques*. ISBN: 978-1-0322-0369-0. Boca Raton, FL: CRC Press.

Diehl, C. and V. Bhise. 2006. Design Concepts for Vehicle Center Stack and Console Areas for Incorporating New Technology Devices. Presented at the Society of Information Display Vehicle and Photons Symposium held in Dearborn, Michigan.

Kang, H., N. Bhat and V. Bhise. 2006. Parametric Approach to the Development of a Front Bucket Seat Frame. SAE Paper no. 2006-01-0336. Presented at the SAE 2006 World Congress, Detroit.

Shulze, Roger (Ed.). 2007. *Designing a Low Mass Vehicle*. Dearborn, MI: College of Engineering and Computer Science, University of Michigan-Dearborn.

Appendix 1
Human Factors Historic Landmarks

TABLE A1.1
Human Factors Engineering Historic Landmarks

Period	Reference Information	Year	Human Factors historic event
Evolutionary Process		Early Man	Hand tools and utensils development
-------------	-----------------------------	1750	------------------------------------
Age of Machines – Steam Power	James Watt developed condenser for steam engine.	1764	
		1828	Civil Engineering was defined to be concerned with – "The art of directing the great sources of power in nature to use and convenience of man". – from the Institute of Civil Engineers.
	3010 steam engines in use in the U.S.	1838	
	First Electric bulb – Edison	1879	
		1880	ASME prepared boiler safety standards.
		1887	Application of Scientific Method to Work in Industry – Fredric Taylor
-------------	------------------------------------	1890	--
Power Revolution	Ford's Model T Roadster	1908	
		1912	Frank Gilbreth presented a paper on Micro-motion study in ASME.

(continued)

TABLE A1.1 (Continued)
Human Factors Engineering Historic Landmarks

Period	Reference Information	Year	Human Factors historic event
	Model Ts accounted for 50% of the world's motor vehicles.	1924	Harvard Fatigue Laboratory established in 1920s.
		1925	Hawthorne studies on relationship between lighting and productivity.
--------------	World War II ends.	1945	Army-Air Force psychology program established. First Chief of Psychology branch was Dr. Paul Fitts.
Knobs and Dials Science		1947	Publication of *Applied Experimental Psychology Journal*. Fitts and Jones published famous paper on Pilot Errors in Operating Controls and Reading Displays.
		1948	Method-Time-Measurement (MTM) developed – Human performance prediction model based on hand/body motion analyses. First Human Factors Laboratory at Bell Labs began under John Karlin.
		1949	Ergonomics – the word was invented by Murrell, British Navy Labs. Ergonomics Society formed in Britain.
		1950	Formation of Human Factors groups in industry – especially related to the air force/military.
		1956	Human Factors Engineering department established in Ford Motor Company.
		1957	Incorporation of the Human Factors Society in the U.S.
Product Liability Cases begin to accelerate		1960	Hennigsen vs. Bloomfield Motors Case – Each product, by being out on the market for sale, bears an implied warranty that it is reasonably safe to use.
Strict Liability		1963	Greenman vs. Yuba Power Products Case – A manufacturer is strictly liable when the article that he places on the market, knowing that it will be sold without inspection for defects, proves to have a defect that causes injury to a human being.
Passage of Safety Laws		1966	National Traffic Safety and Motor Vehicle Safety Act. Highway Safety Act.
		1968	Radiation Control for Health and Safety Act. Fire Research and Safety Act.
		1969	Child Protection and Toy Safety Act.
		1970	Occupational Safety and Health Act.

TABLE A1.1 (Continued)
Human Factors Engineering Historic Landmarks

Period	Reference Information	Year	Human Factors historic event
		1970	Thomas vs. General Motors Case – The manufacturer is liable for any or all foreseeable unintended uses, misuses and abuses of the product and even abnormal uses which were foreseeable and could have been designed against or otherwise safe guarded.
		1971	Federal Boat Safety Act.
		1972	Consumer Product Safety Act.
		1973	Balido vs. Improved Machinery Case – Warnings and directions do not absolve the manufacturer of the liability if there is a defect in design or manufacture.
		1976	NHTSA vs. GM Case – Actual proof of harm is not required for a safety defect to be considered to exist in a vehicle. The government can therefore order defective products to be recalled even if the defect had not caused a collision or injury.
		1979	Three Mile Island Incident.
Information Age	-------------------------------------	1980	--------------------------------------
	WIMPS (window, icons, menus and pointers) interface.		Macintosh Computers, PCs, Boeing 757
		1984	CRTs in Cars
	Ford Taurus/Sable introduced.	1986	
	User-friendly software issues	1990	*Usability* term was introduced in the literature.
	Internet, cellular phones and palmtops.	1990s	Voice-activated controls. Data gloves. Prototyping. Virtual reality displays.
		1995	Chrysler showcases Programmable Vehicle Buck during Super Bowl XXIX commercial.
	New Technologies	2000	Driver Distraction Due to New In-Vehicle Devices Considered as a Major Traffic Safety Problem by NHTSA: Telematics, Bluetooth, Haptics, OLEDs Increasing usages of: Digital cameras, Internet in Palm PC, text pagers, i-pods, cell phones and plug-and-play devices(with USB) in cars.
		2010	Capacitive touch screens in cars and SUVs.

Appendix 2

Anthropometric Dimensions of Driving Populations from Seven Countries

TABLE A2.1
Anthropometric Dimensions of Driving Populations from Seven Different Countries

Anthropometric Dimension	Percentile Value	US Males	US Females	UK Males	UK Females	German Males	German Females
Stature (mm)	95	1870	1730	1855	1710	1845	1750
	50	1755	1625	1740	1610	1745	1635
	5	1640	1520	1625	1505	1645	1520
Sitting Height (mm)	95	975	920	965	910	975	930
	50	915	860	910	850	920	865
	5	855	800	850	795	865	800
Sitting Eye Height (mm)	95	860	810	845	795	850	800
	50	800	750	790	740	800	740
	5	740	690	735	685	750	680
Sitting Shoulder Height (mm)	95	655	620	645	610	640	570
	50	600	565	595	555	595	525
	5	545	510	540	505	550	480
Shoulder Breadth (bideltoid) (mm)	95	515	440	510	435	505	445
	50	470	400	465	395	465	400
	5	425	360	420	355	425	355
Hip Breadth (mm)	95	410	440	405	435	385	445
	50	360	375	360	370	350	375
	5	310	310	310	310	315	305
Buttock-to-Knee Length (mm)	95	650	625	645	620	640	635
	50	600	575	595	570	600	580
	5	550	525	540	520	560	525
Knee Height (mm)	95	605	550	595	540	590	555
	50	550	505	545	500	545	505
	5	495	460	490	455	500	455
Shoulder-to-Elbow Length (mm)	95	400	365	395	360	395	365
	50	365	335	365	330	365	335
	5	330	305	330	300	335	305
Elbow-to-Fingertip Length (mm)	95	515	470	510	460	505	470
	50	480	435	475	430	475	435
	5	445	400	440	400	445	400

Source: Anthropometric Dimensions from: Bodyspace by Stephen Pheasant (2006)

French Males	French Females	Japanese Males	Japanese Females	Hongkong Males	Hongkong Females	Indian Males	Indian Females
1830	1700	1750	1610	1775	1655	1745	1615
1715	1600	1655	1530	1680	1555	1640	1515
1600	1500	1560	1450	1585	1455	1535	1415
970	910	950	890	955	900	905	850
910	860	900	845	900	840	850	790
850	810	850	800	845	780	795	730
855	800	835	780	840	780	795	745
795	750	785	735	780	720	740	690
735	700	735	690	720	660	685	635
670	625	635	600	655	610	690	565
620	580	590	555	605	560	640	515
570	535	545	510	555	510	590	465
515	470	475	425	470	435	455	390
470	425	440	395	425	385	415	350
425	380	405	365	380	335	375	310
410	430	330	340	370	365	365	390
370	380	305	305	335	330	320	330
330	330	280	270	300	295	275	270
640	610	600	575	595	570	610	590
595	565	550	530	550	520	560	540
550	520	500	485	505	470	510	490
575	535	530	480	540	500	565	515
530	495	490	450	495	455	515	470
485	455	450	420	450	410	465	425
395	360	365	330	370	340	370	340
360	330	330	300	340	315	340	310
325	300	295	270	310	290	310	280
505	455	475	430	480	440	485	435
470	425	440	400	445	400	450	405
435	395	405	370	410	360	415	375

Appendix 3

Verification of Hick's Law of Information Processing

An Excel-based computer program is included in the publisher's website to allow the reader to measure his/her choice reaction times under different equally likely alternatives ranging from N = 1 to 8.
The procedure to use the program is as follows:

1) Click on the icon "Reaction Time program" (See Table A3.1 for the screen).
2) Click on "Reaction Time Setup" box (Enable macros, if prompted, and set lowest security level through the Trust Center and via "Excel Options").
3) Select number of choices (N) (choose between 1 to 8).
4) Select repetitions (Number of trials) (Use at least 30 repetitions).
5) Set minimum and maximum delay time (Min = 1s; Max= 3 s).
6) Click on "OK."
7) Click on "Ready."
8) Now, a random number between 1 to the selected number of choices will appear in the white box (to the left of the "Reaction Time Setup" box). Your task is to push the number key corresponding to the displayed number as quickly as possible. Your reaction time (RT) (the time between when the number appeared and when you pressed the key) will be displayed.
9) Press "Ready" button as in step 6 and repeat the procedure until all the repetitions are completed.
10) Note down your data (average and standard deviation).
11) Repeat the above process for a different number of choices (Repeat all above steps).
12) Plot the average reaction time versus Log_2N and fit a straight line through the data points.
13) The y-intercept of the line corresponds to constant "a" and slope of the line is equal to the reciprocal of constant "b".

(Note: Hick's law: $RT = a + b\ Log_2N$).

Figure A3.1 presents reaction time data obtained by using the above program. The data were obtained by asking a practiced and alerted subject to complete 30 trials in each session where the number of choices were 1, 2, 4 and 8. The figure shows mean, mean plus and minus one standard deviations and also a straight line fitted to the data.

TABLE A3.1
Screen Shot of the Reaction Time Program

Reaction Time Demonstration

3

Reaction Time =	**1.047**
Average =	1.025
standard deviation =	0.110
number of errors =	0
Min	0.859
Max	1.266

Reaction Time Log

Trial #		*Trial #*	
1	0.938	*9*	0.984
2	0.934	*10*	1.188
3	1.027	*11*	1.109
4	0.859	*12*	1.047
5	1.266	*13*	
6	0.938	*14*	
7	0.984	*15*	
8	1.031		

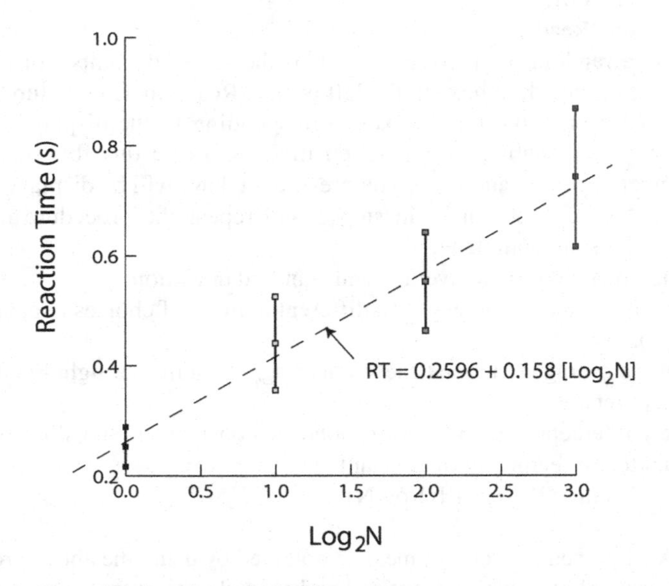

FIGURE A3.1 Reaction time data obtained by using the program.

Appendix 4

Verification of Fitts' Law of Hand Motions

Fitts' Law can be verified by collecting measurements on hand movements between circular targets separated by a given distance. Figures A4.1a and A4.1b present 12 targets labeled as T#1 to T#12 that will be used in this experiment. Each target consists of two circular areas placed apart by a distance. The twelve targets are created by a combination of diameters of the circles (W) and distance between the centers of the circles (A). The "W" is varied between 1/8" to 1" and "A" is varied between 2" to 8".

In this experiment, a subject is asked to sit down in a chair in front of a desk and to place a given target (selected randomly from the 12 targets) on the desk at a comfortable distance and orientation in front. The subject is given a pencil to make "hit" marks on the targets by moving his/her hand back and forth between the circular targets. The subject's task would be to move his/her hands between the circular targets as fast as possible in a given time interval (5 s) and count the number of hit marks in the interval. The hit mark should not fall outside any of the target circles. An experimenter should be used to monitoring the subject and giving the subject "begin" and "end" instructions by using a stopwatch. The subject, thus, is asked to first place his pencil inside one of the circles, and when told by the experimenter to "begin", make hand movements back and forth between the circles and keep track of the number of back and forth movements (i.e. by counting the "number of times the pencil hits the paper"). The hand movement time is calculated by dividing the time interval (5 s) by the number of hits.

The experimenter should give the subject a few targets to practice the task first and then ask the subject to participate in a number of trials. A different target should be used in each successive trial.

The hand movement time for each target should be plotted against the index of difficulty (Log_2 [2A/W]) of the target. If all the points can be fitted by a straight line, then Fitts's Law can be considered to be verified.

Figure A4.2 illustrates data obtained by asking one subject to follow the hand movements procedure for four targets with the index of difficulty ranging from 3 to 6. The figure shows mean and mean plus and minus one standard deviation values of the movement time and a straight line fitting the data.

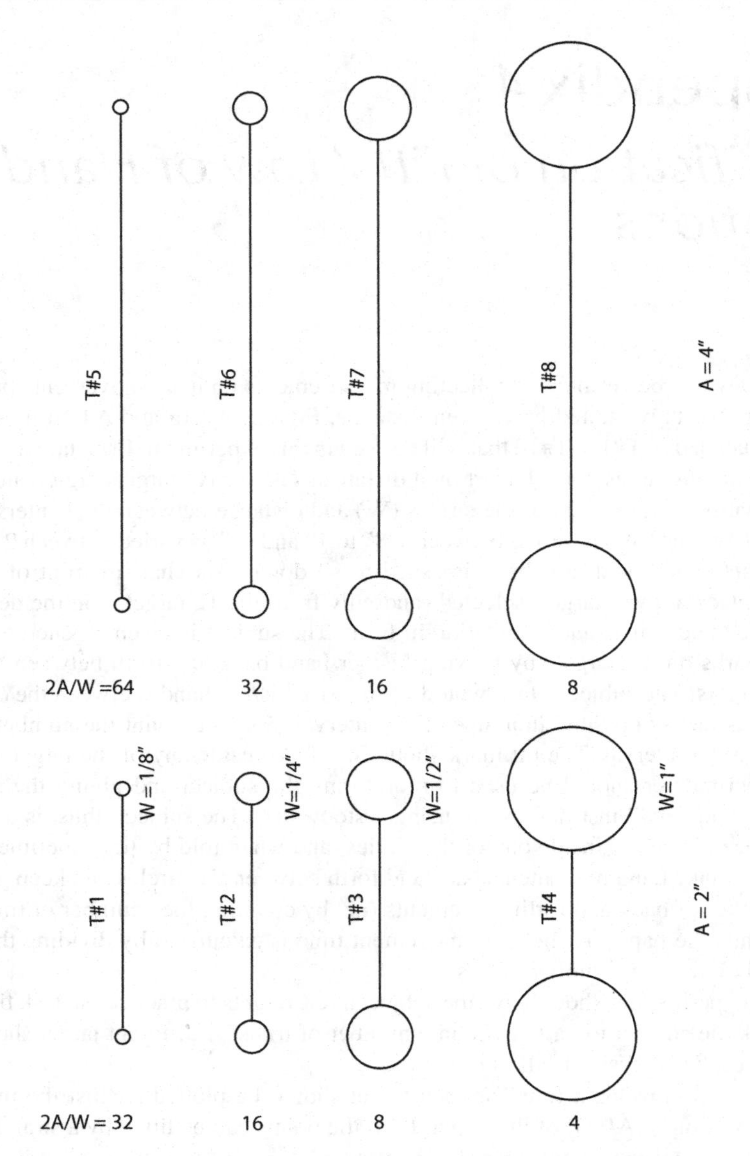

FIGURE A4.1A Targets for verification of Fitts' Law of Hand Motions for targets separated by 2" and 4".

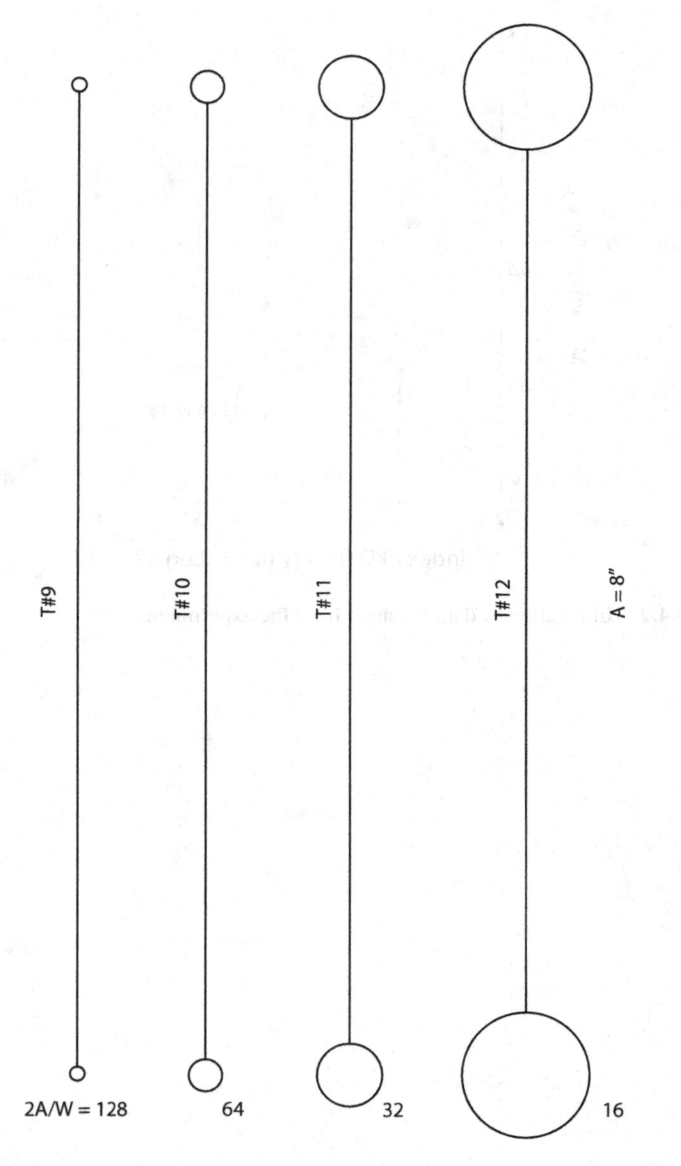

FIGURE A4.1B Targets for verification of Fitts' Law of Hand Motions for targets separated by 8".

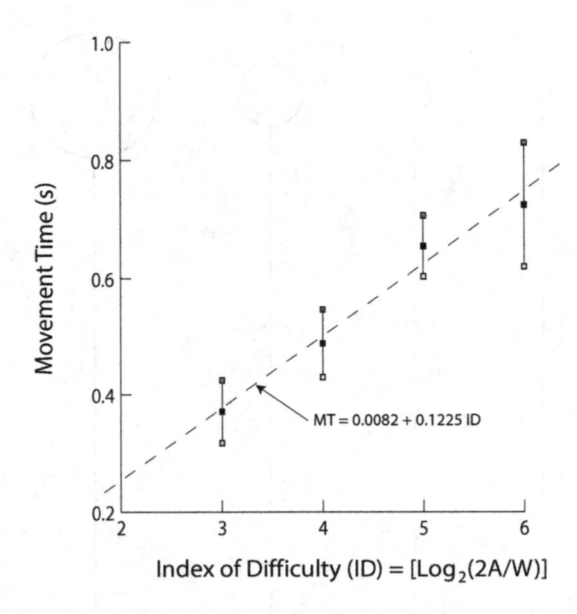

FIGURE A4.2 Illustration of data obtained from the experiment.

Index

Note: Page numbers provide volume number, hyphen and page number (or range of page numbers) in the volume.

E

F

G

H

I

K

L

M

N

O

P

Q

R

S

T

U

V

W

Printed in the United States
by Baker & Taylor Publisher Services

Printed in the United States
by Baker & Taylor Publisher Services